W. Gunn, C. T. Clough

The Geology of Cowal

Including the Part of Argyllshire Between the Clyde and Loch Fine

W. Gunn, C. T. Clough

The Geology of Cowal
Including the Part of Argyllshire Between the Clyde and Loch Fine

ISBN/EAN: 9783337183004

Printed in Europe, USA, Canada, Australia, Japan

Cover: Foto ©berggeist007 / pixelio.de

More available books at **www.hansebooks.com**

Memoirs of the Geological Survey,
SCOTLAND.

THE

GEOLOGY OF COWAL

INCLUDING THAT PART OF

ARGYLLSHIRE BETWEEN THE CLYDE AND LOCH FINE.

By W. GUNN, F.G.S., C. T. CLOUGH, M.A., F.G.S.
AND J. B. HILL, R.N.

(WITH PETROLOGICAL NOTES BY J. J. H. TEALL, M.A., F.R.S.,
SEC. G.S., AND DR HATCH, PH.D., F.G.S.)

EDINBURGH:
PRINTED FOR HER MAJESTY'S STATIONERY OFFICE
BY NEILL & Co., OLD FISHMARKET CLOSE.

And to be purchased, either directly or through any Bookseller, from
JOHN MENZIES & CO., 12 HANOVER STREET, EDINBURGH, and
90 WEST NILE STREET, GLASGOW; or
EYRE & SPOTTISWOODE, EAST HARDING STREET, FLEET STREET, E.C.; or
HODGES, FIGGIS, & CO., LIMITED, 104 GRAFTON STREET, DUBLIN.

1897.

Price Six Shillings.

PREFACE.

The district described in the present memoir includes that part of Argyllshire, generally known as Cowal, which lies between Loch Fyne and the Firth of Clyde. It embraces the south-western extension of the various bands of metamorphic rocks which form the southern edge of the Highlands. Bounded on three sides by coast-lines and penetrated by a number of sea-lochs it affords better and more continuous sections of these rocks than are generally to be met with in the interior of the country. Advantage has been taken of these exceptional features to describe the rocks in much greater detail than would otherwise be desirable.

From the detailed study of this part of the Highlands much information has been obtained by the Geological Survey regarding the structures of the schists and the successive movements by which these structures have been produced. Originally most of the rocks described in the following chapters formed a thick series of sedimentary deposits, the geological age of which still remains to be determined. These strata have been found to have undergone a remarkable series of repeated movements. After being thrown into folds and having been cleaved so as to acquire a first system of deformation, they have again suffered a repetition of the process more than once. They consequently present secondary and tertiary perhaps even quaternary structures, probably due to mechanical movement with accompanying re-crystallisation. The regional metamorphism thus produced is not uniformly distributed, but seems to increase in intensity both from the south-east and north-west towards a nearly central line, ranging about north-east and south-west, which is an anticline of the foliation. It has not been traced to any intrusion of igneous rock, and is so general and diffused that it can hardly be regarded as in any sense a contact phenomenon. Where intrusive masses occur in the district they have given rise to their own accompanying alteration, quite apart from the general metamorphism of the whole area. These interesting and complicated structures, so well displayed in Cowal, are fully discussed in the present Memoir.

The extreme south-western part of the promontory south of an east and west line drawn at Kilfinan was surveyed by Mr Gunn. The region lying to the east of Loch Riddon and Glendaruel and thence northward to Loch Fyne was mapped by Mr Clough, while the rest of the ground lying along the east side of Loch Fyne was the work of Mr J. B. Hill. Each officer describes in the Memoir the tract which he himself has surveyed, their several contributions being indicated by their initials. Mr Clough having mapped by far the largest part of the whole district has had general charge of the Memoir, which is mainly written by him.

The petrographical notes whether embodied in the text or placed in the Appendix have been chiefly supplied by Mr J. J. H. Teall, but some were furnished by D. F. H. Hatch, when he was attached to the Staff of the Survey. In the Appendix will be found a list of all the microscopic slides in the collection of the Survey, which have been prepared from rocks in Cowal; likewise a list of fossils from the "Clyde Beds" of the region and a Bibliography of the chief literature relating to the geology of this part of Argyllshire.

ARCH. GEIKIE,
Director General.

GEOLOGICAL SURVEY OFFICE,
JERMYN STREET, LONDON,
5th April 1897.

CONTENTS.

LIST OF ILLUSTRATIONS.

PLATES—(all at end of Volume).

THE GEOLOGY OF COWAL.

CHAPTER I.

INTRODUCTION.

The district of Cowal includes that part of Argyllshire which is almost isolated from the rest of Scotland by Loch Fine, the Kyles of Bute, and Loch Long. The boundary on the north-east side is not clearly defined physically, and we were for some time in doubt as to the limit of the district in this direction. It appears from our inquiries * that at the present time the north-east boundary is generally understood to extend to the head of Loch Long, and then to keep along the boundaries between Argyllshire on the one side and Dumbartonshire and Perthshire on the other side as far as the parish of Dalmally. It then coincides with the boundary between the parishes of Dalmally and Kilmorich until it meets Loch Fine ½ mile east of Dunderave Castle. The area north-west of Loch Fine and Glen Fine is said to have been added to the parish of Kilmorich and the district of Cowal at a comparatively late period, about the year 1700, in consequence of the change of ownership from the MacNaughtons of Dunderave to the Campbells of Ardkinglas. The area then added is separated from the rest by the well-marked physical boundary of Loch Fine and Glen Fine, and we have not described it in the following pages. Nor have we described the extreme north-east portion of Cowal as above defined, as this portion was surveyed by other officers, and there seems no physical reason for describing it with the rest. The north-east boundary of the area described by us runs approximately along Coilessan Glen (1-inch Map 38), and thence almost along the boundary between 1-inch Maps 37 and 38 up to the margin of the granite by Meall Beag, then along the granite margin to Glen Fine and down this glen to Loch Fine. This area is comprised in 1-inch Maps 29, 37, and 38, the portion in the latter Map being, however, only a few square miles in extent. In the Appendix we have added the list of glacial shells collected by Mr Bennie in Balnakailly Bay, which is in the island of Bute, and outside of Cowal.

The extreme length of the district described by us, from Ardlamont Point on the south-west to the granite edge near Cruach Tuirc on the north-east, is about 44 miles. The breadth, from Toward Point on the south-east to Otter Beacon on the north-west, is about 18 miles. Owing to the irregularity of outline, and the way in which the sea lochs, Loch Riddon, Loch Striven, the Holy Loch, and Loch Goil, run up into the land, it is not easy to estimate the superficial area of the district. It is certainly somewhat less than 432 square miles, the area of the 1-inch Maps of Scotland.

* We have specially to thank Mr Daniel Anderson of Dunoon, and Mr Duncan Whyte of Glasgow, for information on this matter.

A

It seems to be the popular opinion that the district of Cowal derives its name from Cumhal (or Coul), the father of Fionn, the mythical hero of Celtic tradition, who is generally spoken of by old writers as Finnmaccoul. We find him mentioned by Barbour, Dunbar, Gawin Douglas, Hector Boece, and Sir D. Lyndsay.

Professor Rhys, in his *Celtic Britain*, speaks of a Cunedda, the grandson of Coel, who comes from Coelin, which is probably Kyle in Ayrshire, but these names have probably no connection with Cowal, though the district, as J. F. Campbell says, is brimful of Celtic traditions. There is a Tobar an Fhion, or Fionn's Well, at Lachlan Bay on Loch Fine, and there is a Fingal Well near Ardentinny on Loch Long. Mr D. Whyte of Glasgow has kindly furnished us with a long list of farms, streams, and forts, in which the name of Fionn occurs, many of which are not found either on the 6-inch or 1-inch Ordnance Maps of the district; and so long ago as the year 1567 we find Carswell, Bishop of Argyll and the Isles, in the Introduction to his Translation of John Knox's Liturgy into Gaelic,* complaining of those composers and writers and supporters of the Gaelic, who prefer and practise the framing of vain, hurtful, lying, earthly stories about the Tuath de Dhamond, and about the sons of Milesius, and about the heroes and Fionn mac Cumhail with his giants (fhind mhac Cumhail gona fhianaibh).

It is stated by Dr W. F. Skene (*Celtic Scotland*, vol. i. p. 229) that the district of Cowal took its name from Comgall, a grandson of Fergus Mor, who came from Ireland in the end of the fifth century, and laid the foundations of the Scottish kingdom of Dalriada. Domangart, a son of Fergus, had two sons, Gabran and Comgall, and their possessions consisted of the districts of Cowal and Kintyre (Kintyre in those days included Knapdale and the small islands off this coast). It is not stated what the exact boundary of Cowal on the north-east side then was.

The highest hills are in the north-east part of the area. These are: Beinn an Lochain, on the west side of Loch Restil, 2992 feet; Bionein an Fhidleir, 2658 feet; Ben Donich, 2774 feet; Beinn Bheula, 2557 feet. The general elevation of the hills decreases rather uniformly in a south-west direction. Between Loch Eck and Glendaruel the highest points are: Beinn Mhòr, 2433 feet; Clach Beinn, 2109 feet; Cruach nan Capull (183),† 2005 feet; Sgarach Mor, 1972 feet; Cruach Neuran (not named in the 1-inch Map), 1 mile slightly north of west of Corrachaive, 1988 feet. West of Glendaruel the chief hills are: Cruach an Lochain, 1658 feet; Cruach Chuilceachan, 1428 feet; Beinn Bhreac, 1488 feet; and Beinn Capuill, 1419 ‡ feet. The district that lies about half way between the Clyde and the part of Loch Fine above Otter is generally rough and rocky, even when of low elevation: indeed, in some parts of the less elevated south-west part of Cowal, the ruggedness is perhaps increased, owing to the less amount of drift. The shores of the Clyde and upper Loch Fine are smooth and gentle sloping by contrast.

All the chief streams flow south or south-east. Among these are the Goil, the Echaig (the river that flows out of Loch Eck), the Massan, the Little Echaig (Glen Lean river), and the Ruel. The area that drains into Loch Fine forms in most parts but a narrow zone close to the side of this loch. The only large streams that flow into this loch from this

---

* The first book printed in the Gaelic language.

† The number in brackets indicates the 6-inch Map in which the locality is situated. See p. 4.

‡ This is the height given both on the 6-inch and 1-inch Maps, but it must be a mistake. From many measurements made with an aneroid, the conclusion was come to that it was much higher. Probably 1419 is a misprint for 1491.   W.G.

side are the Kinglas, Kilfinan Burn, Acharosson Burn, and Allt Osda. The last three are in the south-west corner of the district, where the watershed approaches near to the Kyles of Bute and Loch Riddon. None of the rivers are large by comparison with those of other districts. This follows necessarily from the way the area is surrounded or indented by sea lochs. But the rivers run in deep-cut glens which intersect the country so completely that no great extent of high-lying ground is left in any part; the hills rise steeply, and there are no broad flats at their tops. The roads, of course, take advantage of these deep glens, and, as the watersheds separating the glens are often low, they rarely ascend to great heights. The roads between Glendaruel and Otter Ferry, Lochgoilhead and St Catherine's, Glencroe and Cairndow, are exceptions. The first of these roads reaches the height of 1026 feet, the second, 727 feet, and the third, 860.

The farms and shepherds' houses nearly all lie in the shelter of the glens, and at low elevations, but sometimes they are so shut in by the mountains that for several months in the year they get no sun. The highest houses are : Butterbridge, 596 feet; High Glencroe (134), a little under 500 feet; Blar Buidhe (183), above 550 feet; Corlarach (192), about 500 feet ; and Caol-ghleann (152), about 450 feet.

The wildest glen views are those in Coire Athaich (173), Corrachaive Glen (172), south-west side of Glen Lean above Corrachaive, west side of Glen Laoigh (172), west side of Glendaruel near Ormidale House, Garrachra Glen under Creag Tharsuinn, south-west side of Gleann Canachadau (142), near Loch Restil, and Glen Fine near Cruach Tuirc. The views from the higher mountains in the north-east show a multitude of other rugged peaks and ranges on the north and east of them, stretching far into the heart of the Highlands. In the south half of Cowal the views from Cruach nan Capull (183) and the Bishop's Seat are extensive, and show well the general physical character of the district : from the former you can see the hills of Mull and the Paps of Jura: from the latter the peaks of Ben Cruachan (1-inch Map 45), through the Loch Eck valley.

As in all the west coast of Scotland, the annual rainfall is heavy. Perhaps taking the average over the whole area it may amount to between 60 and 70 inches. The wettest part is at the head of Loch Fine near Cairndow: here the average is over 90 inches. In the lower areas in the south and south-west the average gradually decreases. At Dunoon it is a little over 60 inches. The months of July and August, which represent the height of the holiday season, are usually broken and unsatisfactory. The spring months, April and May, are generally fine and dry, the prevalent wind then being from the east.

Woods and scattered clumps of trees are less scarce than in many parts of the Highlands and Islands, but in most districts not much is left of the old natural forest growth, excepting along the sides of the rockier burns and in crag faces, where the birch, the alder, and stunted oaks, safe from the nibbling sheep, still add a charm to the landscape. In Caol Ghleann there are unenclosed scattered clumps of Scottish fir which may be self-sown descendants of an earlier forest.

Owing to the mildness of the winter climate many trees lately introduced from warm countries thrive well. The arbutus ripens its scarlet berries in several places, the myrtle is frequently seen in blossom in December and January, and fuchsias not only survive the winter in the open air, but are commonly seen growing as hedges of gardens and cottages (see *New Statistical Account of Argyllshire*, p. 575). A fine specimen of the blue gum tree (Eucalyptus) thrives in the garden of South Hall, Kyles of Bute.

The combined town of Dunoon and Kirn is much the largest in the district. All along the shore from Toward Point to Blairmore, there is an almost continuous fringe of villas connecting Toward, Inellan, Sandbank, Dunoon and Kirn, Kilmun, and Strone. These places, being so easily reached from Glasgow, form a favourite holiday resort for the people of this and the other large towns adjacent. The chief towns or villages outside the fringe mentioned are Carrick Castle, Lochgoilhead, Strachur, and Tighnabruaich. All these are on the coast, and owe most of their growth and prosperity to tourists and summer visitors : in the winter they are half deserted. There are no large fishing or crofter communities.

The trees and the neat appearance of most of the cottage dwellings add a feeling of comfort and home to the landscape which is wanting in many of the more outlying parts of the Highlands. The far-stretching arms of the sea give a charm to the country, and at the same time add immensely to the facilities of travelling. In summer there is a general rush of holiday makers to these erstwhile quiet shores. The sea lochs are then dotted over with fast and well-appointed steamers and the " white wings" of yachts.

There is a beautiful description of Argyllshire scenery which seems to date back to the year 1238, and may be much earlier. It occurs in the lament of Deirdre over Alban, and is thus described by Dr W. F. Skene in his Introduction to the *Book of the Dean of Lismore*, p. lxxxvii. :— " One of the poems in the tale of the Children of Uisneach contains such a tender recollection of and touching allusion to Highland scenery that it is hardly possible to suppose that it was not originally composed by a genuine son of Alban." There are nine quatrains in Dr Skene's translation, of which we quote 1st, 5th, 8th, and 9th.

> " Beloved land that Eastern land,
> Alba, with its wonders.
> O that I might not depart from it,
> But that I go with Naise.

> " Glenmasan ! O Glenmasan :
> High its herbs, fair its boughs.
> Solitary was the place of our repose
> On grassy Invermasan.

> " Glendaruadh ! O Glendaruadh !
> My love each man of its inheritance.
> Sweet the voice of the cuckoo on bending bough,
> On the hill above Glendaruadh.

> " Beloved is Draighen and its sounding shore ;
> Beloved the water o'er pure sand :
> O that I might not depart from the east,
> But that I go with my beloved."

The scenery is all in Argyllshire, and Glenmasan and Glendaruadh will be easily recognised as being in this district. The MS. of the original Gaelic is in the Advocates' Library, Edinburgh, and is believed to have been written in Glen Massan in 1238. Alba or Scotland is Eastern land, in relation to Ireland.

In descriptions of localities the names referred to are those which occur in the 1-inch Maps. The spelling of the names on the Maps has been strictly adhered to, whether this is supposed to be correct or not, in order to help in their identification. With the same view the numbers, and sometimes the quarters, of the 6-inch Maps in which the localities occur, have been given, excepting where the localities are likely to be readily found. On the lower margin of each 1-inch Map there is a small index Map of the 6-inch Maps therein, and by referring to it any one may ascertain in what part of the 1-inch Map the places that are mentioned occur.

The numbers attached to the rock slides referred to are their numbers in the collection of the Geological Survey of Scotland, kept at Jermyn Street, London.

Tabular View of the Rocks.

The following is a list of the different superficial deposits and rocks noticed in the area :—

RECENT AND POST-TERTIARY—

> Blown sand.
> Peat, in well-defined basin-shaped hollows or on the hills.
> Freshwater alluvia.
> Marine alluvia.

GLACIAL—

> Glacial shell beds.
> Moraines.
> Sands and gravels.
> Boulder clay.

UPPER OLD RED SANDSTONE—

> Cornstones and magnesian limestones.
> Red sandstones, breccias, and conglomerates.

METAMORPHIC ROCKS*—

> Schists probably of sedimentary origin.
>> Phyllites, including the two series of Dunoon and Ardrishaig, which consist of phyllites and thin limestones, mixed in the first series with schistose grits and in the second with quartzite schists.
>> Schistose grits and greywackes.
>> Quartzite schist or Quartz schist.*
>> Albite schist.
>> Garnetiferous mica schist.
>> Graphite schist.
>> Schistose limestones, on various horizons, including the Loch Tay limestone.
>> Mica schist.   Areas coloured thus in the Maps may also include unseparated albite schists, sheared grits, and grey wackes, and phyllites.
>> Green beds — chlorite-epidote schists. The group lines may include some mica schists and schistose greywackes.
> Igneous rocks—
>> Epidiorites, hornblende and chlorite schists.
>> Serpentine.

---

* The order in this list is not a stratigraphical one.

† The term quartzite schist has been already applied to these rocks in the published 1-inch Map 29, and has, therefore, been also generally used in the subsequent descriptions in this Memoir.   The rocks in question are thoroughly granulitic in structure.   There seems no doubt about their sedimentary origin, but perhaps in strictness it would be better to call them quartz schist or quartz granulite rather than quartzite schist.

Unfoliated Igneous Rocks—
  Intrusive—
      Basalt and tachylite, including dolerite and augite andesite.
      Trachyte.
      Andesite (? intrusive).
      Lamprophyre group, including camptonite and mica traps.
      Hornblende-porphyrite and felsite.
      Hyperite.
      Granite.

The glacial deposits were separated from the other rocks, but are not shown in the 1-inch Maps, except occasionally by writing.

Excepting for about 1 square mile, near Inellan and Toward, the whole district is composed of schists or rocks intrusive in them. The intrusive rocks generally occur in dykes and thin sheets, and are in places closely crowded together.   W.G., C.T.C.

Summary of Literature.

Very little has hitherto been written about the geology of the district. A list of the works referring to it is added in an Appendix, but we also give below a short statement of the more important conclusions arrived at from time to time by earlier workers.

A hundred years ago the schists of this area were always spoken of by geologists as primary or primitive strata.

Macculloch, in his *Description of the Western Isles* (1819), does not directly describe any part of this district, but he gives in vol. iii. a general Geological Map of the Western Isles which indicates the greater part of the area of Cowal as made up of micaceous schist, with a narrow band of his chlorite series along the upper part of Loch Fine, and a pretty broad band of argillaceous schist near the Clyde. His Geological Map of Scotland (1840), as far as this district is concerned, is a repetition of that of 1819, except that the micaceous schist is now called mica slate, including chlorite slate, and he gives a patch of limestone in the phyllites at Cove on the eastern side of Loch Long. In neither Map does he show any limestone, either at Dunoon, Kilfinan, Otter, or at Toward, though he gives a strip of red sandstone at the latter place. In the small Geological Map prefixed to *Ami Boué's Essai Geologique sur L'Ecosse* (1820), Macculloch's argillaceous band is shown as running from Bute to Stonehaven, but is coloured as rocks chloriteuses and quartzeuses.* The rest is all mica schist, except a small patch at the head of Loch Fine, which seems intended for the Garabal Hill and Meall Breac granite.

The Rev. M. Mackay, in his account of the united parishes of Dunoon and Kilmun, *New Statistical Account of Argyllshire*, p. 581, notices serpentine 4 miles west (south) of Dunoon and a limestone on the lands of Castle Toward near the western extremity of the parish which judges had pronounced to be good marble. This must be one of the limestones in the Dunoon phyllites. He notices other limestones in the sandstones of Toward, p. 580, and the alteration produced by the Dunoon Castle dyke on the surrounding beds, p. 581.

Dr James Bryce, in his *Geology of Clydesdale*, 1859, first called attention to a band of intrusive greenstones, 60 to 100 yards wide, in the country north of Dunoon, ranging from near Huuter's Quay to Dunloskin Hill, as being different from the ordinary intrusive dykes of the coast, its structure being slaty and with hornblende in excess ; and in a subsequent

* See, however, his description, pp. 72-78, where he includes schiste argileux under this head.

edition in 1865, he gives a small sketch Map showing the position of this greenstone and of the Castle dyke. He also describes the limestone at Cove previously noted by Macculloch.

Mr T. F. Jamieson and Professor James Nicol are the two who have hitherto principally contributed to our knowledge of the structure of the district. The former, in his paper in 1861 ("On the Structure of the South-West Highlands of Scotland," *Quart. Journ. Geol. Soc.*, vol. xvii. p. 133), notices the Otter limestones, etc., and the epidiorites which accompany them are described as greenstones, and rightly regarded as intrusive. He describes the course of the anticlinal, which, however, he regards as a true anticline in the beds, and hence he identifies the Dunoon phyllites with those of Ardrishaig. He seems generally to take the foliation planes for true bedding. The greater metamorphism of the beds with production of felspar near the centre of the anticline was noticed by him.

Professor Nicol (*Quart. Journ. Geol. Soc.*, vol. xix. pp. 197–199) showed that the order of the beds on the south side of the anticlinal in Cowal was the reverse of that in the more central districts about Loch Earn and Dunkeld, and he was inclined to think the order in Cowal, of clay slate over mica schist, was the true one, and that the beds had been reversed in the central district with the production of overfolds by a force acting laterally and not vertically. The greater metamorphism of the mica slate in comparison to that of the clay slate he seems to ascribe to its being taken down further into the "interior regions of the earth's crust where the chief laboratories of metamorphic action are situated."

In the first sketch of a New Geological Map of Scotland by Sir R. Murchison and A. Geikie in 1862, an attempt was made to identify the Highland rocks with certain portions of the Southern Uplands, and this

area is coloured and lettered as

b'1 for the Dunoon phyllites, etc. (clay and mica slates).

b'1* for all the rest of the area—in a division which embraced quartzose flagstones, quartz rock, and associated limestones.

The New Geological Map of Scotland by Sir A. Geikie in 1892 contains much of the Survey work of this district in a reduced form.

In Sir A. Geikie's paper on "The History of Volcanic Action during the Tertiary Period in the British Isles" (*Trans. Roy. Soc. Edin.*, vol. xxxv. 21, 1890) there is a Sketch Map showing the numerous dykes in part of this area, with observations on them, from notes supplied by Mr C. T. Clough.　　　　　　　　　　　　　　　　　　　　　　　　　W.G.

CHAPTER II.

INTRODUCTION TO THE SCHISTS. DESCRIPTION OF VARIOUS SECTIONS.

The schistose rocks make up much the largest part of the district, occupying perhaps thirty times the area of all the other rocks combined. Our ignorance of them is, in many respects, still exceedingly great: yet it is worth while to set forth the knowledge at present possessed, in hope that this may serve as a stepping stone in the path of future progress.

No organic structures have been observed in any of the schists, and the geological age of these rocks is at present unknown: as regards most of them, even their age in relation to one another is not certainly known, so that, in making a traverse in any particular direction, we cannot say whether we are going from older rocks to newer, or *vice versa*.

A glance at the Maps shows that the different schist bands traverse the region in a north-east and south-west direction. Starting from the schist boundary fault near Toward and proceeding in a north-west direction we first pass over a group of phyllites and schistose pebbly grits with an average breadth of outcrop of 1 mile. Further north-west we come to the Bull Rock greywacke schist with a breadth varying from a little over ¼ mile to a little under ½ mile. Then comes the Dunoon phyllite series with subordinate schistose grits and limestones : the average breadth of this series is about 1½ miles, but there are some subordinate outcrops further north-west, which owe their existence to folding. Next we find a series of "green beds" (chlorite-epidote schists, generally pebbly), mixed with schistose grits and greywackes and phyllites. North-west of this there is a zone, not less than 7 miles broad, of schistose greywackes, phyllites or phyllitic mica schists, and albite schists. In this zone the rocks are much more highly metamorphosed, brown mica is abundant and a general change takes place in the direction of the dip of the foliation : in the schist bands, already mentioned in this paragraph, and in the south-east part of the zone the foliation dip is almost constantly south-east ; in the north-west part of the zone and in the schist groups to be subsequently mentioned in this paragraph the foliation dip is north-west. The axis of this great anticline of foliation traverses the whole of Cowal and strikes north-east and south-west through Loch Riddon. After this wide zone comes another set of "green beds," mixed with schistose greywackes and albite and phyllitic mica schists. Next north-west of this is the Loch Tay limestone, a bed which probably does not exceed 40 or 50 feet in original thickness, but is the most definite of all the sedimentary schists of the district. After this we get alternations of thin-banded mica schists, generally garnetiferous, with some limestones and "green beds." Then a set of graphitic schists, interbedded with dark graphitic limestones, quartzose schists, and phyllitic mica schists. Finally, in the most north-west part of the area, and composing the greater part of the shores of upper Loch Fine, comes the broad Ardrishaig phyllite series with subordinate quartzose schists and limestones. The characters and positions of these different schist groups are described in detail in subsequent chapters.

The different bands of limestone and of epidiorite or allied rock have, as far as possible, been mapped out, but we have only rarely been able to show the other rocks in the same detailed way, as it is found that they both graduate into one another and alternate with one another in the most minute way. The face of the Maps might, we fear, convey the impression that the phyllites and schistose greywackes, etc., are less mixed with one another than is the fact. There are two large zones separated as phyllite—those of the Dunoon and Ardrishaig phyllites respectively—but it must not be supposed that these consist wholly of phyllite. As a matter of fact, the Dunoon phyllites contain a very considerable admixture of quartzose pebbly grits, and the Ardrishaig phyllites of quartzite schist. Still there is on these horizons a distinctly greater proportion of phyllite than we usually find, and we have mapped them out as groups in order to trace the broad physical structure of the district. So, again, the groups of "green beds" may, particularly the wide one in the north-west half of the district, contain a considerable proportion of schistose greywacke or other mica schist like that outside them.

Numerous sea lochs intersect the area across the strike of the schists, and afford excellent sections of the different horizons. Some of these sections we will now describe in detail, with the aim, as we proceed, of explaining

the characters of the different rock groups met with and the minute
structures in them. On the south-east side of the great anticline of foliation which runs north-east and south-west through Loch Riddon (which,
for brevity, we shall in future speak of as "the anticline") the characters
and structures of the different rock groups are found to vary somewhat
in proportion to their distance from the centre of it. We shall therefore
describe sections which are representative of zones at different distances.

It will be shown from sections examined that there have been various
periods of folding and foliation production. The earliest foliation observable either agrees in direction with the bedding or with the axial planes
of the first folds which affect the bedding. In certain areas and beds
this earliest foliation has been folded again and crossed by "strain slip"
foliations. These strain slip foliations are by no means all of the same
age. One set agrees roughly in hade with that of the first foliation where
this foliation is not itself contorted by the movements which gave rise to
the strain slipping. Another later set is only found in strength near the
centre of the "anticline," and is supposed to have been developed in connection with the ridging up of it. This later set, and the folds which
gave rise to it, can frequently be seen in clear section to affect the strain
slips of earlier age. The later or anticline strain slips differ from the
earlier ones in hardly ever having quartz veins along them. The earlier
foliations, on the other hand, are often accompanied by quartz veins, and
in the region of the "anticline" the numerous folded quartz veins are a
ready indication of the folding of the earlier foliations also. Some phenomena of "stretching" or elongation of particles in the schists will also be
described in this chapter, but further particulars of this subject are re-
served for chapter VII.

1. *Section South of Dunoon.*

A Section representative of the most distant zone occurs on the west
coast of the Clyde, a little south of Dunoon, between Glen Morag and
Clyde View. The rocks here belong to the south-east part of the Dunoon
phyllite series and to the schistose greywacke, which we may call the
Bull Rock greywacke, on the south-east side of this series. The former
includes, besides the bands of phyllite of different colour—pale grey,
grey-green, buff, purple or black—many exposures of schistose grit and
schistose limestone.

The black phyllites are probably somewhat carbonaceous, but they do
not stain the fingers : in places they contain a considerable abundance of
small crystals and nodules of iron pyrites, and, more rarely, thin films of the
same mineral running between the foliation planes of the rock. Pyrites
crystals may also occur in the phyllite of other colour, but not so
commonly. The purple bands are seen in several places to change to
pale buff : the change may clearly take place across the strike, and seems
connected with the proximity of small crush lines. Near basalt dykes
the grey-green phyllites are often altered to dark grey or purple, and the
schists are evidently hardened, so as to stick up in rims by the dyke side.
The colours most nearly akin may occur in such thin stripes, *e.g.*, of $\frac{1}{2}$
inch or so, that the rock has generally a finely ribboned appearance,
which in clean tide-washed exposures is very beautiful. These slight
differences of colour are often associated with slight differences of texture
also, one band being rather less of a true phyllite than another, or even
containing occasional pebbles of quartz, etc. These differences of texture
indicate that the colour striping is itself also essentially an original
structure, but they do not stand alone as a proof of this. Repeatedly

throughout the whole section, secondary planes of schistosity are seen crossing these earlier stripes at marked angles. It is along these secondary planes that the rock now splits. There is often in this section more than one set of secondary planes, and the most prominent, which corresponds to the chief foliation planes of the district generally, may at times run with the bedding of the rock, but we can never go more than a few yards without seeing instances where this is not so: over and over again the bedding is disposed in folds, and the foliation crosses the folds parallel to the axes of fold * : and even when no distinct folding is visible the foliation planes may strike and dip rather differently to the bedding planes—the dip of the former in these cases is generally rather steeper, and the strike rather nearer east and west than that of the latter.

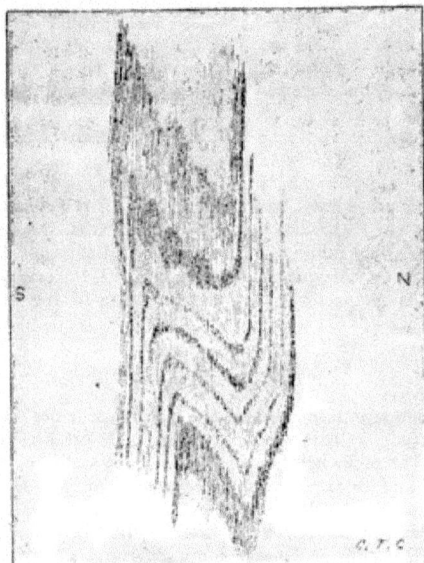

FIG. 1.—× 1. Shore S. of Dunoon. Fold in banded phyllite, as seen on a surface sloping gently E. The direction of shading indicates that of the foliation.

These splitting planes are distinctly lustrous, and can therefore be with accuracy called foliation planes, there having been a mineral rearrangement in the rock which has led to the growth along them of minute flakes of secondary sericitic mica, generally too small to be picked out by the unaided eye. They are also characterised by a set of fine streaks— " rodding " or " stretching " lines. One may not be always able to discern these in the phyllite itself, but they are always conspicuous in the thin quartz veins and pegmatites which occur so abundantly in it, roughly parallel to the foliation. They are also more generally prominent in the grits and greywackes associated with the phyllites. We shall

---

* There are cases, though, in which it is not quite parallel. In this and subsequent passages the word axis is used to denote the section of the axial plane at right angles to the axes, the axial plane being the plane which contains the axes or lines along which the sharp twists of successive bands take place. This double use of the word axis cannot be justified ; but in describing rock exposures and writing field notes the section above referred to has so often to be mentioned that it is advisable to get some brief term to express it, and the word "axis" has thus slipped into use.

shortly (pp. 16, 17) have to mention details of structure in these veins, grits, etc., which are intimately connected with such lines.

The direction of these stretching lines is not always the same : even on one plane they may be occasionally seen to be slightly waved, now in one direction and now in another, and in planes only a few inches separated from one another the general direction may be different. They are but rarely quite at right angles to the strike of the foliation : perhaps most generally they run some 10° or 11° nearer east and west than the foliation dip.

It has already been said that sometimes we may see more than one set of foliation planes in the section. Some subordinate planes especially prominent strike at a considerable angle, often about 80°, to the chief foliation, and bade nearly vertically. The general strike is perhaps 30° to 45° west of north. They are of variable prominence in different localities, and make little or no show in the more massive beds associated with the phyllites. Their surfaces are distinctly lustrous and coated with a minute film of sericite. In one of the best exposures such planes occur at intervals of

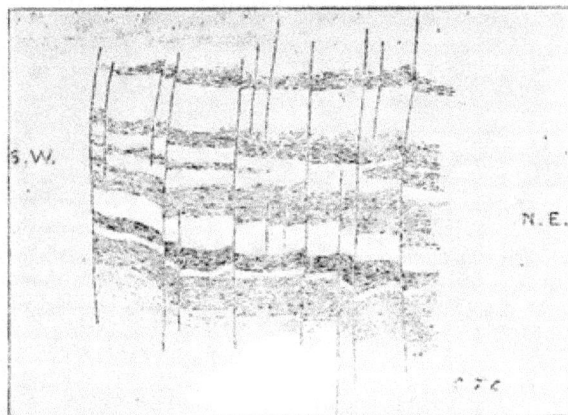

Fig. 2.—× 1. Shore S. of Dunoon. Banded phyllite scar dipping from observer Many, nearly vertical, late foliation planes throw the earlier foliation and bedding.

every $\frac{1}{16}$ inch or so, and in between these there may be minute lines of contortion with axes running parallel.

In age this subordinate foliation is clearly later than the main one, for one often sees that it is accompanied with slight throws which affect the main one and the thin quartz veins which run with it. As far as examination goes it would seem that the throws and contortions accompanying them have a constant tendency to shift the north-east side of any band slightly to the south-east.

The planes of the earlier foliation are also frequently accompanied by distinct throws. Sometimes the amount of this can be estimated by the help of the bands cut, and we find that shifts of several inches are not uncommon : one was observed of not less than 15 inches, and there are others which may exceed even this. These little faults occur along the thinnest lines running parallel to the other foliation planes, and seem of quite the same character as these planes, except that the amount of throw is large enough to be readily observed. In some places they occur so abundantly that the bands representing original bedding are cut up into thin lenticular strips which cannot be satisfactorily correlated with the others near, and so help, together with the folding, to confuse the appearance of the section. The throws are not always in the same direction, and there may be both direct and reversed faults close together crossing the same limb of a fold ; it seems impossible to say which of the two kinds is generally the more common.

FIG. 3.— × ⅓.   Shore ⅜-mile S. of Glen Morag, Dunoon.   Sketches from rock slopes inclining gently E.   The dark bands are purple phyllite ; the paler bands are pale grey rather harder phyllite.   The shading shows the foliation direction ; the darker lines are accompanied with throws.

These foliation faults must not be confused with some later crushes, which also strike and hade almost with the main foliation. These crushes make little nicks or gulleys along the shore line, which may be filled with thin streaks of crushed phyllite, and have clearly been instrumental in destroying the phyllite character, not accompaniments of its production.

In the phyllites the tendency to split along the foliation direction is of the closest possible character, but in the more siliceous and pebbly beds it is not so close, the splitting planes having, for one thing, to go round the pebbles, unless where they occasionally succeed in breaking through one. Hence, in pebbly beds, the distance between the planes may be as much as ¼ inch or even more. Where phyllites and pebbly grits are folded together in clear section, it is seen at once that the number of foliation planes in the former is very much greater than in the same breadth of the latter : it is also observed that the angle of dip of the foliation in the latter is not usually quite the same, being often rather less steep.

Almost universally in this section where the bedding is folded the prominent foliation still keeps straight, with an average strike of about 30° north of east, and a dip to south-east of about 53° ; hence the rocks split along straight lines, and the phyllites admit of being used as slates. But some cases occur where this foliation has itself been folded ; it may be quite gently, it may be so sharply that "strainslip" or "ausweichung" foliation has been developed along the axes of the folds. The direction of hade of these axes is not always the same : in many cases both limbs of the fold hade south-east, but close to this may be others with axes hading slightly north-west : on the whole, however, it seems as if the sharpest and most prominent of these later folds have axes hading south-east. These strain slips always seem to have some

throw, and to have been developed from a series of corresponding limbs
of folded bands overlying one another, along which actual breaks have
occurred.   Neither here nor in other parts of the schist area have quartz
veins been commonly developed along strain slips of this age, though
such are so common along the earlier foliation.   Hence, in this section,
one sees but rarely folded veins, whereas in " the anticline " country they
are excessively common ; and lying on the shore here are some large
boulders of " anticline " schists, which testify at once to the difference
between themselves and the rocks they rest on, both in this and other
respects.

The first foliation is exceedingly uniform in strike and hade, with the
exceptions due to the subsequent folding just referred to.   It is, as
already stated, parallel, or all but parallel, to the axes of the early folds
in the section, and we suppose it may be regarded as having been produced at the same time with, and as a direct consequence of, these folds.

It is a general law here, and throughout the south-east side of "the
anticline," that in these early folds the longer limbs of anticlines are the
south-east limbs.   The effect of this structure is to make the general dip of
the beds less steep than that of foliation.   It is noticeable, too, that the
original bands of bedding vary in thickness in different parts of the folds
according to a general law.   Looking, indeed, at some of the small folds

Fig. 4.—Diagram Vertical Section. To illustrate the laws of folding and thinning
of limbs in the Dunoon horizon.

that we may chance to see, it may be sometimes the south-east limbs
that are the most thinned, and sometimes the north-west, or there may be
no difference.   But if we examine the parts adjacent to such folds we shall
find that the area in which there is no evident law of thinning forms but
a small part, in which the limbs are all comparatively short, coming in
between longer and generally much more thinned limbs, each hading
south-east.   If we regard the series of short limbs as representing together one limb, with a hade in the contrary direction to that of the
longer limbs, we may say that the most thinned limbs are the south-east
limbs of anticlines.   The most thinned limbs are the parts which, if
further pressure had been applied, would ultimately have thinned away
to nothing in one part, and have given rise to "thrust" lines analogous
to those in the Durness and Torridon rocks of the north-west of Scotland ;
and, from analogy, we should expect them to be on the underlying sides
of isoclinal anticlines, and to have been produced by pressure originating in a locality lying in a direction pointed to by the dip of the axes
of fold.   How, then, has it come about that in this Dunoon section it is
the uppermost limbs of anticlines that are the most thinned, and is it
likely that the pressure which produced the early foliation was produced
by earth movements in a locality lying south-east from the present Highland frontier ?   Both questions seem to receive an answer if we bear in
mind the fact that all these beds on the south-east side of "the anticline"

have, by the agency of this plication, been given an inclination which was not originally theirs. The upper limbs of anticlines at Dunoon must once have been under limbs, and the dip direction of their axes of fold must once have been north-west and not south-east.

Three specimens of the phyllites have been examined by Mr Teall (slides 3825, 3826, 3827). Of these the first, of pale grey colour, was found to consist essentially of quartz and sericite, with certain portions crowded with rutile needles. 3826, a dark leaden colour, was also very rich in these needles. 3827, greenish-grey in colour, is composed of quartz, sericite and chlorite, and contains rutile needles in much less abundance.

The limestone outcrops have not in general a greater apparent thickness than 5 or 6 feet. It is rash at present to speculate on the extent to which different bands may have their outcrops thickened by folding, or repeated in other parts of the section. In our 6-inch Field Map no less than 17 different exposures are shown in a length, diagonal to the strike, of $\frac{2}{3}$ mile. They generally weather a rusty colour, but the fresh fracture is dark grey, and the grain fine. The foliation planes are coated with a micaceous lustre, but these planes are not so close or so regular as in the phyllites, and not uncommonly two can be seen to run into one another at a gentle angle.

The limestones seem to have suffered from folding more than any of the other rocks in the section, the first foliation in them being more frequently waved or twisted. A striking instance of this occurs in a broad exposure a little over $\frac{1}{2}$ mile north of the Bull Rock: here not only is the most prominent foliation folded and sharply twisted, and crossed by strain slip foliation, but it seems as if this foliation was itself preceded by another, by the contortion of which it was developed as a strain slip foliation: the earlier thinner foliation is indicated by an alternation of pale grey purer limestone streaks with other darker more micaceous bands. The latest folding in this section may be of "anticline" age, like much of that described in subsequent pages, but it cannot be stated for certain that this is so, Dunoon being so far from the region where the anticline folding occurs in force. At all events this latest* Dunoon folding agrees with the anticline folding in being later than the early foliation and strain slips of the district. In another exposure the prominent foliation planes, clearly corresponding to that in the phyllites, occur at intervals of every $\frac{1}{2}$ inch or so, and, besides the mica in these planes, there is a distinctly lustrous surface along a folded bedding plane which is crossed by the prominent foliation. This suggests that there was an earlier foliation here again, which agreed with the bedding, before the development of the prominent foliation.† But perhaps during the production of foliation the formation of new mica may take place along the old bedding planes, as well as the new foliation planes, in cases of hard beds in which new splitting planes are not readily formed. This might be supposed to be more likely to happen if the angle between bedding and foliation is only a small one. In the phyllites, however, we have not noticed any limit to the angle at which bedding and foliation may lie without actually coinciding: it can certainly be as little as 10°. The prominent foliation in the exposure last mentioned does not look like the later "anticline" strain slip foliation: it is quite parallel to the prominent phyllite foliation, and has quartz veins parallel to it: that it

---

* Latest with the probable exception of the minute contortions accompanying the foliation which strikes 30° to 45° west of north (see p. 11).

† It is not impossible that this may have been the case in the Dunoon phyllites generally. The earlier structures in such micaceous beds as these seem to be comparatively readily obliterated (see p. 21).

is not so close as in the phyllites is only in accordance with the law that prevails in all hard beds, and should we see a phyllite parting in a folded limestone the parting is sure to be crossed by the closest foliation planes, parallel to the rarer ones in the limestone.

In the part of the section included in the phyllite group the pebbly beds are generally distinguished by a considerable abundance of opaque white quartz pebbles, which may attain the size of a sparrow's egg. In the thick band south of this group the more prominent pebbles are of felspar, generally pink, and it is hence called a greywacke—the "Bull Rock" greywacke—rather than a grit. In the same band there are also pebbles of quartz, white and granulitic, or else free from granulitisation, and with an opalescent appearance. In those of the last character a pale shade of blue may often be observed. There are also patches up to over an inch in breadth, of purplish-red hard clayslate, and thin streaks of the same. These streaks are often so long in proportion to their breadth, perhaps 1 foot by less than 1 inch, that we consider them to be essentially cotemporaneous deposits, or derived from such.

The opalescent quartz is like that of many of the quartz streaks in the Lewisian Gneiss, and of pebbles in the Torridonian rocks derived from this. But it must be stated that similar quartz also occurs not uncommonly in the epidiorites of St Catherine's and of the central Highlands. The felspars of the pebbles have not been examined in detail, either in this or the other pebbly schists of the Map. In the grits which occur some distance further north-east, outside of Cowal, along the foliation strike, it has been determined by Mr Teall that they are generally a soda felspar, oligoclase. The felspar of the felspar pebbles in the chocolate and red Torridon sandstones of north-west Scotland is a potash felspar, but in the grey Lower Torridonian rocks of Skye oligoclase is about as abundant as the potash felspar.

We do not think any pebble was noticed in this greywacke which could not be referred to quartz or felspar or clayslate. Pebbles of composite rocks are indeed exceedingly rare throughout the schists of this area. Sometimes, however, compounds of quartz and felspar, much like those of many pegmatites, have been noticed : and some pulled-out streaks of chloritic rock occur in a schist $\frac{1}{3}$ mile east of Inellan Hill (194 south-east), for the description of which see p. 30. There are also some pieces of green-grey hornstone-like rock in the band mapped south of Gairletter (174).

On the shore there is a continuous section of the greywacke for a length, not quite at right angles to the strike, of over $\frac{1}{3}$ mile, and a little of the north-west end of it is probably covered under drift. Throughout this, bands of phyllite are rare, and never exceed 4 or 5 feet in breadth, so that it hardly seems likely that they are folded up inliers of the mapped phyllite series. There are frequent alternations in the size of the pebbles in different bands, and the original bedding is not generally hard to make out. There are not so many prominent folds as in the phyllites, and perhaps the greywacke, being so massive, may to some extent have escaped the folding to which the softer beds yielded, as is often seen to be the case in the later foldings of "anticline" age.

The colour of the greywacke is often somewhat greenish. This is probably due to the chlorite which the microscope shows to occur in it in association with the sericite. Mr Teall describes the microscopic character of a specimen as follows (slide 3823): "Micaceous (silvery) lustre on cleavage planes. Relics of clastic grains of quartz and felspar very conspicuous. These lie in a fine-grained granulitic matrix containing wavy strings of chlorite and sericite. Iron ores and small opaque white grains (? leucoxene) sparingly represented. Structure—microflaser. It seems probable that this is the type of rock termed sericite gneiss on the continent." On either side of the quartz or felspar pebbles in the direction of " stretching," there are often special accumulations, or tails, of granulitic quartz, and it is suggested that these have been formed by the breaking down and recrystallising of the original edges. The felspar pebbles often show cleavage faces across their whole breadth, and on the average have perhaps suffered less from granulitisation than those of quartz. In some bands of greywacke schist, e.g., in the thin band mapped 1 mile E.S.E. of Lephinkill (172 south), the felspar substance (slide 3828) has a dusty aspect near the centre of the pebbles, but it is clear at the outside : this clear part is probably a product of recrystallisation, but it is in optical continuity with the centre. The sericite flakes are perhaps larger and more conspicuous than in the phyllites, but there are fewer of them, there being fewer foliation planes.

Fig. 5.—×2. Shore S. of Dunoon. Surface of pebble exposed on a foliation plane in Bull Rock greywacke. The felspar substance is shaded ; the clear parts are granulitic quartz. to counterbalance this foliation planes.

The direction of "stretching" is shown in many places by the parallel elongation of the pebbles, but particularly well at a spot just below the quarry in the greywacke. The direction runs a few degrees nearer east and west than the direction of dip of the foliation. The pebbles are often stretched to a length of over $\frac{1}{2}$ inch, while the breadth does not amount to $\frac{1}{8}$ inch, and they are repeatedly cracked and the

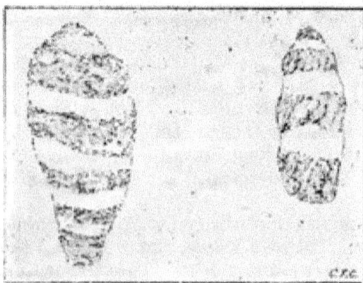

Fig. 6.—×1. Pebbles out of greywacke schist $\frac{1}{4}$-mile N.E. of the "k" of "Dalleik," 1 mile E.S.E. of Lephinkill (172). The felspar substance is crossed by lines of granulitic quartz in a direction at right angles to their length. In the right hand pebble the cleavage of the felspar is distinct.

Fig. 7.— ×2. From a boulder in Glen Lean. Polished section of lenticular pebble-form patch in greywacke schist, cut at right angles to foliation and in the direction of stretching. The shaded areas are felspar ; the rest granulitic quartz.

cracks filled with granulitic quartz. In one pebble, not over $\frac{1}{2}$ inch long, ten such cracks were counted. Similar cracks are also to be observed in the opalescent quartz, but these are not so conspicuous, probably from the want of sufficient colour contrast. The cracks do not usually run quite at right angles to the length of the pebbles or the dip of the foliation : we could not satisfy ourselves that there was a law guiding their directions : sometimes they incline more steeply than the plane at right angles to the foliation, and sometimes less so. Besides the cracks nearly at right angles to the lengths of the pebbles, there are sometimes other thinner ones roughly parallel to the lengths, and these last may slightly throw the former.

In the Bull Rock greywacke, and also in some grit bands near the north end of it, one sees in places that the prominent foliation is crossing another less continuous set of planes. There are certainly films of sericite lying along the less continuous planes as well as along the others. By the dyke side ½ mile S.S.E. of Glen Morag it seems clear that there must have been an earlier foliation preceding that which is usually the most prominent, the pebbles in a quartzose pebbly band being distinctly elongated along planes which are cut by the prominent foliation.

The stretching of these pebbles is supposed to have taken place at the time of the production of the rodding or streakiness seen on the foliation planes of the adjoining phyllites, and, probably, both were accompaniments of the production of foliation. The rock was not only rendered fissile and to some extent mineralogically changed along these foliation planes, but the particles on these planes were elongated or stretched in a definite direction along it. This elongation seems quite comparable to the well-known distortion of fossils on the cleavage planes of slate.* In no part of Cowal have we been able to distinguish on foliation surfaces an elongation due to the "anticline" movements apart from that due to the movements of earlier date. Sometimes, though, one sees in vertical sections that the pebble-like specks in pebbly schists are more elongated in the limbs of folds of "anticline" age than in their centres, so that some stretching or elongation must have taken place at the time of the formation of these folds. At Dunoon the schists are, with the possible exception of some small exposures (chiefly of limestone, see p. 14), unaffected by the anticline movements, and the stretching noticed there must consequently be essentially of pre-anticline date.

The veins and pegmatites in the section rarely exceed a few inches in breadth, and occur much more commonly in the phyllites, etc., than the more massive beds. The felspar in them is generally pink, and in small proportion to the quartz. Lumps and strings of green chlorite are common in them, or at their sides, and flakes of it are often enclosed in the outer parts of quartz bands, giving them a green colour. Thin strings and lenticles composed almost entirely of chlorite are also common. In general, the veins run roughly with the foliation, and show on their sides prominent "stretching" or "rodding" structure, the quartz and felspar running in rod forms in a definite direction, that of the stretching of the pebbles in the pebbly beds: in some cases the rod structure extends through the thickness of a vein several inches wide. Many of the veins have been cracked across, and the cracks filled with a new growth of quartz: such later growths are very prominent in the polished face of a specimen obtained from 50 yards south of the Bull Rock, and on examining a slide of it under the microscope we see that the new quartz is much clearer than the old, being much more free from minute dust-like enclosures: probably these enclosures are of secondary origin, and their greater abundance in the older quartz depends on the greater number of strains to which it has been subjected. A ferriferous carbonate is frequently associated with the veins, and may occur in masses several inches across: in some places this mineral is certainly of earlier date than some of the quartz, for strings of the latter may be seen breaking across

* See, for instance, Daniel Sharpe, "On Slaty Cleavage," *Quarterly Journal, Geol. Soc. London*, vol. 3, p. 74.

its cleavage planes. When veins of quartz and carbonate run across the foliation their edges frequently show casts of its planes, and also a close alternation internally of quartz and carbonate layers, which may approximately agree in direction with the foliation planes. Among the veins which run neither with the foliation nor at right angles to it there seems a tendency to an inclination rather more steep than that of the foliation, and to a strike less near east and west. The veins of this class are at times affected by the vertical foliation already described, and so must be very old. They are often accompanied by slight throws: the up-throw side is often the south-east, but not always. Some of the veins which run at right angles to the prominent foliation are full of minute twists, the axes of which are roughly parallel to it. It might be supposed at first that these veins are of earlier date than the foliation, and have been contorted at the time of its production. On this supposition, the total length of the vein, calculated through all the twists, would indicate the original thickness of the rock traversed by it, and thus afford a means of finding the amount of compression subsequently undergone. But it is doubtful whether the supposition is a safe one: perhaps the pre-existence of foliation may itself be a cause for a later formed vein to alter its course in directions which have reference to it.

A thin slice (7314) of the pegmatite about 50 yards south of the Bull Rock has been examined microscopically by Mr Teall. The specimen sliced seemed quite the same in character as the other thin pegmatites which occur on the shore both in the greywackes and phyllites. Mr Teall states that the slice is composed of quartz, albite, and a little carbonate, and shows signs of crushing. He continues:—"This is no doubt a segregation vein, and is therefore different from the quartz-microcline pegmatites which are often associated with granites. Similar quartz-albite veins occur in the schists near Start Point in Devonshire and in an area of dynamic metamorphism in the Hartz as described by Lossen." The albite of these Dunoon pegmatites differs from the albite of the albite schists (pp. 39, 40), and of the pegmatites in these schists, in being pink macroscopically and in showing distinct rodding. The albite in the Start Point pegmatites is also pink. In another pegmatite (slide 4198) obtained from a schistose greywacke ½ mile north-east of Mark (153) the felspar has also been determined as albite by Mr Teall: the colour is quite the same as that of the albite in the Dunoon pegmatites, but the specimen does not show whether it is rodded or not. We are also informed by our colleague Mr Barrow that he has sent from twenty to thirty specimens of pegmatite from the greywacke schist of the Stonehaven district to Mr W. M. Hutchings, who has determined the felspar to be albite in all cases.

2. Section on the West Side of Loch Long, South of Ardentinny.

The next Section we will describe lies on a horizon 3 or 4 miles nearer "the anticline" than that south of Dunoon. This occurs on the west side of Loch Long, from ½ mile north of Gairletter Point to about ½ mile south of Ardentinny. Here there is a much less proportion of phyllite, or of soft phyllitic mica-schist, than there is at Dunoon, and nothing we can call limestone. A great portion consists of schist in which granulitic quartz and sericitic mica occur in thin distinct folia, but this schist is mixed with more massive greywackes and hard quartzose bands, which are much less perfectly foliated, and there are also many calcareous, ochreous weathering, bands and lenticles, rarely exceeding a few inches in breadth and a few yards in length, which show no clear foliation at all. These calcareous bands are great helps in making out the original bedding.

As compared with the Dunoon section the number of foliation planes in a similar breadth of exposure is less, these planes being generally further apart than at Dunoon. This is not due to a diminution in the strength of the force that produced foliation, but to original differences in the rocks acted on,—to the fact that the rocks here are generally harder, more quartzose and massive than at Dunoon. For in places we can see hard quartzose bands folded with more micaceous schists or phyllites, the

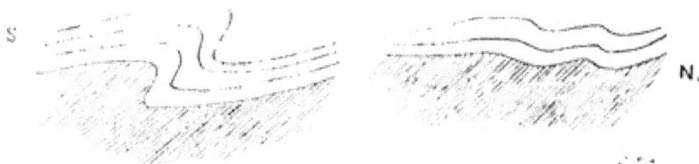

Fig. 8.—Rather more than ½-mile N.W. of Gairletter Point. Hard quartzose schist, with hardly any cross foliation, overlying finely foliated green-grey phyllitic schist. The shading in the latter shows the direction of foliation.

latter being foliated in the finest possible way, while the former are only affected by strain slip foliation at intervals of ½ inch, or even occasionally of 1 inch : these strain slips run parallel to the fine foliation of the phyllites and to the axes of fold in the exposure. One very massive quartzose band, for a distance of more than 3 feet across the axis of fold, shows no strain slips, except 3 or 4 short discontinuous lines. Yet just below it, in more micaceous schists, the very finest foliation is seen. Such interstratification of readily and less readily foliated beds may give rise to appearances closely simulating false-bedding. Not uncommonly the foliation is so poorly developed that the rock splits more readily along the original bedding, now repeatedly folded, than along the foliation, and for this reason the possibilities of acquiring roofing slates are less than at Dunoon.

Fig. 9.—× 1. Shore N. of Gairletter Point, Loch Long. Shaded parts are the more micaceous. Direction of shading shows the lie of the mica plates. The micas which lie on the strain-slip foliation plane are folded, though folding cannot be discerned in the more quartzose parts.

The chief system of fold is the same as at Dunoon, and the same law regulates the length and the thinning of the different limbs of anticlines. The chief foliation planes are parallel to the axes of fold, and correspond roughly in strike and dip with those at Dunoon. They are on the whole, too, about as steady and unfolded as these, the apparent less regularity of splitting planes being due, not to an extra folding of the foliation, but to the fact that some of the rocks split by preference along their bedding planes, which may be folded to any extent. But as we traverse the section from south to north the waving of the prominent foliation becomes gradually more pronounced : in most of it this waving can only be noticed in the more micaceous schists, as a succession of minute folds, probably averaging hardly ½ inch in breadth, or of strain slips parallel to the axes of these folds. But near the north end such folding is common even in massive quartzose schists, and is also of one predominant type, with axes hading north-west : still, how-

ever, the folding has not, in quartzose rocks, been strong enough to
give rise to strain slips. Pretty hand specimens can be acquired which
show the late folding and strain slips in micaceous layers which are
themselves of secondary origin and lie along the lines of earlier strain slip.
The folia of mica in the earlier strain slips are arranged in the same
general direction as the strain slips, and so, too, are often those of the
micaceous layers cut through by the slips. At times there are indications
of an arrangement in the same direction even in the quartzose layers : this
is shown by the elongation of small ferriferous carbonate specks within
them. But if examined more minutely it is seen that, independently of
the later folding, the mica in the slips is not always quite parallel to the
sides of the slips, but lies at a small angle, approximately that at which
the earlier banding approaches the slips. For this reason as one moves
one's finger along the surface of a slip one feels a slight roughness when

C.T.C

Fig. 10. — × 1.   Coast N. of Gairletter Point.   Burn ½-mile E. of Gortnamhuinn
(173).   "Ausweichung" or "strain-slip" foliation crossing earlier planes.   The
shaded parts are the more micaceous : the direction of shading indicates the
general direction of the mica flakes.   The black spots represent hollows left
by the weathering out of a ferriferous carbonate.

the direction of movement is contrary to the direction of inclination of
these earlier bands.

The micaceous layers developed along the early strain slips sometimes
attain a breadth of ⅛ inch or more, and hence secondary alternations of
mineral character of such a marked degree are acquired that they may
rival or exceed in prominence those along the bedding, or along an earlier
foliation. The last words are added, because it is clear in some part of
the section that, besides the foliation which corresponds with the prominent
one at Dunoon, there is also a mineral rearrangement along the bedding
or other planes cut by it. For instance, about ⅛ mile north of Gairletter
Point, a hard quartzose schist is seen folded over phyllitic schist : the
phyllitic schist has a foliation like that at Dunoon, but the quartzose band
is hardly affected by this foliation, and breaks more readily along the bedding, and the curved surfaces thus obtained are covered with secondary mica.
Again, at the north end of the section, there is a schist with many pebbles,
both of quartz and of white and pink felspar. The bedding is shown by
the line between this schist and a much more micaceous dark grey schist, and
crossing this bedding and the whole exposure, a set of foliation planes,

runs closely together in the more micaceous and less so in the pebbly schist. But in addition there is in the pebbly schist another set of planes, which are cut by the more prominent ones. These last mentioned planes agree in direction approximately with the bedding : but they do not represent bedding alone, for the pebbles are distinctly elongated along them. There are other sections indicating the existence of a foliation, which did not agree with bedding, prior to the production of a prominent set of strain slips hading south-east and striking the same as the prominent foliation of the district. In some of these sections the angle at which the bedding is cut by the earlier foliation is much the same at all points of the fold, and it is therefore probable that it was lying evenly at the time the first foliation was set up, and has been subsequently crumpled together with it. But in other cases the angle between the bedding and the earlier foliation is not constant, and we may conclude that the bedding was folded before, or at the same time with, the production of the foliation, as is so constantly the case at Dunoon.

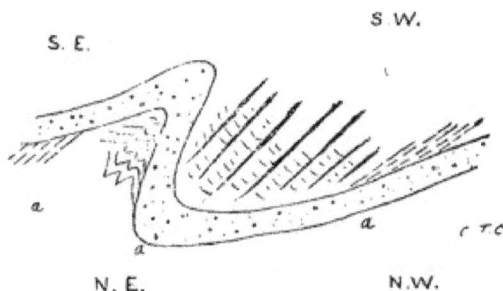

FIG. 11.—Not quite ½-mile N.N.W. of Gairletter Point. Plan of a rock surface sloping E. The spotted band is an ochreous weathering bed which shows hardly any foliation. Rest of rock is a well-foliated quartzose schist : the fine lines in it represent first foliation : the darker lines crossing these are strain-slip foliation. At points over the letter "a" the angle between the first foliation and the bedding is approximately the same in each case.

It seems clear that in alternations of phyllitic and quartzose schists, the old structures of the former may be so obscured by renewed movement and strain slipping as to be no longer recognisable, though the corresponding structures in the quartzose schists may remain quite distinct. In some of the sections mentioned in the preceding paragraph, we suppose there was at one time a distinct foliation in the more micaceous schist running parallel to the bedding or to the foliation still seen in the quartzose beds, but that by a renewal of folding the particles of the former schist have got their directions rearranged parallel to the axes of the new folds. In the crags ¾ mile N.N.W. of Glenmassan (163), in alternations of more and less micaceous schists, one sees that in the former the strain slips may sometimes be exceedingly close—within $\frac{1}{20}$ or even $\frac{1}{10}$ inch of one another—so close that had it not been for the more quartzose bands of the section we should certainly have carried away the idea that the strain-slip foliation was the only foliation present, though, as a matter of fact, there was an earlier one during the contortion of which the strain-slip foliation was produced. The folds in connection with which the above strain-slip foliation was produced are not very acute, and it is probable that by a little more intensity of movement the earlier structures would be still more obscured, until in certain bands they could not be recognised. This last result seems to have been arrived at in parts of the Ardentinny section.

In places where bedding and foliation do not coincide, the pebbles in pebbly beds are elongated along the foliation planes, their greater lengths sticking up at times at right angles to the original planes on which they were deposited.

Quartz veins are common both along the main foliation and the bedding : they seem more common along the latter than they are at Dunoon.

If we compare the amount of mineral change in this section with that at Dunoon we are struck by the occurrence here of brown mica occasionally along the foliation planes, besides sericite ; whereas at Dunoon brown mica does not occur. The sericite also commonly occurs in larger and more distinct flakes.* Brown mica has not been noticed at the south end of the section : as we proceed from south to north it is first seen in a quartzose band a little over ½ mile from the south end. In this band it is not conspicuous : it occurs only in small flakes, and on many foliation surfaces of some extent it could not be found at all, in spite of close search. North from this, brown mica becomes gradually more prominent, and in the massive greywacke lying east of the word "Driseig" is fairly abundant. Even yet the amount of change is not so great as in the schists lying a mile or two further north-west, for, among other things, the number and size of the brown and black mica flakes in these last are greater, and the quartz veins along the bedding or earlier foliation are more prominent : this can be seen from the boulders of these rocks which lie scattered about on the shore. In this section, then, we seem to have an intermediate stage between the types of greatest and least alteration of the district. The amount of folding, as has been said, is not appreciably different from that at Dunoon ; we have not yet advanced far enough north-west to be within range of the more prominent movements connected with the development of "the anticline." Therefore it does not seem likely that folding or physical movement, to be seen in the section itself, can be the cause of the greater metamorphism.

Still it is clear that folding does effect a metamorphism, or, at all events, that greater metamorphism is shown along lines which have been specially subject to mechanical disturbance. There is hardly a space of a yard length that does not show the metamorphism to vary locally and sharply, according to the lie of the rocks and the amount of physical strain they have undergone. Again and again we see broad bands, whether of bedding or early foliation need not here concern us, which are lying nearly flat in the centres of small folds, and in this position they are, on the average, two or three times as thick as they are in the limbs. On the other hand, the more micaceous layers are, at least, as thick in the limbs as in the centres, so that the bands are, as a group, of much darker appearance, and more micaceous than in the centres. This difference must imply a molecular change throughout. The new mica which occurs along the strain slips is also an example of the appearance of mechanical movement and metamorphism together. We may look upon a strain slip as a limb thinned to the utmost extent, and the presence of new mica along it as due to an extension of the law that in thinned limbs the quartzose parts of bands diminish in quantity, while the micaceous remain the same or increase.

* We had the great advantage of Mr Teall's company in going over the section.

3. *Section along East Shore of the Kyles of Bute.*

The next Section we will describe extends along the east shore of the Kyles of Bute, from ½ mile north-east of Colintraive to Port an Eilein. The rocks here consist of an alternating series of schistose greywackes, and more thinly foliated very micaceous or chloritic schists, in many of which small specks of albite are abundant. Near the north-west end of the section the greywackes form a larger proportion of the rock than usual, and have been mapped out as a zone.

The south-east end of the section lies about 1½ mile nearer "the anticline" than the north end of the section last described does, and there is a difference in the amount of mineral change in the rocks of the two localities : the brown and black mica in the Colintraive greywackes forming larger flakes, and a generally more uniform layer along foliation surfaces. But the mechanical movements to which the rocks have been subjected are still essentially the same, the evidence for the later movements connected with "the anticline" not becoming pronounced until we advance north some little way in the section. The first distinct albite crystals that we noticed in traversing the section from the south-east occur about 100 yards south-east of the foot of Allt Glac na maill, in bands of very micaceous schist which alternate with other quartzose bands, which seem free from albite. The albites here are so small that they could easily be overlooked, and it is probable that similar crystals in bands further south-east have escaped notice. North-west from this locality almost any exposure of the more micaceous or chloritic layers will, on careful search, disclose albites, becoming, as a rule, larger and more prominent the further north-west we go. Further details about these albite schists will be given on p. 39. They are possibly a further indication of the greater mineral metamorphism of the rocks.

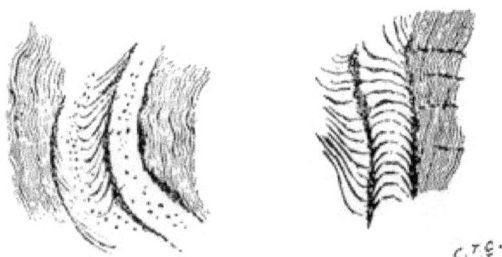

FIG. 12.— × 1. Fearna Bagh. Sketches from specimens of quartzose mica schist.

Folded "pre-anticline" strain slips in middle of sketch. The black spots represent hollows left by the weathering out of a ferriferous carbonate. Nothing to indicate bedding.

"Pre-anticline" strain slips, slightly waved, in left part of sketch. The earlier foliation approaches these at a distinct angle and is cut by them. In right hand of sketch weakly developed strain slips of "anticline" age are crossing the first foliation.

As we proceed north-west the evidence for the movements connected with "the anticline" becomes stronger, and is particularly well seen at the point north of the Burnt Islands, and that north-west of Fearna Bagh. Every few yards we see places where an early foliation has been crossed by early strain slips, of the type of the prominent ones in the Ardentinny section, but these again are crumpled most variously by later "anticline"

movements.   Plate IV., near the end of the volume, shows a natural rock
exposure on the north side of Fearna Bagh.   In this the strain slips are
seen distinctly folded by the later anticline folds.   Plate III., also from
the north side of Fearna Bagh, and Plate V., from the south side of Coylet
Inn, Loch Eck, show folds, which also are probably of anticline age, in
still greater intensity.

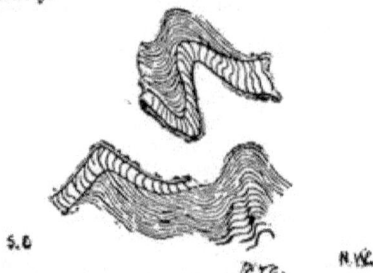

S.E.   N.W.

Fɪɢ. 13.— × 1. From a vertical section of pebbly schist near the top of Beinn Bheag
    (152).   The parts drawn are placed in their natural relation.   The lines all
    indicate foliations.   Folded strain-slip foliation planes, crossing the earlier
    planes, are seen in the top portion, and at left hand of bottom portion.   At the
    right hand of bottom portion there is an attempt to develop strain slips along
    axis of same fold which folds the early strain slips in top portion.

The strain slips of anticline or pre-anticline age may be only a few
yards or less in length, passing into lines of sharp contortion from which
they have been developed, and these lines of contortion may grow much
less conspicuous, or die out altogether, in the massive beds which have
been less readily crumpled.

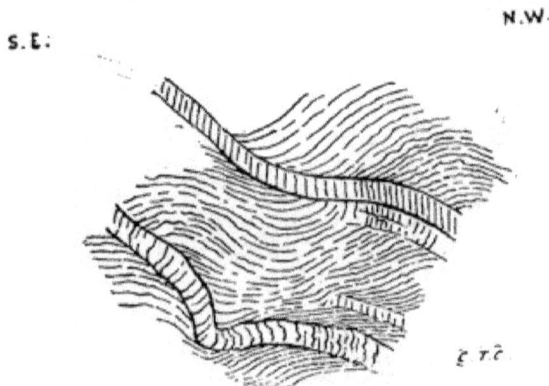

N.W.

S.E.

Fɪɢ. 14.— × 1/10. Vertical section in quartzose schist 250 yards S. of Coirantee (163).
    Strain bands hading N.W. have been subsequently folded along an axis inclin-
    ing steeply S.E.   All the lines indicate foliations. Nothing to show the
    bedding.

The narrow rock strips lying between two closely adjacent strain slips
we will for convenience call "strain bands."   The earlier foliation crossing
these bands is often almost obliterated by the development of numerous
small spots of ferriferous carbonate weathering into hollows, the greatest
lengths of which lie parallel to the strain slips.   It was originally, too, probably not so much developed as in the rock at the sides of the strain bands,

owing to the strain slips having usually occurred near the centres of folds, and so in areas somewhat protected from the drag and thinning experienced in the limbs. From these causes the earliest foliation layers may have less distinct differences between them than the strain bands and the rock at their sides. Hence, in a first traverse, one is apt to miss the earliest foliation altogether, and take strain bands as indication of bedding; whereas, in fact, these bands only occur crossing, and contorting or breaking,

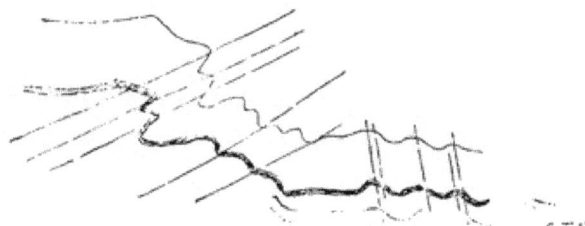

Fig. 15.—Half a mile N.N.E. of Glenmassan (163). Folded pebbly greywacke schist. The lines across the bedding show the directions of two sets of strain-slip foliations: the relations of these are not known, the more vertical and weaker set dying away before the other is reached.

an earlier foliation, and even this earlier foliation need not coincide with bedding. The want of parallelism between the early strain slips and the earlier foliation is often so slight that it may, at first, be taken as merely a sign of slight false-bedding, and it may, too, be less readily noticed owing to the later folds and strain slips which affect them both. We cannot help thinking that many of the supposed instances of false-bedding

Fig. 16.— × ¼. Headland S. of Fearna Bagh (182). Appearance in a vertical section in thin alternating schists. The crumpled line represents a foliation plane. The lines converging towards right hand of sketch are strain slips and axes of folds: these probably belong to some of the last "anticline" movements.

in mica schists which have been mentioned by early observers are really due to structures of this kind.

The later folds and strain slips are not all of one type, nor of one age, but they are only found in strength near "the anticline," and were no doubt produced in connection with it. The axes of one set of these folds, and the direction of the strain slips produced with them, hade north-west, and these are generally the most prominent, even if not the most abundant, through the angles of fold being the most acute, and their strain slips most abundant. But there are also other late folds with axes hading south-east or vertical. The relation in age of these with the others whose axes hade north-west cannot always be determined, but sometimes they are both seen in one section, together with strain lines of " pre-anticline " age. In some cases the late folds with axes hading south-east must be later than those with axes hading north-west, the strain slips belonging to the latter being folded by the former. In others, folds with axes of very varying hades may have been produced together.

It is possible to get specimens which show four different systems of fold or foliation. In order of age these are :—

1. Folding previous to, or accompanying the development of the first foliation.
2. Folding of the above first foliation, accompanied with early strain slip foliation.
3. Folding and strain slipping of "anticline" age, with axes hading north-west.
4. Folding still later than 3, with axes hading south-east.

1 and 2 are both of " pre-anticline " age. 4 was probably produced near the close of the ridging up of "the anticline," when the resistance

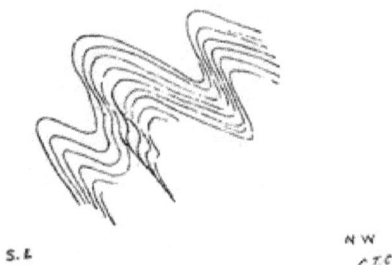

S. E.

N W

FIG. 17.—Diagram Vertical Section to illustrate the form of, and law of thinning of limbs in, the most prominent set of "anticline" folds. The folded lines represent an early foliation merely, not necessarily bedding. Compare fig. 4.

on its south-east side was becoming equal in strength to the pressure producing the anticline.

" The anticline " itself, in those areas where folds of type 3 are observed in strength, may be regarded as a very great enlargement of these folds, for its axis too hades north-west, the foliation dip on the north-west side being generally low, perhaps 20° to 30°, compared to that on the south-east side.*

The connection between movements 1 and 2 and the amounts of alteration is the same as that already described in the Ardentinny section. Similarly, movements of types 3 and 4 are connected with a further alteration in the rocks affected by them. Where affected strongly, the average thickness of the earliest foliation laminæ in the quartzose bands is perhaps $\frac{1}{10}$ inch (though in some very much more), which is distinctly less than the average thickness in unaffected rocks. The amount of mica in the former is also greater. This is connected with the fact that when "anticline" foldings are abundant there are but few of the earlier laminæ of the rock that can remain unaffected by folding,—nearly all having been further stretched and thinned, and, with this thinning, made more micaceous. In this section one sees but few wide centres of folds of comparatively unaltered rocks, like those near Ardentinny, and one

* West of Loch Riddon there is no marked difference.

SECTION IN KYLES OF BUTE.

reason for this is, that any portions that were so left by the "pre-anticline" movements have been further affected by "the anticline" movements at a later time. Some bands in the limbs of "anticline" folds are hardly half the thickness they are in the centres.

The strain bands and the centres of folds generally show the rock in its least altered character. If we want to make out what the unaltered rock was like, whether, for instance, it is likely to have been igneous or sedimentary, it is to the examination of these parts that we have to turn. If, in comparing them with the more thinned parts, we find a more sedimentary aspect in the former, more numerous distinct pebbles, etc., we may conclude that the rocks in question represent sediments. This conclusion seems confirmed by the frequent alternation, in some parts of the section, of bands of slightly different character, as regards size and abundance of pebbles, just as we should expect in pebbly sediments. In the headland on the north-west of Fearna Bagh it is surprising to see how many of the felspar pebbles have escaped granulitisation: frequently we can see such pebbles, generally pink in colour, with cleavage planes occupying the whole area of section. Opalescent quartz pebbles are also often seen, and without any evident distortion. The microscopical structure of these rocks is described fully by Mr Teall on p. 296.

The same parts of the rock that are the least affected by the "pre-anticline" movements have a tendency to continue the least altered through the subsequent movements also, for, being more massive parts, they have not been so readily broken through by subsequent strains, or so much altered in internal character. Still even these parts, the early strain bands, are often distinctly thinned along the limbs of later folds, are broken across by strain slips parallel to the axes of these folds, and may have the minerals in them for a little distance arranged parallel to these strains. The breadth of mica introduced along "the anticline" strain slips has sometimes been observed to exceed $\frac{1}{2}$ inch. "Rodding" or "stretching" indications may appear on the surfaces of the "anticline" strain slips, as on those of "pre-anticline" age, but we observed no, or only exceedingly few, quartz veins running with them. In this respect there is undoubtedly an important difference between the strain slips of "anticline" and "pre-anticline" age.

The "anticline" strain bands are not so likely to be mistaken for bedding as the "pre-anticline" bands, for the first foliation lying at an angle to the former is sure, in some part or other, to have been further accentuated by extra drag and thinning, and so forms more fissile conspicuous laminæ.

South-east of Port an Eilein the general dip of the prominent "pre-anticline" foliation is still south-east. Half a mile or so further north we begin to observe this foliation flat in places, when looked at on a large scale, though it is full of small folds, the axes of which strike north-east and south-west and hade usually north-west. A little further on, the general dip is south-east again; further still, flat on the large scale; and so we may proceed with alternations of this kind for some distance. There is no distinct line which we can say is the centre of "the anticline." So, too, on the north-west side of "the anticline" there are alternating areas of north-west dipping and essentially flat foliation. It is not until a mile or more north-west of Port an Eilein that the general dip of foliation is to the north-west, and, for several miles beyond, exposures are not uncommon in which it is again south-east. This arrangement is seen well in the crags on the hillside north-east of Upper Stronafian, and between here and Lephinkill.

In the same locality it is evident that "the anticline" folding of type

3, with axes hading north-west, is more prominently developed than
it is south of Port an Eilein. Here the closeness of the new strain
bands, and the differences developed on either side of the sharp twists
accompanying them, are quite comparable to those of the "pre-anticline"
movements, excepting that quartz veins running along the strains are
rare, though one or two have been noted. We think, too, that the
beds are generally more thinned and altered than they are south of
Port an Eilein. One notices less commonly, and less distinctly, than in
the last mentioned locality, evidence of those "pre-anticline" strain
bands which have existed previously to, and been folded by, the later
folds with north-west axes. Why is this? Can it be that the greater
thinning, etc., effected during the later movements, has now almost
entirely obscured the earlier structures, or were these latter never so
strong in this locality? We incline to the former supposition. We
know that even south of Port an Eilein the early strain bands are often
markedly thinned by the folds of later date. On this hillside, "the
anticline" movements being stronger, the general thinning and drag

NW   S.E.

Fig. 18.—Diagram to illustrate the direction of "pre-anticline" folds and strain
slips in different parts of the anticline, and their appearance compared with
those of "anticline" age. The single dotted line represents roughly, neglecting
earlier folds, the position of a "pre-anticline" foliation plane as it is traced
across the anticline. The coupled dotted lines represent the strain slips and
axes of fold of "anticline" age. The single continuous lines may represent
either bedding or foliation planes folded by the prominent type of "pre-
anticline" folds. The coupled continuous lines are the strain slips of "pre-
anticline" age.

should be still greater, and the earliest foliation planes, those lying across
the early strain bands, might be made to cross these bands at such a
small angle that it would with difficulty be perceived. Bands which
may represent such extremely pulled out strain bands are not uncommon.
They appear in many cases almost like thin quartz veins, but differ from
these in not standing out in relief on weathered faces, and in being
granulitised, and so not in possession of such even surfaces: there are
indications, too, of flakes of mica lying somewhat frequently within them,
and of spots of ferriferous carbonate, like those so commonly noticed in
the strain bands of other districts. In some parts of the hillside, too,
there are early strain bands of "pre-anticline" age, which still remain
perfectly clear, themselves sharply folded with axes of fold hading
north-west.

In parts where the main foliation is now dipping north-west, these
early strain bands may either incline north-west also, but at a distinctly lower angle than the strain bands connected with the later
"anticline" folding, or they may be approximately horizontal, or even
incline south-east at a gentle angle. Which of these positions is found
depends on the present dip of early foliation; the angle made by the

early strain bands to the general dip of the early foliation remaining nearly the same. The strain bands are supposed to be parallel to the axes of fold accompanying them, as all those observed in clearer sections are, and therefore they testify to the existence of a set of early folds with axes hading less steeply to the north-west, or, what comes to the same thing, more steeply to the south-east, than the early foliation, which they affect, did. These early folds are in fact of the same type as those near Dunoon and Ardentinny, etc. Their axes on either side of "the anticline" now lie differently with respect to the horizon, because this "anticline" has given to those on one side of it a direction of inclination which they did not originally possess. The same law respecting the amount of thinning in the different limbs is observed in the early folds on either side of "the anticline." At Dunoon it will be remembered it was the upper limbs of anticlines that were the most thinned. Here it is the under limbs, these corresponding with the upper series on the south-east side of " the anticline."

In "the anticline" folds with axes hading north-west, it is the under limbs of anticlines that have a tendency to be the most thinned, whether we are on the south-east or north-west side of the centre of the anticline. Hence, if we regard the early " pre-anticline " folds as having originally had axes hading north-west, the same law of the greater thinning of under limbs of anticlines prevails in both : and we may conclude that the source of the pressure which produced them both, lay to the north-west of the area being described, and that the pressure was outwards from the Highlands in a south-east direction. The evidence in the north-west of Scotland is now well known to show that there were there, partly at all events in post-Cambrian times, mountain-making forces pressing outwards from the Highlands in a W.N.W. direction. Hence, the central Highlands represent an area from which earth moving forces have pressed outwards, on the one side in a west-north-westerly and on the other side in a south-easterly direction.   C.T.C.

CHAPTER III.

SCHISTS PROBABLY OF SEDIMENTARY ORIGIN.

We shall describe the different groups of sedimentary schist as nearly as possible in the order in which they are mentioned on p. 8, at the beginning of chapter II.

Schists South-East of the Bull Rock Greywacke Schist.

The most south-easterly zone of schists in Cowal, between the Bull Rock greywacke and the schist boundary fault, consists of a series of phyllites and schistose pebbly grits. The average breadth of outcrop is 1 mile. The pebbles in the grits differ from those in the Bull Rock greywacke in being chiefly of white opaque quartz with comparatively few of felspar. The proportion of phyllite is less than in the Dunoon phyllite series, and there are fewer limestone outcrops. The different bands of grit and phyllite have not been separated out, except for short distances. The best sections are those on the shore at and east of Toward Quay and between Inellan Pier and Glenacre.

The characters and structures of most of the types of schist are like those of corresponding types in the Dunoon series, as seen in the coast section south of the West Bay of Dunoon, described on p. 9. The phyllites make as good slates as those of the Dunoon horizon, or perhaps slightly better. They have been worked on the west side of the Tom, near Inellan : the colour here is black and purple. There are also large old quarries in purple and green phyllite on the east side of the burn to the north of the "a" of "Chapelton" * : a little below the band quarried, comes another in which the colour is purple and black.

A good example of the massive pebbly schists with pebbles of quartz, prominent both in size and number, occurs on the north side of the Tom and runs from the "a" of "Glenacre" for rather over $\frac{1}{2}$ mile W.S.W.

Only three or four limestone outcrops have been seen, and these are thin and of poor quality. Various thin bands and lumps occur in black phyllite on the Clyde shore $\frac{1}{3}$ mile south-east of the last "e" of "Glenacre." Some calcareous bands are seen in a peculiar gritty phyllite on the coast west of Toward Point, and in a black phyllite still further west. On the coast close to and 100 yards east of Toward Quay there are thin calcareous stripes in black phyllite.

In a few places near the south-east margin of the zone there are exposures of a peculiar schist which deserves special mention. This schist is characterised by the association of very fine and very coarse sediment. In the Map the exposures have been coloured the same as the schistose grits and greywackes, but this is rather misleading. The most northerly exposure of such beds occurs in a little burn $\frac{1}{4}$ mile south-west of the top of the Tom : it contains pebble-like lumps of quartzose pebbly schist in a phyllite matrix : the lumps are sometimes 1 inch in length. A little further south-west, south of the word "Free," two other much clearer exposures of peculiar pebbly beds occur. These have at first sight a very ashy aspect. They are of a green or purple tint, effervesce freely in many parts with dilute hydrochloric acid, and show on fresh fracture many small specks of ferriferous carbonate, up to the size of a pea. The calcareous character causes the pebbly portions to stick out on weathered surfaces, and gives a rough appearance to the rock. There are many small roundish specks of water-clear quartz contained in the rock : also narrow lenticular wedges of black and purple phyllite, which do not always continue as much as 1 inch in the direction of the stretching lines, and small chloritic patches. The phyllite patches are distinctly lustrous, and like the phyllite seen at the sides of the exposures : we suppose them to be "galls," portions derived from essentially cotemporaneous beds. The specks of carbonate often occur at the sides of quartz pebbles, and may have originated in the way described on p. 80. A specimen from one of the outcrops was sliced (slide 4840), and is thus described by Mr Teall : "Dark greenish fine-grained schistose grit. Quartz grains, green patches mainly composed of chlorite. Matrix in places opaque in consequence of minute grains or particles of iron ores or carbonaceous material. Schistose chloritic grit." The chloritic patches appear to represent pebbles of some old igneous rock or elongated lenticles which may have been formed by the pulling out of pebbles. The splitting planes of the rock are not chloritic except where these chloritic patches occur.

On the shore about $1\frac{1}{4}$ mile west of Toward Point we get a black phyllite which contains many pebble-like lumps of quartzose pebbly grit, and other lumps, fewer in number, of a calcareous rock or limestone. The exposure is crossed by many north-east crushes, and is probably repeatedly folded, but it hardly seems that the pebble-like shape of the lumps

---

* In describing localities it has often been found convenient to mention their distances from certain letters or words engraved on the 1-inch Maps.

can have been acquired by crushing and folding only : we see sharply outlined pieces of grit again and again and can hammer them out, and thus ascertain that they sometimes end sharply in all directions. But there are other pieces which seem more like galls than true pebbles, for they have thin streaks of phyllite penetrating into them for some distance. The phyllite matrix of the rock is as lustrous as that of the Dunoon phyllites. The grit patches contain abundant pebbles of water-clear quartz, which are about the size of small shot: these are not noticeably pulled out and no mica is seen in the matrix of these patches, so that they have a less altered aspect than the associated phyllite. Such massive portions would naturally suffer the least during mechanical movements of the rock. A specimen from this bed was sliced (slide 4841), and is thus described by Mr Teall :—" Phyllite and grit (? interfolded). Irregular grains and patches of quartz in a fine-grained matrix essentially composed of the same substance. The outlines of the original grains are now only preserved in a few cases. Crushed grit."

Two exposures of much the same character are seen in the little burn near the " a " and " p " of " Chapelton." In the section on the south side of Toward Quay also, there are beds with an approximation to the same appearance, massive pebbly schists being irregularly mixed with bands and irregular pieces of green-grey phyllite : some of the pieces are nearly as broad as long.

The Bull Rock Greywacke Schist.

The character of the Bull Rock greywacke schist is described on p. 15. Besides the coast section there described, there are other good exposures of the bed on the Gantcocks, in the burn going through the " r " of " Tor Aluinn," in the wood ¾ mile W.S.W. of Inellan Hill, and in various places isolated within the raised beach deposits north-east of Ardyne Point. The bed is largely quarried for building stone near the coast a little north of the Bull Rock. The outcrop varies in breadth from a little over ¼ to a little under ½ mile. Hillside exposures generally give one the impression that the band is nearly free from phyllite. In the section south of Dunoon there are no bands of phyllite within it which exceed 4 to 5 feet in breadth. There are no clear sections of the margins except one, of the south-east margin, 200 yards south-east of the Bull Rock : here there is a fairly sharp line between the greywacke and the purple and green phyllite on the south-east side.

The Dunoon Phyllite Series.

The section of the Dunoon phyllites on the coast south of Dunoon is already described, p. 9, together with their minute structures. This may be taken as representative of the series generally. Fortunately we have on either side of it, through most of the area, well characterised bands of schist, a green bed on the north-west, and the Bull Rock greywacke on the south-east. Were it not for these bands it would not be so easy to separate the series from the adjacent rocks, for on either side of it, beyond the green bed and the greywacke respectively, phyllites of the same character are found, excepting that the purple tint very rarely appears on the north-west side; and there are pebbly grits in fair abundance within it. Still the horizon between the green bed and the greywacke contains a distinctly larger proportion of phyllite and a greater number of limestone outcrops than we find on the other sides of them. The pebbles in the schistose grits within the phyllite zone are also more frequently of quartz than is the case in the schists on the north-west of it; but this character is common also in the schists south-east of the Bull Rock greywacke.

Examples of black phyllites, indistinguishable from some of those within the phyllite series, occur just above the top of the most north-westerly "green bed" in the burn ½ mile N.N.E. of the "o" of "Bishop's Seat," on the east side of the west of the two faults that run north to south by the "K" of "Giant's Knowe," in outcrops between the top of the north-west green bed and the south-east band of hornblende schist. A little on the north-west side of the above green bed black phyllite also commonly occurs, e.g., in the burn ½ mile south-west of the "G" of "Giant's Knowe," ¾ mile E.S.E. of the last "n" of "Inverchaolain Glen," and in the burn that enters this glen by the second "n" of "Inverchaolain." Other good exposures are seen in the burn 200 yards south of the "l" of "Tighnuilt," ⅛ mile south of the "h" of the same word, on the shore of Loch Striven 1 mile and ¾ mile north of Strone Point (194), and on the west side of the fault ⅜ mile north of the "S" of "Strone Point" (194).

The green-grey phyllites are still more abundant outside the Dunoon phyllite series, and it is unnecessary to mention particular exposures.

Besides the section south of Dunoon, there are other good sections of the phyllite series on the south side of the East Bay of Dunoon, from ⅜ mile south-west of Kirn pier to ¼ mile south of Hunter's Quay, and for some distance on either side of Creag na Cailliche (194). There are also good sections in the different burns that run into the West Bay of Dunoon, in Glen Fyne burn (194) between Knockdow and the "F" of "Fyne," and many other places. Many outcrops of limestone are seen in all the coast sections mentioned. There is an unusually broad exposure of limestone ¼ mile south of the "C" of "Corlarach Hill" (183). In only one locality does limestone appear to have been quarried, and this is on a small scale, by the junction of the little burns ⅛ mile W.N.W. of the "D" of "Dunoon" (town). Most of the bands have a dark grey fracture, but some are buff or creamy. Examples of the last occur immediately north of Dunoon Pier, ¼ mile N.N.E. of this pier, and ⅛ mile north of Kirn Pier.

Near both margins of the zone the purple phyllites are conspicuous, but especially so near the south-east margin. In some places, near thin crushes, and where there are indications of faulting, the purple tint changes along the strike into buff or yellow with streaks of red. This is seen on the shore 100 yards S.S.E., and not quite ⅛ mile south, of Ardfillayne.

The phyllites are not now quarried for slates, but some bands were worked in former times: e.g., at the edge of the 100 foot beach by the first "o" of "Dunoon" (town), in the deep ravine of the burn ¼ mile W.S.W. of this letter, at the burn side a little over ⅛ mile E.S.E. of Corlarach, and on the hillside ¼ mile north-east of Corlarach. None of these workings are on a large scale, perhaps not even so large as the old quarries in the phyllite on the south-east side of the Bull Rock greywacke. The colours of the phyllites so used are dark grey, black, and purple. There are many other quarries within the phyllite series, particularly in the close neighbourhood of Dunoon, but the rock quarried is pebbly schist in most cases, and is used for building stone.

Schistose grits included within the phyllite zone occur, among other places, at Creag na Cailliche (194), ¾ mile north of Creag na Cailliche, ⅛ mile north of Kirn Pier, at the foot of the burn that runs into the East Bay of Dunoon, various places between Ardfillayne and the second "l" of

"Bull Wood," on the south side of the Dunoon reservoir, and on the hillside ¼ mile south-west of Ardtillayne.

On the shore a little north of Ardfillayne there are a few lenticular bands, only an inch or so thick, of a heavy dark grey hard rock which is only slightly schistose. Dilute hydrochloric acid does not cause effervescence, but after digesting with this acid and then adding ammonia to the liquid, a considerable quantity of ferrous oxide is precipitated.

The "Green Beds."

The "green beds" are, as implied by the name, a set of more or less green coloured schists. They are often intersected, too, by thin quartzose veins coloured green by epidote. Some of them are of such fine grain that, microscopically, very little could be learnt of their true character, but these are always mixed with others in which conspicuous grains of quartz and felspar occur, often in bands running parallel to one another and to the pebbly bands in the adjoining greywackes, so that the first presumption is that the green beds, like the greywackes, are also sediments, but of an especial kind. This presumption seems confirmed by the quite insensible gradations by which we may pass from a green bed into a greywacke. A typical green bed, whether pebbly or not, can be identified at once, but there is no sharp line between such a green bed and one of the schistose grits or greywackes with a grey matrix, the green tint passing into grey quite insensibly. Also, green beds with a green matrix may alternate so repeatedly with other beds with a grey matrix that it has not been practicable to map them all out separately. Instead of this the green beds have been mapped as groups, each group representing typical green beds mixed with a certain proportion of others. On the north-west side of "the anticline" it is estimated that about half of the green bed group consists of other schists : on the south-east side the different outcrops mapped are comparatively unmixed.

The green beds are usually tough to the hammer, and give rise to a number of scars which weather into blocks that are often rather rounded at the edges. Smooth hill slopes, when free of drift, have a slightly greener and drier appearance than slopes composed of the quartzose schists and phyllite. This colour is noticeable on the hill south-east of Ardkinglas woods, and near Cairndow and Upper Clasheoin (126).

During the processes of folding to which the district has been subjected the green beds seem to have behaved in a similar way to the other massive beds, the schistose greywackes, etc., for they have evidently not been crumpled so much as the phyllitic mica schists.

The green beds on the south-east side of "the anticline" are not exposed in any satisfactory coast section, but burn and hill-side sections are numerous. Among the best are those in the burns south of the Badd (184), and in the burns 1 mile north of the Badd (184), and on the raised beach margin a little north of Strone Point (194). The green beds on the north-west side of "the anticline" are exposed in Kinglas Water, from a point ½ mile E.N.E. of Ardkinglas House for a little over ¾ mile up the burn, the burns that run north-east from Stob an Eas, the river Cur above its junction with the Cab, and Strath nau Lub below the "S." In all these sections from the north-west side of "the anticline," bands of grey pebbly schist and phyllitic schist alternate with the green beds proper, as already mentioned.

In some of the coarser pebbly green beds, e.g., in those in the wood 200 yards south-east of Ardnadam Farm (174 south-west), it can be seen that many of the pebble-like pieces, in fact nearly all of them which do not con-

C

sist of quartz (which here is usually opalescent), have a pale yellowish-green tint, and that the abundance of these pieces gives the rock its colour. These green pieces are seen on microscopical examination (slide 3793) to be very rich in grains of epidote, and it is these grains that give the colour. In all probability these pieces represent altered felspar, for in a slide of somewhat similar rock (slide 3791) from 150 yards west of Conchra (162 south-east), there are signs of the development of epidote in the felspar fragments. The evidence to be obtained in this area does not enable one to decide whether the epidote of the fragments was formed before or after their inclusion in the bed as pebbles.

The green tint of this Ardnadam rock is clearly due to these epidotised fragments, but in others of the green beds, perhaps we should say most of them, the shade of green is much darker, and is chiefly due to chlorite. Two of these on the south-east side of "the anticline" have been examined by Mr Teall, and their microscopical appearance is described below. Slide 3793, from ¼ mile south of Ardnadam Farm (174 south-west) : "Dark green, fine-grained slightly schistose rock. Composed of chlorite, felspar, quartz, white mica, iron-ores. Chlorite-mica schist." Slide 3795, from Tighnuilt, Inverchaolain (183) : "A green schistose chloritic grit. Contains large fragments of quartz and felspar more or less granulitised, embedded in a schistose matrix composed of epidote, chlorite, white mica, sphene, quartz, and felspar." Dr Hatch has also described two from the same area. His notes run as follows :—Slide 2830, from 1 mile south-west of Ardnadam Farm : "Epidote-chlorite schist. Marked parallel structure produced by the mechanical extension of the constituents. Chlorite in rudely parallel lamellar bands, between which are lenticular grains (eyes) of quartz and felspar, the latter slightly turbid in the central portions of the crystal, always clear at the periphery, with here and there included spicules and arrow-headed twins of rutile. Scattered abundantly through the section are grains of epidote—some of considerable size—showing the characteristic high double refraction and pleochroism (in yellow tints). Here and there lie isolated plates of a green strongly absorptive mica. Iron ore (magnetite or ilmenite) in isolated grains." Slide 2831, from the burn ½ mile north of Bishop's Seat (183 north-east) : "Chlorite-mica schist. Marked eye-structure, the lenticles consisting mostly of quartz, sometimes a mosaic of differently orientated granules, sometimes one crystalline individual. A few of the smaller eyes are turbid felspar. Chlorite in scales and lamellæ. Mica in strongly absorptive plates. Epidote in small granules—not so abundant as in 2831."

In the green beds on the north-west side of "the anticline," chlorite and black mica associated with it are, however, much more abundant, and in larger flakes. There is, too, a great prominence of albite, both as isolated individuals throughout the mass of a bed, and collected into pegmatite veins like those of the albite schist. For the description of the albites refer to p. 39. In the green beds on the south-east side of "the anticline," such albite is hardly ever observed. In slide 3794, from a green bed near the top of Blar Buidhe (183), a mineral referred doubtfully to albite occurs, and we have also seen a few thin pegmatites, about 250 yards south-west of Kilmarnock Hill (194 north), which seemed like those of the albite schist. These occurrences, though so rare, are perhaps an indication that on this horizon albite of this kind is more common than in the other schists. The felspar in the pegmatites of the Dunoon shore section (p. 18) is also albite, but these pegmatites usually show distinct rodding, and we are not inclined to connect them in age with the albites of the albite schist.

Small magnetite grains and crystals, up to a small pea in size, are very abundant in some of the green beds on both sides of "the anticline," though the beds associated with the green beds on the south-east side do not usually show them in any prominence. These magnetites are referred to again, p. 80. Spots, up to ½ inch in length, of ferriferous carbonate giving rise to rusty-coated hollows, are also common on both sides of "the anticline." Not uncommonly it may be seen, e.g., on the east side of Glendaruel near Lephinkill (172 west), that the spots have their greater lengths orientated in one direction—that of the stretching common in the neighbourhood. In slide 4726, from ¼ mile north-west of Mullach Choire a' Chuir (142), the spots of carbonate contain inclusions which show that the rock was a foliated rock before the time of their development : in spite of this, they are developed with their greatest lengths in one direction. Veins which consist largely of ferriferous carbonate are also unusually abundant in the green beds, particularly, we think, on the north-west side of "the anticline."

In the north-west bank of Strath nan Lub (163), ½ mile south-west of the letter " S," there is a band, 1 to 6 inches thick, of white coarsely crystalline limestone, in which the calcite crystals reach $\frac{1}{10}$ inch in diameter. This is in tough pebbly green beds. Just at the side of the band the rock is rather calcareous. The band is repeatedly folded with both limbs hading north-west. We know of no other such band in the green beds, and it is perhaps possible that the one described may represent an early segregation vein.

As stated elsewhere, p. 149, the green beds have a habit of changing colour to dark grey or black near basalt dykes. Instances and particulars of the change are given in the reference.

Many microscopic sections of the beds on the north-west side of "the anticline" have been examined. Slide 3831, from about ⅔ mile slightly north of east of Lephinkill (172), is thus described by Mr Teall : "Dark green schist, with brilliant lustre on the cleavage surfaces. Albite, chlorite, white mica, rutile, and dark narrow lath-shaped sections, probably pseudomorphs after some other mineral, also tourmaline. The albite occurs in idiomorphic crystals and crystalline groups. It contains single prisms and knee-formed twins of rutile as inclusions. Sometimes a distinct fluxional arrangement of inclusions may be seen in the albite. The matrix of the rock in which the albite crystals occur as porphyritic constituents is essentially composed of chlorite and white mica, probably paragonite. The tourmaline is present as beautifully formed prismatic crystals, often showing perfect rhombohedral terminations : E pale brown, O opaque. In addition to the constituents above mentioned, a yellow mineral occurring in flat square tables, and giving a negative uniaxial figure, may be observed in the tailings formed by vanning the powder of the rock. The double refraction of this mineral appears too low for anatase, and there are no indications of the pyramidal faces which bound the square tables of that mineral."

In some quartz veins in the green beds not quite ½ mile W.S.W. of An Socach (162 south-east) there are needles of black tourmaline attaining a length of over 1 inch ; they form sheath-like aggregates, sometimes with a breadth of ¼ inch.

We suppose that the green beds on both sides of "the anticline" were essentially of the same original character : and that the differences now seen in them result from the greater metamorphism experienced by those on the north-west. These differences are only of the same kind as those exhibited by the other groups of rocks in the respective areas. It does not, of course, follow from this that the green beds on either side of "the anticline" are stratigraphically equivalent. This is a question which we

discuss elsewhere, p. 86.   What is their original character likely to
have been ?   During our early examination of the district we supposed
them to have been formed by an admixture of volcanic and normal sediments—to be beds in the formation of which submarine volcanoes had probably a share.   The abundant chlorite might be supposed to be an alteration
product, derived from original hornblendic rocks, and the magnetite
indicated that these rocks were of basic character.   But we have never
found, in any green bed, pebbles which could be referred to a cotemporaneous igneous rock, either ashy or vesicular, nor are there any bands
which could with probability be attributed to interbedded lava flows.
Where schistose igneous rocks are associated with them these are always
of the same class as those found in the schist areas adjoining—epidiorites,
hornblende or chlorite schists—bands which, indeed, have a habit of
running as sills with the bedding, but most probably represent old
intrusions, not cotemporaneous rocks.

It has been known for some years that the old surfaces of the Lewisian
Gneiss in the north-west of Scotland, where they underlie the Torridon
rocks, are often largely epidotised, and quite lately thick beds of green grit
of Lower Torridon age have been found, which have clearly been formed
by the detritus from such gneiss surfaces.   Grits of this kind are well
shown, among other places, on the west side of the Sound of Skye,
between Isle Ornsay and Kylerhea.   These grits have a striking resemblance to the green beds now being described, though they are not so
highly altered.   The resemblance extends to the microscopic appearances.
Epidote grains occur abundantly, mixed with others of quartz and felspar,
and pebbles occur which seem to have had their epidotic character
before inclusion in the rock.   Small flakes of secondary sericitic mica
and chlorite also occur, and small grains of iron ore.   It is curious, too,
that these green rocks of Skye often change their tint to dark grey or
black near basalt dykes, just as these Argyllshire green beds do.   We
cannot say that the two groups of rock are to be correlated in time, but
their resemblance is certainly to be insisted on.

In no part of Cowal have hornblende needles, either macroscopically or
under the microscope, been noticed in the green beds.   In this respect
they are noticeably different from the green beds which occur further north-
east.   In the glens of Forfarshire examined by Mr Barrow, the occurrence of hornblende within the green beds is usual, and there are rocks in
which parallel black hornblende needles occur in abundance, together
with pebble-like specks of quartz and felspar, as large as a pea : Mr
Barrow considers these needles to have been formed by thermal metamorphism in beds which were previously chloritic.   Mr Hill finds that
hornblende begins to appear in some of the green beds near the north-east
corner of 1-inch Map 37, and that it rapidly increases in prominence to
the north-east.   In the exposures south of Achadunan we looked for signs
of hornblende, but saw none.

In a little scar ⅛ mile south-east of Cruach nam Mult (134) a pebbly band
of the green beds contains a number of small oval or round shapes from the
size of a marble up to that of a duck's egg.   They are slightly harder than
the rock in which they occur, and separate from it more or less readily
on hammering.   They seem also more epidotic than the inclosing rock,
and contain larger and less thinned pebbles, which consist in part of
opalescent quartz.   The shearing or foliation goes through the shapes to
some extent, but they also act as "eyes," the fine foliation of the matrix
curving round them : at the sides they merge, in some places, insensibly
into the surrounding more foliated rock.   Both the shapes and the rest
of the rock contain abundance of black mica.   Some of the shapes show

indications of concentric structure: on the weathered face we may see an outer shell projecting, then a zone which weathers in depression, then an inner projecting shell, and lastly a hollow central core.

It is probable that these shapes were in existence before the foliation of the rock was produced, and that the less sheared character of the pebbles in them is due to the protection afforded by the hardness of the shapes. Some of the shapes are distinctly longer in the direction of the foliation than across it, so it is clear they have not wholly escaped the shearing action. Slides 4728 and 4729 are cut from these shapes. Mr Teall describes the microscopic aspect of the former as follows: "Clear patches of quartz, either formed of simple individuals or of granular aggregates. These may represent original clastic grains. The green matrix is an aggregate of epidote, calcite, and quartz. The epidote occurs in small granules closely crowded together: the calcite, which is often closely associated with the epidote, extinguishes uniformly over large areas. A few scattered grains of iron-ore, partially changed to leucoxene, are scattered through the slide."

The south-east margin of the green beds on the north-west side of "the anticline" is not a well-defined line. For a little space below the line taken for the base of the main zone there are in many places thin bands of the same character. These are mentioned in the detailed account of each area. Such bands may possibly be due to infolds of the higher lying bed, but this has not been made out with certainty in any case. In the north part of Cowal we have taken the north-west margin of the zone some distance below the Loch Tay limestone. Between it and the limestone comes usually a considerable thickness of massive quartzose schist. But green beds do not stop at this line. In several places, e.g., 1 mile S.S.W. of "h" of "Creag Dhubh," and ½ mile north-east of the same letter, a thin band rests directly on the main outcrop of limestone, and there are two other bands which have often been mapped in the space between this limestone and the graphitic schists. As the limestone does not make a double outcrop in this area it is not possible to regard the green beds north-west of the limestone as repetitions by folding of the green beds on the south-east. The bands in question occur within a broad zone of garnetiferous mica schist, and garnets may occur in the green bands also within this zone, e.g., on the shore ¼ mile south of "R" of "Rudha Bathaich Bhain" (133). Still further north-west, on the shore of Loch Fine on either side of Mid Letter, green beds occur again. This is within the Ardrishaig phyllite series.

In a few places there has been noticed within the green beds a structure not easy of explanation. We do not think the structure is confined to these beds; in fact, neither of the beds in which the best examples occur is a typical green bed, though both are within the green beds zone. The impression received on first seeing the structure is, that we are dealing with a case in which the foliation is folded, while the bedding keeps straight. It is seen best of all in a greenish albite-bearing bed in the north bank of Garvie burn (162) ½ mile south of "S" of "Strondavain." The section is not easily accessible, and cannot be examined properly without wading. There is a distinct alternation of more or less pebbly beds dipping W.N.W. at 43°, and there are thin quartz veins running parallel to them: albites are common in some of the bands. Crossing the bedding are more or less intermittent lines which are specially rich in biotite, and appear darker than the rest of the rock. These run parallel to one another: they are in some places straight, in others they have the appearance of sharp folds with the axes parallel to the bedding. When these dark lines are examined more carefully they are seen to be made up of flakes of biotite, the flat sides of which lie across the general direction of their lines of occurrence, but parallel to the bedding. Also in the greener rock at the sides of the dark lines, the greenish flakes are lying approximately parallel to the bedding. The dark lines are not accompanied with any throw. They are most conspicuous in the soft more micaceous schists, and are not noticed in some of the pebbly bands at all. Where a pebbly band comes against one that is more micaceous, the dark lines come up at a more or less marked angle to the junction of the bands, and then suddenly stop. Indications of some structure running in the same directions as the dark lines are also sometimes seen in the quartz veins. On the bedding planes of the more micaceous bands it is often seen that the surface is slightly waved, and that the lines of black mica have the individual flakes lying at a slightly different angle to the flakes of mica or chlorite in the adjoining area. Of the bedding there is no doubt. The parallelism of the mica flakes to the bedding is doubtless an instance of the parallelism of the early foliation and the bedding. The dark lines of biotite may perhaps be considered to accompany small strains of later date (perhaps of the same age as the later anticline movements described on p. 26), along which strains the pale micas, etc., have been altered to biotite. But what about the folding of these lines? Has there been a later movement along the bedding and foliation planes, which has had the effect of contorting them?

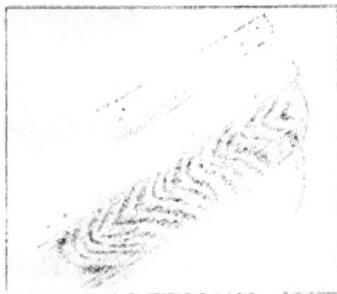

Fig. 19.—× ½. North bank of Garvie burn. Section in alternating pebbly green beds and more micaceous schists. The direction of shading shows the direction of foliation and the lie of the mica flakes. Where the shading is dark the mica is chiefly black.

Schists between the South-East group of "Green Beds" and the Albite Schists, etc.

The schists that come north-west of the green beds on the south-east side of "the anticline" are well represented in the Loch Striven coast section between the Couston high terrace and a point ¾ mile north of the second "o" of "Strone Point" (194), on the east side of this loch between Finnart Point and Inverchaolain, and on the south-west side of the Holy Loch near Hunter's Quay. The proportion of quartzose schists in these sections is considerable, and massive pebbly bands are not uncommon. Good examples of these last are seen 100 yards S.S.E. and 170 yards N.N.W. of Hunter's Quay.

The general colour of the phyllites is green-grey, but black bands are also common, particularly near the base of the green beds zone. For the localities of some of the exposures, refer to p. 32.

Thin ochreous-weathering calcareous bands and lenticles are common. These are like those described, p. 18, in the section on the west side of Loch Long from ½ mile north of Gairletter Point to about ½ mile south of Ardentinny. The thickness very rarely exceeds an inch or two. Good examples occur on the coast ½ mile N.N.E. of Inverchaolain Church, and in Inverchaolain burn ½ mile north of the "n" of "Glen."

South-east of the horizon of the Loch Long section just alluded to, we have noticed no black or brown mica, no albites excepting the few instances in the green beds mentioned on p. 34, and those in the rodded pegmatites (p. 18), nor any tourmaline. It is not likely, however (see p. 35), that the absence of these minerals implies any original difference between these beds and the others a little nearer the anticline. We take it rather to indicate that the metamorphism was less intense.

The Albite Schists and other Schists that occur with these near "the Anticline."

The albite schists of the Kyles of Bute, and "the anticline area," etc., appear to play the same rôle in the section as the phyllites, and other schists which are mainly made up of mica and chlorite, in the districts lying further to the south-east of "the anticline." But whether they were originally of quite the same character as these is not clear. If they were, it must be admitted that a considerable chemical change, and a large impregnation with soda, have subsequently taken place in them, and they might be quoted as additional evidence of the greater metamorphism of the areas in which they occur. Two chemical analyses have been made by Mr Teall to illustrate this point : one (I.) of green phyllite from 100 yards north of Blairmore pier, which we think may be taken as representative of the phyllites generally : and another (II.) of albite schist from the head of Stuck burn (152 south-east). The following are the analyses :—

					I.	II.
Silica,	.	.	.	.	43·3	63·4
Titanic acid,	.	.	.	.	1·2	trace
Alumina,	.	.	.	.	26·2	18·1
Ferric oxide,	.	.	.	.	13·6	6·7
Lime,	.	.	.	.	0·5	0·9
Magnesia,	.	.	.	.	3·8	1·9
Potash,	.	.	.	.	4·6	3·2
Soda,	.	.	.	.	1·8	3·2
Loss on ignition,	.	.	.	.	4·5	2·8
					99·5	100·2

Mr Teall adds : " If the whole of the soda in II. were used to form albite it would make 28 per cent. ; that is, 28 per cent. of the rock would be formed of this mineral." It is curious that the potash and soda taken together form the same percentage in both rocks.

The albite spots are almost confined to the more micaceous or chloritic beds : it is doubtful whether they occur at all in the quartzose pebbly beds. Even in the former rocks they are of very variable occurrence : in some parts of a band they hardly appear, while in others they may form the chief constituent. They seem generally of fairly equal dimensions in all directions, rarely exceeding $\frac{1}{10}$ inch in length, and occasionally they show an approach to idiomorphic outlines. The usual colour is pale brown : this colour is probably due to the inclusions, which we know from microscopical examination are so common in them. In rock exposures they generally project somewhat from the micaceous matrix.

They never show any appearance of stretching or clear deformation, and are thus easily distinguished microscopically from original pebbles. From this fact it might be inferred that they are of later age than the mass of the movements which affect the rocks in which they occur.

There is abundant further evidence to the same effect. In field sections we notice that they have a habit of occurring in greatest abundance in the more micaceous layers which have been developed in strain slips : they may occur in these layers even when the rock traversed by them is a quartzose grit, and shows no albites. Many of the strain slips along which the albites occur can be made out to be of "anticline" age, of types both 3 and 4 (see p. 26), by the fact that they affect the abundant early quartz veins of a section, and the earlier strain bands of "pre-anticline" age. It is therefore clear that they have been developed, either simultaneously with or subsequently to "the anticline" movements. They occur, too, in the micaceous layers along the "pre-anticline" strains, and, as these earlier strains are more abundant than the later, the albites are more often seen in them.

The microscope affords still more conclusive evidence of the movements being earlier in age than the albites, the puckered foliation planes of the rock being traceable through the albites by the help of frequent inclusions of magnetite, quartz, and some nearly colourless prismatic microlites.* Inclusions of white mica never occur. On the other hand, the magnetite specks may form a larger proportion of the albite space than they do of the schist outside it. Mr Teall has made a detailed microscopical examination of a specimen of the albite schist on the hillside ½ mile E.N.E. of Ard a' Chapuill (182 north-east), and his full description is given in Appendix. Mr Teall has also taken photographs from this slide and from slides of albite schists of two other localities, viz., slide 4733 from Beinn Tharsuinn (142), and slide 5215 from near the head of Stuck Burn (152). Reproductions of these photographs are given in Plates VI. and VII., at the end of the volume.

In these schists it is usual to find veins or pegmatites which consist of albite, or albite and chlorite, mixed with quartz. Occasionally such veins traverse the grits associated with these schists, but this is not very common. In the veins albite crystals occasionally attain a large size, cleavage faces sometimes exceeding ½ inch in breadth. One-sixth of a mile N.N.E. of Craigendaive, head of Loch Striven, the veins reach a breadth of several feet, and are traceable several yards, but more usually a breadth of several inches is the most that is seen. On the hillside east of Upper Stronafian (182 north-east) there are albite pegmatites, as much as 2 feet broad and several yards long, which contain black tourmaline needles 1 inch in length. Smaller tourmalines are common in the matrix of the albite schists generally. Good examples of them occur in the schist by the road side ¼ mile north-east of Ard a' Chapuill, and on A' Chruach (182 north-east). These tourmalines are referred to again, p. 77. Tourmaline prisms are included within albite in slide 5215, from near the head of Stuck burn (152), and so must be of earlier date than the albite. The albite pegmatites of the albite schists never show rodding or stretching at their sides, and thus differ markedly from those with pink or opaque white albite, which are common in some of the schists, e.g., in those already described at Dunoon,

The albite schists have not been mapped in any part of 1-inch Map 29, and their character as albite schists escaped observation for some time after the commencement of the survey. As a result, however, of various sub-

____

* It seems to follow from this that these included minerals are of earlier formation than the albites. The albite schists of the Hoosac and Greylock Mountains, described by Messrs Wolff and Dale ("On the Geology of the Green Mountains," vol. xxiii. of the *Monographs of the U.S. Geological Survey*), seem in many ways comparable to those of Cowal, but the albites are considered by these observers to have "evidently crystallised contemporaneously with the other minerals in the rock." See p. 297 for further reference to these and other albite schists.

sequent traverses, we may conclude that albites do not occur conspicuously in any beds south-east of a line drawn between Ardentinny and Colintraive. Only two or three occurrences have been noticed south-east of this, and all these were in "green beds." On the north-west of this line they are abundant, up to nearly as far as the Glendaruel or Loch Tay limestone, and there are certain bands on the north-west side of this limestone, within the garnetiferous mica schist zone, e.g., $\frac{3}{8}$ mile slightly north of east of Mid Letter (141), $\frac{1}{2}$ mile north-east, and $\frac{1}{4}$ mile W.S.W., of Sith an t-Sluain (152).

In the north part of Cowal various zones of albite schist, or of schist in which albite is in some places very abundant, have been traced. They are intimately mixed with more quartzose pebbly schists, and seem, as at the Kyles of Bute, to play the same part in the section as the phyllites south-east of their area of occurrence. Near "the anticline" centre at Lochgoilhead all the more micaceous schists contain albites: in some sections of a bed they are rare, and may escape observation: in others only a little removed they may be very abundant. This is not surprising when one

Fig. 20.—Slightly enlarged. One mile W. of Drimsyniebeg, Lochgoilhead. Plan of a foliation plane of albite schist. The black lines are lines of special occurrence of albite. The almost horizontal lines are the crests of small ridges by which the foliation is affected. The almost vertical lines show apparent direction of stretching.

remembers, as has been already remarked, that the albites frequently occur in strings which may cross the bedding and early foliation, and in micaceous strips which have segregated along lines of strain in the rock.

There are good coast sections of albite schists on the west side of Loch Goil, near Woodside Lodge, and extending rather more than 1 mile to the south, and on the west side of Loch Long near Mark. Most of the hilltop from $\frac{1}{2}$ mile south-east of Cruach a' Bhuic (153) to the south slopes of Beinn Bheula is composed of the same rock. The boundary of this mass of albite schist has been traced from Cruach a' Bhuic to Glen Kinglas in one direction, and down to the valley of the Cur in another. Below this comes an apparently thick mass of pebbly schists, and, below this again, more albite schist. The upper limit of the hill-top albite schist is not well defined. Many thin bands of more quartzose schist are mixed with it as we pass into apparently higher beds, and to separate them would involve much labour. This close intermixing continues generally to prevail up to near the base of the green beds. Albites have not been found in all the more micaceous bands, but we have met with many cases in which, after searching in vain for albites in some exposures of a band,

we have suddenly come across them in abundance in another exposure
of the same band at no great distance.   Probably more prolonged search
would show albites in all the more micaceous schists from " the anticline "
centre up to the base of the green beds on the north-west side.

In the albite schist 1 mile west of Drimsyniebeg (142) the little ridges
caused by the puckering of the early foliation planes by " the anticline "
movements are very conspicuous, and, crossing these ridges almost at right
angles, are obscure indications of rodding or stretching.   Crossing both
the puckering and the stretching at a marked angle are short discontinuous
vein-like outcrops of schist in which albites are very abundant, while
the rock at the sides of these outcrops is almost free from albites.
Sections nearly at right angles to the foliation planes show the lines rich
in albite crossing a succession of foliation planes : the planes may be quite
even, and yet the lines of albite may show an appearance as of sharp contortion.   Sometimes the axes of the apparent contortions show no clear
relation to the direction of the foliation, but in other cases they are
approximately parallel to it.   In the last cases it might at first sight be
supposed that the lines of albites had been folded, but this idea is
negatived by the occasional presence of other albite lines which cross
the foliation without any twisting.

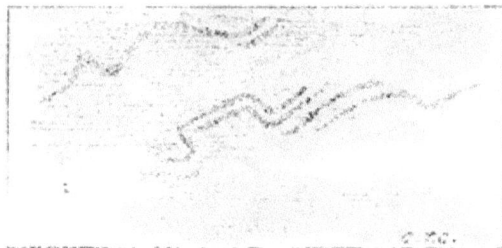

FIG. 21.— × 3.   One mile W. of Drimsyniebeg, Lochgoilhead.   Section of albite
     schist at right angles to foliation.   Shading shows the direction of foliation.
     The black lines are lines of special occurrence of albite.

There are good hillside sections of albite schist on the west side of
the glen between Lochgoilhead and Drimsyniebeg.   This is at a lower
level than the Cruach a' Bhuic band, but we cannot be certain that it is
not the same band repeated by folds with approximately horizontal axes.
The best burn sections are in Allt Glinne Mhoir, above the word " Mhoir,"
and in Kinglas Water from 200 yards west to ⅔ mile west of Butterbridge.
Perhaps both these sections are in the same band as the one between
Lochgoilhead and Drimsyniebeg.   The section in Kinglas Water shows
various interstratified quartzose pebbly schists in its east part, and there
are many thin bands of albite schist still further east.   It rather seems
as if in the albite schists of these sections—beds which at the time
of the formation of "the anticline," must have been among the lowest
beds of the district, whatever their position originally—albites are more
abundant and in larger shape than in the higher lying beds.   In a burn
½ mile north-east of Butterbridge the bigger ones attain the size of peas.
In the higher beds near the base of the green beds the average size is
about that of small shot.   It is certain that as we go along the west
side of Loch Goil from near Woodside Lodge to the mouth of Glen
Finart, the albites become gradually smaller, until in sections south of
Rudha nan Eoin we cannot, without the aid of the microscope, feel

certain of their existence. The albite pegmatites also get rarer as we go south along this coast. A special examination for albite on the coast south of Rudha nan Eoin was made in company with Mr J. B. Hill: we saw no albite pegmatites, and no schists that we could be quite certain with the naked eye were albite bearing. But frequently we met with thin chloritic bands, generally only a few inches thick, which we suspected might contain albites of exceedingly small size—so small that we could not be sure of their character. A specimen of such a doubtful albite schist, close by the lamprophyre dyke, $\frac{1}{8}$ mile north of "ll" of "Toll a' Bhuic," was sliced and proved to be an albite schist. Similar bands with minute albites certainly extend as far south as the foot of Knap burn, and probably further. Now, in walking on the side of Loch Goil from Woodside Lodge south, we are advancing from beds which at the time of the formation of "the anticline" were lower on to beds which were higher. The lower beds contain the larger albites, and in the higher beds the albites gradually decrease in size.

Proceeding north-east along "the anticline," it seems, too, that the albites increase in size. We do not think that in the Kyles of Bute district such large albites occur as in Kinglas Water.

It has already been stated that the albites are of later age than the mass of the movements which affect the schists in which they occur, and various proofs of this have been given. It is not, certain, however, that they are later than all these movements. It is not uncommon for the rows of inclusions in the albites to run in a different direction from the foliation of the rock enclosing them. The inclusions show the direction of foliation at the time of formation of the albites, but there is a suggestion that since then the albites may have received a twist, and that thus a discordance in direction has been brought about between the foliation preserved in them and the foliation now seen at their sides. In slide 4733, from near the side of the east to west basalt dyke on Beinn Tharsuinn (142), the rows of inclusions in the albites are at times at right angles to the foliation at their sides. Albites have, as already stated, a habit of occurring along lines of special strain or twist, but it hardly seems as if this were an adequate explanation of all that is observed in the slide: at the sides of, and in the part between, two adjacent albites placed diagonally to the foliation, the foliation may be quite even, and yet the rows of inclusions are inclined 45° to this. In various places on the south-west side of Glen Fine albites are frequently bordered by quartz grains of larger size than any seen elsewhere in the specimen. It is uncertain whether this is due to the incipient formation of an albite pegmatite, or whether the quartz may not be of later age than the albites, and developed at their sides after the rock matrix has partly been dragged away in the course of earth movements, in the same way as it has from the sides of magnetite crystals, etc. (p. 80). Slide 6102, from Garbh Allt Mor, $\frac{1}{2}$ mile south-east of Achadunan, is cut from a rock of this type. The appearance under a hand lens rather suggested that the albites had been granulitised, but the opaque white aspect is in reality due to the quartz grains at their sides.

It is not clear that the albites are absolutely the last of the constituents formed in the schists. In slide 3416 (described fully in Appendix II.) Mr Teall states that the white mica ends off abruptly when it comes against the albite, and never occurs as inclusions in it. Large white mica flakes cross the foliation of the albite schists not unfrequently, e.g., at the head of Stuck burn (152), and there are sometimes bands, about $\frac{1}{2}$ inch thick, in these schists, which are almost entirely made of mica flakes of this kind. Mr Teall describes the microscopic character of one of these bands from a little over $\frac{1}{3}$ mile N.N.E. of Beinn Lochain (141), as seen in slide 4732, as follows:—"The substratum of this rock is an aggregate of well-crystallised white mica. The different individuals interlock with each other just as the grains of quartz in a quartzite, and there is no orientation of the plates parallel with the general foliation. Embedded in the groundmass of mica are numerous minute opaque grains which are black by reflected light and not attracted by a magnetite needle. These are therefore probably ilmenite. They define by their arrangement the foliation of the rock. The other constituents which can be identified are prisms of tourmaline, plates of chlorite and biotite, and possibly a few grains of quartz. There are also small hexagonal prisms of a nearly colourless mineral having a much lower double refraction than tourmaline. These may be apatite. The crystallisation of the mica has obviously taken place after all movements had ceased, as this mineral forms the matrix in which the other minerals are uniformly distributed." The biotite flakes are arranged parallel to the ilmenite. The biotite is not usually included in the white mica: it often ends suddenly at the sides of this mica, as if, perhaps, it had been used up in making the latter mineral. The tourmaline prisms sometimes have their long axes parallel to the foliation, and sometimes across.

In other places brown mica flakes occur lying across the foliation of the albite schists, e.g., on the hill top $\frac{2}{3}$ mile slightly east of north of Binnein an Fhidleir (134).

The matrix of the albite schists may be either green, pale greenish-grey, or leaden coloured. Near the head of Stuck burn the two last colours are in close association. The pale greenish-grey occurs on the outside of the numerous little quartz veins in the schist, in the flakes enclosed in these veins, and in thin bands mixed with the harder quartzose schists. The leaden coloured rock is very fine in grain, and no separable flakes can be made out by the eye or hand lens: it breaks up less readily under the hammer than the greenish-grey type. Besides a close interbanding of the two types, there is also apparently a gradual passage between them: it is often uncertain to which type a particular specimen belongs. In the Stuck burn, albites seem most abundant in the leaden coloured variety.

The alteration of albite schists by various of the unfoliated intrusive rocks is described on pp. 98, 106, 114, 149.

Bands of typical albite schists are included within the green bed zone on the north-west side of "the anticline." Besides these, there are other beds which, from their toughness and richness in epidote, we should prefer to class with the green beds, which yet contain albite of the same kind as the typical albite schists. A green bed, the next below the base of the main zone of green beds, $\frac{1}{4}$ mile north-west of Mullach Choire a' Chuir (142), may be mentioned as an example. Slide 4726 is cut from a specimen of this. Probably there is no distinct line between these rocks and typical albite schists, any more than there is between pebbly green beds and pebbly quartzose schists. On the shore 1 mile slightly south of east of Dundarave Castle there is a green bed with abundant epidote both in grains and prisms, and chlorite and biotite in equal abundance; this also contains albite specks, but the specks are hardly like those of the typical albite schists: some are very free from inclusions.* This occurrence is within the garnetiferous schist zone. Another outcrop of green beds with albites is seen in the same zone, $\frac{1}{4}$ mile slightly north of west of Meall Reamhar (141).

* A reproduction of a photograph taken by Mr Teall from the slide (4725) of this rock is given in Plate VII. near the end of the volume.

Quartz veins of irregular occurrence and thickness, but not generally more than an inch or two thick, are abundant in the albite and phyllitic mica schists. They are certainly more numerous in these schists than in the more quartzose schists.

The quartzose and pebbly schists associated with the albite schists in "the anticline" district, generally do not differ in any marked way from the corresponding schists in the Kyles of Bute section (see pp. 23–27, 296). A relative abundance of black and brown mica is characteristic of them when compared with the less altered grits to the south-east, in the area where there are no visible albites. Some portions of bands are much more full of pebbles than others, and when seen in good sections the alternations from more to less pebbly portions closely resemble alternations due to original deposition. The shapes of the pebbles are generally more deformed in the limbs of folds than in the axes (see pp. 27, 296), but there is no area where it is probable that the absence of pebbles is due to complete destruction of them in a band that was coarsely pebbly at first. It is more likely that the finer-grained quartzose schists free from pebbly specks were from the beginning marked by a comparative absence of large pebbles. There are also many bands of schist which are neither genuine phyllitic schists, nor yet very quartzose. Their original representatives may have been sandy shales.

Tourmaline prisms are more common in the albite and more micaceous schists than in the quartzose, but they do occasionally occur in the latter. A quartzite-like schist from a burn 1 mile N.N.W. of Ben Donich shows on the foliation planes a set of closely parallel rods, which are made up of small prisms of tourmaline, each running parallel to the direction of the rods. Slide 4731 is cut from a specimen of this rock.

Ochreous weathering lenticles and short bands of calcareous schistose grit, 1 or 2 inches thick, are common, just as they are among the schistose grits further south-east (see p. 38), in areas where no albites have been recognised. Good examples occur on A' Chruach (182), on the south-west side of Tom nan Con (164), and on the hillside east of Clunie Wood (164).

C.T.C.

CHAPTER IV.

SCHISTS PROBABLY OF SEDIMENTARY ORIGIN—*(continued)*

The Loch Tay Limestone.

The Glendaruel or Loch Tay limestone is perhaps the most readily mapped and best defined schist in the district. It does not usually make swallow holes, though here and there, e.g., on the watershed a little more than ½ mile slightly south of east of Cruach nan Capull (133), and ¾ mile north of Cruach an Lochain (151), such do occur, but there is a distinctness of character in it, and the associated beds, which usually enables it to be recognised in the different outcrops. It is also probably thicker than any other limestone in the district in its unfolded state. It is seldom seen without a band of hornblende schist either in or just at the sides of it, or in both positions.

Among the best sections are those in the burn ½ mile north-east of the "g" of "Creag Dhubh" (141), the nearest burn south-east of this stream, the burns ½ mile south-east, and ¾ mile S.S.E., of the same letter, at Glensluan, various burns on the south-east side of Sith an t-Sluain (152), Kilbridemore burn (162), and various burns on the west side of Glendaruel as far south as Glendaruel House.

In the north part of Cowal the part of the limestone above the included hornblende schist is commonly about 20 to 30 feet thick ; and below the hornblende schist limestone or calcareous quartzite schist occurs again with an average thickness of about 20 feet. There may in some places be greater thicknesses than these, owing to folding. Both parts of the limestone, but especially the lower part, are sometimes mixed with bands and lenticles of calcareous quartzite. Below the limestone comes often a few feet of leaden coloured phyllitic schist, and below this a considerable thickness of massive quartzose schist. Above the limestone is soft phyllitic schist which may contain small garnets, and above this often another outcrop of limestone, usually about 6 to 7 feet thick, and above this again is soft garnetiferous schist. The interval between the upper and the main limestone is in places 30 feet, in other places only 12 feet, and in others, again, we see nothing of the upper outcrop. Perhaps this upper limestone is only a repetition of part of the lower one by folding. Further details of the section in different localities are given on pp. 98, 106, 114, 149.

The calcite crystals which make up the mass of the limestone show cleavage faces varying in size from a small mustard seed up to ½ inch in length : perhaps the average length is about $\frac{1}{20}$ inch, but it may vary rapidly. In some bands the general colour is nearly black, in others pale grey, and in others there is a mottled appearance due to the mixing of these colours. In the mottled bands the black parts are formed of calcite crystals of larger size than those in the pale grey parts. The black crystals, either in combination or singly, form island-like areas which are separated by narrower portions made up of smaller pale grey or white crystals. Both black and white parts effervesce freely with dilute hydrochloric acid. The outlines of the black parts are sometimes markedly angular, and the white parts may penetrate into them along bays or thin crack-like lines, and then suddenly end. In sections across the foliation it is seen that the black parts have often a greater length parallel to the foliation than across it. It is suggested that the white parts have been formed by a kind of crushing, and recrystallisation in smaller crystals, of the black parts. The section in the burn below the road 1 mile W.S.W. of Cruach nan Capull (151 and 152) shows in one place a succession of dip slopes of mottled limestone, and it is clear that the longer axes of the black crystals are parallel to one another. This is so whatever the size of the black areas : sometimes they are 2 or 3 inches long. The direction of elongation is the same as that of the stretching or rodding in the neighbourhood, as seen in some surfaces of calcareous quartzite schist in the same burn.

It is not clear what is the cause of the colour in the black calcite. In slide 5136, from black limestone not quite ⅓ mile south of Cruach nan Capull (141), the calcite is often dusty, with minute inclusions, and possibly it is these that give the colour.

The section at the quarry at Glensluan is described by Mr W. Ivison Macadam ("On the Chemical Composition of certain Limestone Rocks from Ballimore, Argyllshire," *Trans. Edin. Geol. Soc.*, vol. iv. 1883, p. 101), who gives a series of chemical analyses of the limestone and of vein and crush rocks which occur in it. Analysis No. 3, of blue limestone of the common type of the quarry, is as follows :—

Ferric oxide, Fe_2O_3,	.	1·14
Aluminic oxide, Al_2O_3,	.	0·18
Calcic oxide, CaO, .	.	37·21
Magnesic oxide, MgO,	. .	0·35
Silica and silicates, SiO_2,	. .	29·52
Carbonic anhydride, CO_2,	. .	29·63
Sulphuric anhydride, SO_3,	. . .	1·56
Organic matter and moisture, .	. .	0·41
		100·00

This corresponds to a percentage of 66·44 of carbonate of lime, and 0·75 of carbonate of magnesia.

Analysis No. 7, of a red coloured band, shows a considerably higher percentage, viz., 84·86, of carbonate of lime. The details of this analysis are as follows :—

Ferric oxide, Fe_2O_3,	. .	1·24
Aluminic oxide, Al_2O_3,	. .	0·42
Calcic oxide, CaO, .	. .	47·51
Magnesic oxide, MgO,	. .	0·25
Silica and silicates, .	.	12·28
Carbonic anhydride, CO_2,		37·64
Sulphuric anhydride, SO_3,		0·23
Organic matter and moisture,		0·43
		100·00

We have not elsewhere noticed any red coloured band in the limestone, and perhaps this colour is confined to the proximity of a crush line or vein.

C.T.C.

Small pale coloured micas are sometimes seen, but in no great abundance, lying flat on the foliation planes. Weathered faces frequently show small specks of quartz, about the size of small shot. These are often opalescent, e.g., in the burn ½ mile south-west of Glenshuan, and probably represent pebbles. They are not generally elongated in any particular direction, but perhaps the calcareous parts of the rock, being readily soluble, would, during the shearing, flow easily from the sides of the pebbles, and allow them to remain undistorted. They are often somewhat rusty coloured on the outside, but this may be owing to residues left from the weathered limestone. In the limestone quarry at Otter Ferry pebbles may be seen as large as peas. Some of these are of clear glassy felspar with partially rounded outline, some are of a dull pink coloured felspar, and others are clear blue quartz. C.T.C., J.B.H.

Weathered faces of some of the impure limestone bands also show in certain places crowds of minute pale straw coloured or white needles, often about $\frac{1}{40}$ inch in length. These needles have a general parallelism of direction. Slide 5547 is from a band of this kind from the limestone just above the hornblende schist in the burn ¾ mile south-west of the "g" of "Creag Dhubh" (141). Mr Teall identifies the needles as idocrase : he says, "The prisms of idocrase are seen to be eight-sided, due to the forms (100) and (101). Terminal faces appear to be rare, but a combination of the pyramid (111) and basal plane (001) has been observed." There are also small hemispherical knots, about the size of peas, projecting on some of the weathered surfaces, and somewhat resembling the tops of worm tubes. These spots are distinct in the slide, and are finer grained than the rest of the rock : the calcite in each spot may extinguish simultaneously : the foliation has a tendency to go round the spots. The purer limestone from the same locality also shows slender prisms of idocrase projecting on the weathered face. Mr Teall says there is probably also a mineral of the epidote zoisite group present.

A calcareous quartzite schist from the limestone below the hornblende schist ¼ mile E.N.E. of Bathaich ban Cottage (134) contains hemispherical projections which seem to the unaided eye very like those in specimen 5547. This was sliced (slides 4730 and 5133), and disclosed a great number of minute colourless garnets, forming in some parts more than half the bulk of the rock. They are occasionally collected into spots and streaks, and the spots probably form the projections noticed on the weathered face. There are abundant small pyrites specks in the same rock.

In the calcareous quartzite below the hornblende schist, ½ mile south-west of Glensluan, there are bedding surfaces which show oval projections sometimes more than 1 inch in length, but these have not been examined microscopically.

The Garnetiferous Mica Schist.

North-west of the Loch Tay limestone comes a broad zone wherein nearly all the rocks, whether green beds, epidiorites, thin fissile mica schists, or quartzose schists, may contain garnets, often in great abundance. This zone is not supposed to represent a stratigraphical horizon, but in a rough description of the boundaries, or in an attempt to represent them graphically, it is convenient to take the Loch Tay limestone as the south-east margin, and the most north-west of the graphite schists, which often runs at a distance of about 1 mile north-west of this limestone, as the north-west margin. But, as stated already, white garnets are in places abundant within the Loch Tay limestone itself. A band of hornblende schist within the limestone in the burn ¾ mile south-east of Ardno also contains small pink garnets (see p. 63); and garnets of the same size and colour as those in the defined zone, and many of which have undergone the same changes occur also in places outside the zone. There seems to be a tendency for these occurrences gradually to increase in number, and to lie at a further distance south-east of the Loch Tay limestone, the further we go north-east, so that a little way up Glen Fine it is doubtful whether the limestone continues to be the best boundary for the main zone. In the river Cur, ¼ mile west of Creag Dhubh, garnets occur in a green bed some distance below the north-west boundary of the green beds zone: at the junction of the Cur and the Cab they are also seen in a green bed: in the Cab, ¼ mile above its foot, they are in a thin fissile mica schist near the base of the green beds: in Leamhanin they are in quartzose schist and phyllite east of the letter "m," near the base of the green beds. Further north-east still, garnets are common in the albite schists below the green beds. The first we noticed occur in small outcrops, and their size is but small—perhaps $\frac{1}{16}$ inch in diameter. This is on the high ground near Binnein an Fhidleir: there is one small outcrop 16 yards west of the Ordnance Station 2658, another 150 yards W.N.W. of the same station, and another $\frac{7}{12}$ mile slightly north of east of it. Further north they become more abundant. They are prominent in the crags ½ mile W.N.W. of Lochan Mill Bhig, and by the "Ri" of "Eas Riachain," and in a little burn running parallel to Eas Riachain, on its north-east side, at a distance of about 160 yards above the delta of the burn.

In the same way the occurrences of garnet do not stop sharply at the graphite schist band mentioned, but here and there we find them on the north-west side of it, e.g., in the schist at the edge of a hornblende schist $\frac{5}{12}$

mile south-east of Newton, on the shore at the south side of Strachur Bay, and in the branches of the burn that enters Loch Fine at St Catherine's.

In the burn south-east of Leanach (151) the garnets are about as prominent and large as in the areas further along the strike to the north-east, but south-west of the burn they seem gradually to decrease in size. In the south part of Cowal near Kilfinan, the schists on the north-west of the Loch Tay limestone were surveyed before those on the same horizon further north-east, and it was not observed at the time of the survey that garnets were specially characteristic of this horizon. It has been subsequently noted, however, that small garnets, not exceeding the size of a pin's head, are not uncommon on or a little south-east of this horizon, e.g., in the hornblende schist about 50 yards north-west of the Loch Tay limestone on the north side of Auchalick Bay (181), and in the green beds just west of Ballochandrain (172), and they were noticed in slide 3830 from the green beds near Lephinkill (172).

Among the best sections within the main zone are these,—on the shore from $\frac{1}{8}$ mile to $\frac{7}{16}$ mile south-west of Rudha Bathaich Bhain (133), in the different burns that run N.N.W. from Cruach nan Capull (133); in the lower part of Eas Dubh (141), and in the burn south-east of Leanach (151), up to a point $\frac{1}{2}$ mile south-west of the "n" of Leanach." The most common type of rock within the zone is a thin fissile phyllitic mica schist, on the foliation planes of which there is sometimes a good deal of chlorite irregularly distributed. The rock breaks readily across the chief foliation, and shows usually a great number of close folds with axes parallel to this foliation, and the impression is conveyed that the folding is generally more repeated than in the rock zones further south-east. Intermixed with this type are more quartzose schists, with thin rusty-weathering calcareous lenticles, and many thin quartz veins. The quartzose bands and veins are comparatively free from garnets, but in one place, not quite $\frac{1}{8}$ mile south-west of Cruach nan Capull (133), there are quartz veins which are nearly half composed of garnets. Frequently, e.g., in Eas Dubh $\frac{1}{8}$ mile east of Inverglen, the vein-quartz is in the form of short rods which run parallel to one another and to the other indications of stretching in the neighbourhood. Often, too, in the more quartzose schists, there is a rod structure which is more prominent than the foliation : the rock splits readily into a series of parallel rods, but it is hard to get a large plane slab. This is shown well in the scars of An Carr (141 and 142). There is a constant repetition of sharp folds in the scars. It has been suggested by Mr G. Barrow, that in cases like these, where the rodding is prominent but the foliation is not, the rock may have been folded before the force which produced the rodding commenced to act.

Two chemical analyses of mica schist which possibly belong to this zone are given on p. 55.

The garnets are of a port-wine colour, and vary in size from large peas or small marbles down to specks which are hardly visible. The small sizes occur near the margins of the zone, or in areas outside it. The most common is that of a small pea. Perhaps the largest noticed are those in Eas Dubh (141) $\frac{1}{8}$ mile S.S.E. of the letter "h" of "Creagan an Eich." A first glance at some exposures may fail to show any clear distortion or indication of movement since the formation of the garnets. But more careful search nearly always shows distortion : the angles of the garnets are slightly rounded, or the shape elongated in one direction, or there is a rim of chlorite around them. This rim usually extends along the foliation to a greater distance in one direction than in others, or rather in two directions opposite to one another, forming two tail-like extensions. The ex-

D

tensions on the same foliation plane are parallel to one another, and also to the rodding in the quartz veins. The direction on adjoining slabs is not necessarily quite the same, but neither are the directions of rodding. The most general direction is north-west or N.N.W. As seen in sections at right angles to the foliation, the rim is extremely thin at right angles to the stretching, or it may hardly be observed at all. These are just the directions that would show the least thickness of any material formed from the garnets by rubbing or crushing along the foliation. If the larger garnets are examined they are not uncommonly seen to be traversed by strings of chlorite, and to be distorted, or even broken into pieces. In the most westerly burn

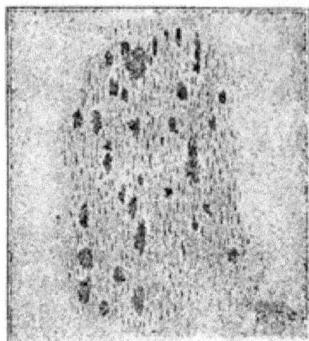

FIG. 22.— 1. Garnetiferous mica schist ½ mile S.W. of Bathaich ban Cottage. Surface of a foliation plane with patches of dark green chlorite, supposed to be replacing or edging garnets.

of two parallel burns nearly $\frac{5}{16}$ mile slightly east of south of Laglingarten (133), the garnets are pulled out into streaks ½ inch long. In a little scar nearly ¼ mile south-west of Cruach nan Capull (133), ⅓ mile north of the "C" of "Creag Dhubh" (141), and at a point 70 yards west of Meall Reamhar (141), the garnets are crossed by minute veins of quartz, running nearly at right angles to the foliation, and accompanied with slight throws: there may be three such in one garnet. In the most westerly of three neighbouring burns ¼ mile slightly south of west of Ardno, there are thin bands of leaden coloured

C. T. C

FIG. 23.— × 2. One quarter of a mile S.W. of Cruach nan Capull (133). Garnets crossed by cracks filled with granulitic quartz.

schist in which the garnets are sometimes surrounded by rims of quartz which have a special extension in two opposite directions along the foliation. No doubt these rims are of the same nature as those around the magnetite crystals, etc., described on p. 80, and were formed in the same way. Some of the garnets with quartz rims are distorted: round those with rims of quartz we did not notice any chlorite, but others with chlorite rims occur within 2 feet of them.

All stages seem to occur in the relative prominence of the chlorite rim compared with the garnet centre, the latter getting smaller as the former enlarges. In many localities we see no garnets at all, but many small specks and streaks of chlorite which probably represent them.

On the shore ½ mile south-west of Rudha Bathaich Bhain there are, besides the chlorite areas surrounding the garnets, spots and crystals of black mica. We discerned no common orientation in these, and no garnet centres, and it is not probable they have any connection with the garnets.

In some places the garnet schists contain also distinct radiate shapes, probably after actinolite. The best localities are the burn ¼ mile slightly west of south of Ardno, and another burn (on either side of a little east to west crush) nearly ⅛ mile south-west of Ardno. These two places are on much the same horizon. The rays of the shapes may attain a length of 3 to 4 inches. The substance composing them is quite soft, and presumably chlorite. Spots of similar substance occur on the same foliation planes replacing garnets. The faint stretching lines and wave lines on the foliation planes which contain the shapes, cross the surfaces of the shapes also, without projecting beyond the general level of the surface. Similar shapes also occur in the more micaceous parts of some rather quartzose schist blocks tumbled from the west bank

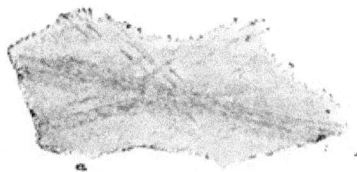

FIG. 24.— × 1. Chlorite pseudomorphs after actinolite, in the garnetiferous mica schist ¼ mile slightly W. of S. of Ardno : as seen on a foliation plane. The fine shading indicates the direction of the crests of the small waves affecting the foliation plane. At a place vertically over "a" in the sketch these waves are clearly seen to affect the pseudomorphs.

of the burn not quite ½ mile south of Ardno, in the burn 130 yards south-east of the farm buildings at Ardno, in Eas Dubh ¾ mile east of the "h" of "Creagan an Eich" (141)—in thin bands which are rather more quartzose than the normal type of garnet schist—¼ mile north-west of Meall Reamhar (again in a rather quartzose schist), and in a scar 1/12 mile south-west of An Carr in a greenish fissile schist.

Besides these, there are other places where the shape and arrangement of many of the chlorite areas rather remind one of deformed actinolite rays. It is possible that most of the irregular patches of chlorite may have originated in this way, or from the occurrence of garnets and actinolites in close association.

It is to be noted that the garnets in the albite schists of Glen Fine are also at times surrounded with a distinct chlorite rim. This is seen in a burn (not marked in the 1-inch Map) ½ mile south-east of the big "R" of "KILMORICH" (126), and in a little burn running parallel to Eas Riachain on its north-east side, at a distance of 160 yards from the delta of the burn. In slide 6103, from a specimen obtained ⅝ mile slightly south of west of the Eagles Fall, the garnets are irregular in shape, and chlorite veins cross them, and there are large chlorite flakes at their sides. The garnets are sometimes partially or wholly replaced by carbonates, and it is to this that the white rim noticed in the hand specimen is due.

In a few places within the garnet schist area small white knots have been observed, of about the same size and frequency as the garnets in the adjoining schists. In slide 5054, from a dark slightly graphitic schist ⅚ mile slightly north of west of Cruach nan Capull (133), the knots are composed of aggregates of quartz, mica, and an opaque brown substance : in one of them the mica flakes are arranged somewhat radially, starting from a nearly opaque mass near the centre and passing through the clearer quartz at the edge. In the same knot there seemed to be minute specks of garnet. Possibly the knots are pseudomorphs after garnets. White knots of an apparently similar kind also occur in the burn ½ mile north-west of the "r" of "Creag Dhubh" (141).

Though the garnets have suffered much alteration and physical distortion, it is not likely they have passed through the same amount of movement and change as the schist in which they occur. Probably the rock was a schist before the formation of the garnets. In slide 4739, from a specimen obtained from the shore ¼ mile west of Bathaich ban Cottage (133 and 134), the garnets contain inclusions, part of which consist of some brightly polarising needle-shaped substance, and part of some black opaque grains, probably ilmenite. These inclusions run in rows across the garnet: the rows are slightly waved, and run in much the same direction as the schist foliation. The same slide shows two or three portions of vein quartz, the interiors of which are nearly free from chlorite or garnet, but contain a good deal of white mica : the chlorite is specially abundant at the sides of the garnets, but is not confined thereto—it may also occur in small isolated flakes.

An albite schist on the hillside east of Mid Letter, from within the garnet schist area, contains also elongated chloritic streaks which probably represent garnets. A slide was prepared from this schist (slide 5432), but it failed to give any clear evidence of the relation in age of the albite and chlorite. Considering how commonly the garnets are deformed, while there is no satisfactory evidence of distortion in the albites, it is probable that the metamorphism which produced the garnets is of earlier date than that which produced the albites—at all events it is not of later date.

An instance of the presence of garnets in green beds in the garnet schist zone is given on p. 37. Garnet bearing hornblende schists within the same zone are seen ½ mile slightly north of west of the first " C " of " Cruach nan Capull " (133), ⅝ mile north-west of the same letter, and, as already mentioned (p. 48), in a band within the Loch Tay limestone.

In a burn ¼ mile north-east of Sith an t-Sluain (152) there are large brown mica flakes in a phyllite of the garnet schist zone. The flat sides of the flakes cross the foliation at a considerable angle, and are evidently of later formation than it.

The north-west band of graphite schist may be considered as the north-west boundary of the area of chief occurrence of garnets. Apart from the presence of garnets the schist for some distance on the south-east side of this band is in most respects very like that on the north-west side. There is, however, one difference. Most of the quartzose bands intermixed with the phyllites on the north-west side are fine grained, and very compact—are, in fact, a kind of quartzite schist—while the harder bands within the garnet schist zone are generally pebbly, and rarely so compact or purely siliceous. For this reason, apart from others, it does not seem probable that the mass of the garnet schist zone can be a repetition of the beds at the base of the Ardrishaig phyllite series in a different state of alteration.

In the area between Strachnr and Ardno there is a limestone outcrop a little south-east of the graphite schists. A band of hornblende schist usually occurs in or below this, or there may be more than one band. In the burn ⅛ mile S.S.W. of Ardno the upper exposure of limestone is 6 to 8 feet thick, and contains conspicuous quartz pebbles : below this is a hornblende schist about 6 feet thick : then limestone again, mixed with a larger proportion of calcareous quartzite, the thickness of the two together being perhaps 12 feet. Below this is about 6 feet of hornblende schist resting on shivery garnet schist mixed with occasional quartzose bands containing quartz pebbles. On the hillside ½ mile N.N.W. of the " C " of " Creag Dhubh " the limestone is divided into four parts by three bands of hornblende schist : the upper part is rather a calcareous

quartzite than limestone.   In a little burn 1 mile W.S.W. of the
" C " of " Creag Dhubh " some parts are quite a coarsely crystalline pale
grey limestone, others are more a calcareous quartzite : the weathered
faces are a deep ochreous colour, deeper than those usual in the Loch Tay
limestone, and many of the splitting planes are black, as if partly graphitic :
we saw no mottled bands. There is no corresponding limestone outcrop
seen on the south-west of Strachur, excepting in a little burn ¼ mile
north-west of the " n " of " An Carr " (141 and 152).

If this limestone be regarded as a repetition of the Loch Tay limestone
by folding, it must occur in a sharp syncline or anticline, both sides of
the outcrop representing the same margin of the bed, for the Loch Tay
limestone on the south-east side does not make a double outcrop.   It is
difficult on this supposition to see why the hornblende schist seen close on
the south-east side in some of the sections, e.g., in the burns ⅙ mile S.S.W.
of Ardno, ⅙ mile south-east of Laglingarten, ⅔ mile north-west of the first
" C " of " Cruach nan Capull " (133), and other places, is not also repeated
on the north-west side.   The hornblende schist is supposed to be intrusive
and not necessarily continuous, but we know of no reason why it should
die out more quickly across the strike than along it.

The Graphitic Schists.

In the north part of Cowal there are various parallel outcrops of
graphite schist, impure graphitic limestone, and calcareous quartzite, at a
distance of about 1 mile north-west of the Loch Tay limestone.   As stated
on p. 48, the most north-westerly of these has been adopted as the
north-west boundary of the garnet schist zone.   There are not always the
same number of outcrops in the different localities : in some places we can
only see one or two, in other places four or more, e.g., on the south-west
side of Strachur Bay.   Probably these variations in number are due to
variations in the amount of folding in the different localities.   Among
the best sections of these beds are these,—at the head of Eas Dubh
(141), in the burn ½ mile south-west of Laglingarten (133), in various
burns by the shore between Mid Letter and Strachur Bay, and in the burn
¼ mile north-east of Leanach (151).

All the bands show a great amount of twisting and crushing, there
being special lines of disturbance running along with them in a general
N.N.E. direction.   These lines are of later date than the folding which
accompanied the schist manufacture, for they twist this folding without
producing any further foliation, and the actual breakage lines are filled
with streaks of clay formed from the crushing of the crystalline schist.
But they are in part of earlier date than some of the lamprophyre
intrusions (see p. 109).

Owing to the movements it is hard to get good specimens of the softer
schists, and the blackening of the fingers on handling specimens is great,
owing to abundant thin streaks of crush rock.   The bands of limestone
and calcareous quartzite mixed with the graphite schist are themselves dark,
graphitic, and thin bedded, and in most places they are so mixed with
the schist by contortions and crushes, that it is hard to separate them.
It is possible, too, that in the original succession the beds were closely
intermixed.   At the head of Eas Dubh there is a moderately good limestone at least 6 to 8 feet thick, dark grey in colour, and fine grained.
There is graphite schist on both sides of this outcrop, and other impure
calcareous bands : the graphitic horizon is probably not less than 50 yards
wide.   In the burn ¼ mile north-east of Leanach (151) there is a graphitic
limestone with an apparent thickness of about 10 feet.

Slide 5053, from a rather quartzose variety of graphite schist $1\frac{1}{4}$ mile south-east of St Catherine's pier, shows that the rock is composed of micro-crystalline quartz, white mica, and much opaque dust. Many areas are comparatively clear, consisting mainly of granulitic quartz with but little opaque matter, while other areas are entirely opaque : the clear and the opaque areas being roughly parallel to one another. The white mica is conspicuous in the clear areas. Some of the micas seem to have accumulated in themselves a good deal of the opaque matter. There are indications of sharp folding in some layers : the folds may end at strain or slide-like lines which are parallel with the foliation of the rock. In slide 5054, of the dark knotted schist already mentioned on p. 51, the opaque dust is not so abundant as in 5053 : in addition to the opaque dust there are also opaque folia stained brown.

In the bend of the burn $\frac{1}{2}$ mile north-west of the "C" of "Creag Dhubh" a quartzite schist, under a dark limestone and graphite schist, contains quartz pebbles as large as peas. In Eas Dubh, 1 mile south-east of Ardnagowan Cottage, thin quartzite schists are intimately mixed with leaden coloured garnetiferous schists and thin bands of graphite schist: in one of the quartzite bands there are various black spots, about the size of peas, with prominent cleavage planes. Microscopic examination (slide 5134) shows that the rock consists essentially of quartz, felspar, and white mica. The felspars generally contain opaque particles, which Mr Teall considers are probably carbonaceous, and it is these which form the black spots in the hand specimen : some opaque matter commonly occurs also on the foliation planes.

In the burn $\frac{1}{4}$ mile north-east of Leanach the graphite schist appears as dark as usual, and is associated with a graphitic limestone already mentioned. Graphite schist is seen again on the south-east side of the same burn, near the limekiln 200 yards north-east of Leanach. We cannot say for certain that the graphite schist continues south-west of Leanach. The district to the south-west was surveyed before that to the north-east, and no graphite schist was noted in it.

The Ardrishaig Phyllite Series.

The Ardrishaig phyllites or "calcareous sericite schists" consist of greenish-grey soft phyllitic mica schists, in many parts of which there are abundant calcareous lenticles, often hardly $\frac{1}{2}$ inch thick. Intermixed with these are occasional thicker outcrops of limestone which can be mapped, and bands of hard compact quartzite schist or greywacke schist. The greater part of the zone is extremely soft, and the rock, on hammering, breaks almost as readily across the foliation as along it : for this reason it cannot be used for slates, and the burns traversing the area are unusually deep and narrow. Good sections occur on the shore between Creag a' Phuill and a point $\frac{1}{2}$ mile north-east, between Creggan's Point and Aird Cottage, and between Mid Letter and a point $\frac{4}{5}$ mile south-west.

Only the south-east portion of this series comes into Cowal, and there is a greater breadth of exposure, at right angles to the strike, on the north-west side of upper Loch Fine than on the south-east side. In the part of Cowal north-east of Strathlachlan (151) the most north-westerly band of graphite schist has been adopted as the south-east margin of the series (see p. 53), but the graphite schists have not been noticed in the area south-west of Leanach, and there is therefore a possibility that the margin adopted in this area is not exactly the same as that adopted to the north-east. In the south-west area the bands of quartzite schist near the south-east margin gradually increase in prominence to the south-west, and finally are in excess of the intermixed soft phyllitic mica schists. In the north-east area, on the other hand, the more prominent quartzite schists do not usually occur within $\frac{1}{3}$ mile of the south-east margin. In such a greatly folded series, a bed may have its apparent thickness greatly altered in different localities owing to the varying relations of the ground surface to the different parts of the folds. But it is found that not only in Cowal but also on the west and north-west of this district, the quartzose bands throughout the series, and not alone near the south-east margin, generally increase in prominence to the south-west : for example, they make a great show on the west shore of Loch Fine near Erins and Stronchullin, but away to the north-east, on the north-west side of Glen Fine, they are very thin and inconspicuous. It seems possible that the more quartzose parts of the series were originally thicker and more numerous in the north-east area than they were to the south west.

In his "Preliminary Notice of a Clay Shell-bed between Newton and Strachur, Loch Fynne, Argyllshire" (*Trans. Edin. Geol. Soc.*, vol. iv. 1883, p. 94), Mr W. Ivison Macadam gives a chemical analysis of the mica slate of the district. It is not stated from what locality the specimen analysed was obtained. If obtained from near the shell-bed, it must belong either to the Ardrishaig phyllites or the garnetiferous mica schists. The analysis is as follows :—

Ferric oxide, Fe_2O_3, .	16·91
Aluminic oxide, Al_2O_3,	22·12
Calcic oxide, CaO, .	0·65
Magnesic oxide, MgO,	1·44
Silica, etc., SiO_2, .	58·88
	100·00

In another paper, "Notice of Veins of Specular Iron Ore at Strachur, Argyllshire" (*Trans. Edin. Geol. Soc.*, vol. iv. 1883, p. 95), Mr Macadam gives two analyses of mica schist near the veins described in the paper. The exact locality is not given. The analysis of the schist 6 feet from the veins is as follows :—

Fe_2O_3, . . .	15·18
CaO, . . .	0·36
MgO, . . .	0·84
SO_3, . . .	0·09
Silicates, . . .	83·53
	100·00

The insoluble silicates, after being fused, yielded the following results :—

Fe_2O_3, . . .	1·73
Al_2O_3, . .	22·12
CaO, . .	0·19
MgO, . .	0·61
SiO_2, . .	58·88
Soluble in acids, .	16·47
	100·00

Here and there we get close alternations of colour in the phyllitic schist. Such are seen on the south side of the quartzite at Creag a' Phuill, and in the two burns, at 130 to 270 yards from high-water mark, which run into Loch Fine on either side of St Catherine's pier. These striped schists remind us of the striped phyllites of the Dunoon series, but in the Ardrishaig phyllites there is not the same wide range of colour, there being hardly any purple or black. The Ardrishaig phyllites are also softer and more calcareous than those of Dunoon : many parts in which the eye can distinguish no clear carbonate effervesce freely with dilute hydrochloric acid. A specimen of the phyllite on the shore ⅛ mile south-west of Aird Cottage (slide 5433) was found to consist essentially of carbonates, quartz, sericite, and chlorite ; this is probably more than usually calcareous.

The white mica on the foliation planes seems generally in smaller and less distinct flakes than in the schists nearer " the anticline," and brown mica is less common. The latter is, however, still seen in places. It is, e.g., conspicuous in a pebbly schist within the phyllite zone ½ mile south-west of Stucreach (151). It may be remembered that in the Dunoon phyllites no brown mica has yet been observed (p. 22). Probably, too, the white mica on the shore between Newton Bay and Strachur is in larger flakes than at Dunoon.

The calcareous lenticles weather with an ochreous tint. Similar lenticles also occur, but less frequently, in the bands of quartzite schist. The weathered surfaces of the latter lenticles often show a number of small projecting specks of opalescent quartz, even when none are observed in the quartzite itself. There are all variations in the thickness of the lenticles, from the thinnest streaks up to bands that can be mapped. In the area north-east of Newton Bay there is no limestone outcrop which exceeds 4 feet in thickness, and none have been traced except on the shore ; the thickest occurs ¼ mile north-east of Creag a' Phuill, but it is not a good limestone. Most of the limestones are individually only a few inches thick, but they come so near to one another that they form a considerable proportion of the section. In an 18-inch band nearly ½ mile north-east of Creag a' Phuill the calcite grains are clearly longer in a particular direction along the foliation than in directions across it, and this no doubt is generally so, the direction of elongation corresponding to that of the other indications of stretching. The band referred to has phyllite overlying it, but quartzite schist below.

The quartzites or quartzite schists mixed with the phyllites vary in width from a few inches or less to ¼ mile,—as on the north-east side of Ardchyline (133). In every section of intermixed phyllite and quartzite repeated folding is seen, and there is no doubt that the thicknesses of these rocks have been enormously increased in this way. Scars and fresh fractures are of a pale buff or straw colour. The wider bands split into fairly definite layers which are often from 3 to 4 inches to a foot in thickness, and these layers are sometimes crossed by joints at right angles to the strike. The grain of the rock is usually fine, and shows no indications of pebbles.

On the surfaces of many of the quartzite layers there are scattered small depressions about the size of peas. These are of the same type as the markings on the quartzites on the north-west side of Loch Fine, near Inveraray, which the Duke of Argyll has called attention to ("On certain Bodies, apparently of Organic Origin, from a Quartzite Bed near Inveraray," *Proc. Roy. Soc. Edin.*, vol. xxi. p. 39). He supposes these markings may represent the ends of worm tubes, such as those in the pipe-rock or upper quartzite of Sutherland, etc. Good examples of such markings may be seen in the burn a little over ⅛ mile south-east of St Catherine's, on the shore a little over ½ mile north-east of Ard na Gailich, and ¼ mile north-east of Creag a' Phuill. In the last place there is a thin purplish phyllite between the different posts of quartzite, and the markings weather with a rusty colour. This colour suggests the presence of weathered pyrites, and we have often found either small specks of this mineral, or its decomposition products, in or close by the marks in other localities.

In the burn ⅓ mile south-west of St Catherine's some of the markings on one of the planes of bedding project slightly, others form depressions with slightly raised rims. It seems as if the rock substance composing the marks was a little paler than that outside. We could see no pyrites or ochre in this bed. On another bedding plane near the above there are ochreous spots which weather into hollows : they are sometimes 1 inch long in the direction of stretching, ½ inch in breadth, and rather less than ⅓ inch in depth at right angles to the foliation planes. These last markings are certainly due to the weathering out of pyrites specks. C. T. C.

On the shore 1 mile west of Newton there are somewhat similar bodies both in the quartzite schists and in the micaceous schists which accompany them. They are very numerous, and are found along foliation planes which are oblique to the dip. Most of them are ovate, without any very clear outline. A few of them, however, are partially rectilinear, and contain portions of crystallised pyrites. In others, where the material is brown and rusty, the outline points to its having been derived from pyrites. The shearing of beds containing pyrites has been responsible for, at any rate, a large proportion of these markings. J. B. H.

The microscopic character of slide 5434, from a quartzite schist band ¾ mile S.S.E. of Aird Cottage, is described by Mr Teall as follows :— " Quartz, felspar (scarce), white mica, and chlorite. The last two in detached scales and not very abundant. There are some small grains of sphene."

Pebbly schists may be seen within the series on the shore, 150 yards south-west, and ½ mile south-west, of Creag uam Faoileau ; ₅⁄₈ mile north-east of Stucreach ; at Stucreach, and extending thence for ¼ mile south-west along the shore. In the last exposure brown mica in moderately large flakes is abundant on many of the foliation planes : some of the quartz pebbles are nearly 2 inches long in the direction of stretching, though only ½ inch broad.

As already mentioned on p. 37, green beds of the same type as those further south-east may occur within this horizon, and garnets also are not uncommon. The garnets may be clothed with a rim of chlorite, like those in the adjoining garnet schist zone.

Crystals and specks of pyrites and of magnetite are also not uncommonly seen, surrounded with chlorite rims, e.g., ½ mile south of the " y " of " Strachur Bay," and 200 yards north-east of Tigh na Criche. In the latter locality the rim is sometimes of chlorite and quartz combined, and is prominently extended in two opposite directions along the foliation plane. The magnetite streaks are in places 1 inch long and ¼ inch broad, but when as long as this we could discern no crystal form. On the shore 160 yards N.N.E. of Ardnagowan Cottage, pyrites crystals reaching the size of a pea are abundant, forming nearly half the bulk of some rather quartzose bands, and they seem specially abundant along lines parallel to the direction of extension. The corners of some of them are slightly rounded, and there are growths of quartz and chlorite extending from them in parallel directions.

On the shore 70 yards N.N.W. of M'Phun's Cairn there is an exposure of schist, generally quartzose, which is unusually rich in pyrites. There are some bands, occasionally 6 feet thick, which consist almost entirely of pyrites mixed with some galena and blende. In other bands extremely thin streaks of these minerals run parallel with the foliation, or there are scattered specks of pyrites. The specks often give us the impression of being deformed crystals : they reach the size of a large pea. The bands specially. rich in pyrites seem quite part of the schist, and are sometimes folded with axes parallel to the strike of the schist foliation. The total breadth of the pyritous rock on the foreshore is about 8 yards. It is not seen in the little bank at high-water mark, and probably there is a fault running north-west, between the exposure and this bank. We saw nothing of this rock except on the shore. The exposure is in places overlain by a hard surface-breccia, with a ferruginous cement derived from the weathering of the pyrites; and in other places it is covered by a white soluble efflorescence, which has a strong taste of sulphate of iron. A specimen of the ore that contained more than the average of galena and blende was sent to South Kensington for analysis, and was reported on as follows by Mr Ernest A. Smith.

ORE FROM ARGYLLSHIRE.

	oz.	dwts.	grs.	
Silver,	0	11	18	} per ton of 2240 lbs.
Gold,	0	1	. 7	

Insoluble residue, mainly silica with a little mica, 24·50% ; zinc 3·5% ; iron 30·0% ; lead 3·0% ; sulphur 30·5%. Together with alumina and oxygen. Sample in one lump. Total weight 13,700 grains.

Outcrops of pyritous schist also occur on the shore south-east of Creggans Point, beyond the hornblende schists. These are much like that near M'Phun's Cairn, but we noticed no galena or blende. The exposures extend for about 100 yards along the shore to the south, but there are barren intervals here and there. The rock containing the pyrites is chiefly a quartzose schist, but in the phyllite also there are streaks and nodules of pyrites as large as a pigeon's egg. The pyrites specks are often elongated in the same direction, and there are sometimes quartz growths extending from their sides in this direction. A smaller exposure of similar rock is seen in the bank of the raised beach, along the strike of the rock on the shore.

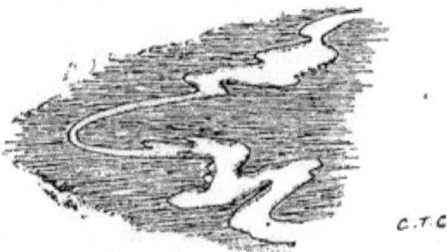

C.T.C

FIG. 25.— × ₄₇. Shore near Newton Bay, Loch Fine. Ground plan of folded phyllite and quartzite schist. Shaded areas are phyllite : direction of shading shows direction of foliation.

The great contrast in colour and hardness between the sericite schist and the quartzite schist shows off the bedding in the sea-coast sections distinctly, and enables one to follow it through many complicated folds. The thickness of any quartzite band varies according to the part of the fold in which it occurs. It is always thinnest in the limbs of folds. At the axes a band may be 7 to 8 inches thick, and yet in part of the limb may be less than 1 inch, or may even be pulled out to nothing.

A peculiarity of the folds here as compared with those at Dunoon, and near Gairletter Point, etc., is that the limbs have very little extension along the strike of foliation, the crests or axes (using the word axis now in its strict sense—see footnote on p. 10) being often steeply inclined and almost parallel to the dip of the foliation. They strike and dip much with the foliation of the sericite schist, but their extension along the strike is so small that the outcrop of any particular band is often more at right angles to the foliation than along it. This is seen on rather a large scale on the shore 100 yards west of Leak (151). The folds are so close that a band only a few inches thick in one place may in another be several feet. If it had not been for the quartzites most of the folds would have escaped notice, as distinct bedding is rare in the phyllites. In the neighbourhood of the folded quartzites there is a close foliation in the phyllites mixed with them, parallel to the axes of the fold; in the quartzites themselves there is usually but little distinct foliation. The close foliation in the phyllites to a large extent obscures original differences of bedding, but on careful examination the phyllites themselves may show indications of folds; sometimes there are slight differences of colour along the bedding, and these different colours cross the foliation. It is possible that the prominent foliation is itself a strain slip foliation, and the earlier colour banding may represent a still earlier foliation running with the bedding. The limbs of folds are sometimes so closely appressed that the soft phyllite is squeezed out from between the quartzite limbs for nearly the whole depth of the fold.

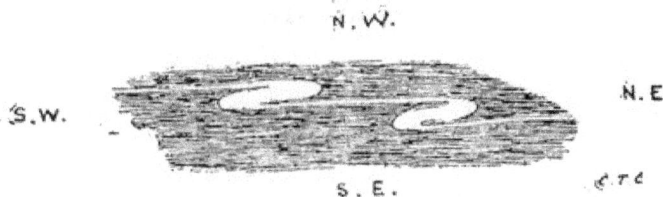

Fig. 26.—Shore ¼ mile S.S.E. of Aird Cottage, Strachur. Ground plan of folded phyllite and quartzite schist. Shaded areas are phyllite: direction of shading shows direction of foliation. Shows the attenuation of the long limbs in the quartzite, and how the phyllite may be squeezed away from between quartzite limbs.

Many of the veins of quartz, or quartz mixed with carbonates, are also repeatedly folded, e.g., on the shore ₁⁵₂ mile south of Aird Cottage.

If we examine vertical sections as nearly at right angles to the foliation strike as possible, we can often see that it is the under limbs of anticlines with axes hading north-west that are the longest and most thinned. But one cannot say that this is so universally, for sometimes the reverse is the case. In a burn ⅜ mile E.S.E. of St Catherine's Pier there is a good section at right angles to the foliation, in a deeper cut than we usually see on the shore, which shows the under limbs to be the longest. On the south side of Creag a' Phuill the well-striped green and grey phyllites on the south side of the quartzite show clear folds with the finest foliation crossing the folds : there are often slight throws, of about ½ inch, accompanying the foliation, in the same way as described in the Dunoon section (pp. 11, 12).

Even near "the anticline" centre there is seldom any doubt which foliations are of "pre-anticline" age, and these are generally much more pronounced than the later ones. On Loch Fine we cannot doubt that the prominent foliation, and the abundant folding which accompanies it, are also of "pre-anticline" age. The folds and foliation of anticline age seem gradually to get less strong in a north-west direction from the anticline centre, just as they do in a south-east direction. Even at the base of the green beds the sharper folds of "anticline" age are not very prominent.

The coast sections give one the idea that the folding is even more repeated than at Dunoon, and the sudden ending of various bands when traced along the foliation helps to confirm this idea. The hornblende schists near Leak and the quartzite near Ardchyline show good examples of this.

Besides the folding of the same age as the prominent foliation, we not uncommonly see sets of folds, of later date, of much less depth and amplitude. These plicate the early foliation without effecting much change in it, and they rarely exceed an inch or two in depth. Generally speaking, the strike of their crests is not far off east to west, and the crests succeed one another in very close succession. They do not seem to affect the more massive beds, quartzites, etc. In the pebbly schist and phyllite on the shore ½ mile or more north-east of Stucreach there is a set of close folds, the crests of which strike rather nearer east to west than the foliation, and the axes hade north-west more steeply. One-third mile north-west of Leak there are small folds with axes hading north-west less steeply than the foliation, and their under limbs are slightly thinned. Between ⅛ and ¼ mile S.S.W. of Aird Cottage some close small folds strike 16° to 23° north of west, and have axes which are almost vertical, or hading S.S.W. In the same locality there are other folds, with nearly vertical axes, which cross the preceding set nearly at right angles : apparently the last mentioned set is the later. At the point ¼ mile S.S.E. of Aird Cottage some prominent small folds have crests striking 29° south of west, and axes hading south-east. These are stronger than most of the later folds in the neighbourhood, and are accompanied by strain slips and thinning of the under limbs of anticlines. The axes hade south-east, but sometimes at so low an angle that they are almost horizontal.

In the albite schists near Achadunan (126) it is also common to see sets of small late folds with axes hading south-east, and generally striking nearer east and west than the strike of foliation. Perhaps these are of approximately the same age as the late folds in the Ardrishaig phyllites. They may both represent the latest "anticline" folds mentioned on p. 26.   C.T.C.

CHAPTER V.

EPIDIORITES, HORNBLENDE SCHISTS, AND RELATED CHLORITE SCHISTS.

The epidiorites, and hornblende and chlorite schists connected with them, are much more abundant in the north-west third of the area than in other parts. Even the most schistose of these rocks are generally harder than the other schists, and so often stick up in field exposures; they are traced with comparative facility and afford a great help in working out the physical structure.

They are never vesicular or scoriaceous, and seem in places to have given rise to a small but perceptible amount of contact metamorphism in the schists at their sides. Hence it is believed that they represent old intrusions, and not lava flows. On the shore of Loch Fine ½ mile west of Newton (151) the quartzite schist on either side of the hornblende schist is baked and altered. A similar alteration is noticed ½ mile south-west of Castle Lachlan in the schist next to a felspar-chlorite schist, ⅛ mile west of Tom na h' Iolaire (171) in a calcareous schist next to a hornblende schist, and on the north side of Auchalick Bay (181).

We cannot in Cowal point to any clearly visible transgression of bedding by these schists. The north extremity of the hornblende schist on the north side of the burn that flows into Loch Loskin (184) is, however, of a very irregular shape and strongly suggestive of an intrusive rock. One of the hornblende schists both on the west and on the north of the Bishop's Seat seems also gradually to change its stratigraphical horizon (see p. 174). The hornblende schists are seen to be repeatedly folded with the sedimentary schists (see, e.g., pp. 174, 176, 213), but it is possible, notwithstanding this, that these sedimentary rocks were partly folded and sheared before the intrusion of the rocks now represented by the hornblende schists, etc. There is no evidence, however, to show that this was the case.

They do, however, in the main, keep to the bedding of the schists of sedimentary origin, and they have undergone the same foldings as these. In the west and north-west parts of Cowal they occur on the horizons of the Glendaruel or Loch Tay limestone, and of some other beds which lie on the north-west side of this. Between this limestone and the green beds on the south-east side of "the anticline" very few bands of epidiorite, etc., have been mapped, but it is believed that some may have escaped notice, the thickness of the bands near Ardentinny, and between there and Uig, being generally not more than a few yards.　　　W.G., C.T.C., J.B.H.

There is a special danger of confusing some of these old igneous rocks with green beds. In both, chlorite may be a prominent constituent; in both, ferriferous carbonate spots, and quartz strings stained with epidote, are common. In hand specimens and microscope slides it is sometimes impossible to decide with confidence whether a certain rock is a green bed or a schist derived from an originally massive igneous rock. Isolated field sections may even leave one in doubt about this question, but after an examination of all the exposures belonging to the doubtful bed it is generally possible to give an answer with a fair amount of confidence. We may, for instance, find indications of pebbles, or bands of pebbles, in some of the exposures, or find the rock passing, especially as we proceed inwards from the margins of the band, into one of which the original igneous character cannot be doubted. Occasionally small specks and streaks of opalescent quartz may occur in the epidiorites, but these specks are, fortunately, in rocks which tell their own tale of igneous origin clearly. The specks of carbonate in these old igneous rocks, and also those in the green beds, are not confined to the surface of the rock, and they have not originated as a consequence of weathering. They occur throughout thick masses of rock, and are an essential part of the soundest varieties. We suppose them to have been formed as a result of the alteration of hornblende into chlorite or black mica. In some bands the specks of carbonate are so numerous and large that the beds containing them acquire the aspect of limestones. This is seen in the band $\frac{1}{2}$ mile south-west of Stronchullin Hill (163 south-west), and in that on the south side of Coire Athaich (173 north-east). The last mentioned band was, indeed, originally mapped as a limestone.

Opaque white patches of leucoxene (decomposition product after ilmenite) are at times large enough to be distinguished by the unaided eye, and may have ilmenite kernels.

In the bands which occur on the south-east side of "the anticline" the hornblende is represented by pale green shreds of fibrous actinolite. This can generally be made out with a hand lens. The felspar is represented by streaks and patches of saussurite, generally pale grey by reflected but almost opaque to transmitted light. These streaks may occur either closely parallel, or in a mesh, according as the rock is more or less perfectly foliated : in either case, they are a help in distinguishing these rocks from green beds. Slide 2832, from the band exposed on the coast 230 yards N.N.W. of Hunter's Quay, is thus described by Dr Hatch : "Flasers diabase or epidiorite (?). Well-marked flaser structure. Lenticular masses and layers of green fibrous hornblende (actinolite) alternating with dirty coloured almost opaque masses of very altered felspar (probably saussurite), which show here and there a tendency to the development of eyes. Leucoxene in isolated turbid grains. Probably a sheared and metamorphosed diabase or epidiorite." We noticed here and there in the coast section from which the rock described above was procured, some squarish sections of saussuritized felspar, reaching to $\frac{1}{8}$ inch in length, but these are not common. In the exposure west of Blairmore pier the hornblende, again pale green, occurs in somewhat stouter and more conspicuous forms, often about $\frac{1}{30}$ inch in breadth.

The first hornblende schists noticed after passing north-west of "the anticline" are some thin outcrops—never apparently more than 16 feet thick and not often so much—which occur near the base of the green beds, sometimes a few feet above and sometimes a little below. Such are seen at the top of Creag Bhaoigh (152), on the north and east of Cruach nam Mult (134), and on the west side of Binnein an Fhidleir (134). These thin beds are, as we might expect, well foliated throughout. The hornblende in them has not been replaced by chlorite in any marked degree : it is dark green in hand specimens and less fibrous than the variety which exists in some of the Cowal hornblende schists. The same general type continues in the different bands of hornblende schist we meet with further north-west as far as the graphite schists, and is seen best of all in the sills which almost constantly accompany the Loch Tay limestone. We may therefore speak of this type as the Loch Tay type.

In a wide band in the green beds $\frac{1}{2}$ mile east of the "1" of "Maol Odhar" (162) the foliation is sometimes less perfect. The hornblende appears in shapes up to $\frac{1}{12}$ inch in breadth, which have a general parallelism of direction, but do not coalesce together into lines. Some scars of this exposure are very blocky. In the probable continuation of this band south-west, yellow mica may occur in abundance, e.g., $\frac{3}{4}$ mile south of the "i" of "Strondavain" and $\frac{1}{4}$ mile south of An Cruachan.

The sill in the Loch Tay limestone in the burn a little more than $\frac{1}{4}$ mile south of the "u" of "Creag Dhubh" appears to be about 30 feet thick. This is perhaps about the average thickness, but near Ardkinglas it seems to be thicker, the outcrop being unusually wide even for the dip slope. In the burn $\frac{1}{2}$ mile south-west of Glensluan there are veins of epidote, quartz, and a little carbonate between the hornblende schist and the overlying portion of limestone. The epidote is in prisms of a pale yellowish-green colour; these occasionally attain a length of $1\frac{1}{2}$ inches, and are sometimes slightly bent.

In the burn $\frac{7}{8}$ mile south-east of Ardno there is a sill a little below the base of the Loch Tay limestone which has an apparently greater thickness than the sill within the limestone; but outcrops on this horizon are not often seen in the north part of Cowal. C.T.C.

In the area west of Glendaruel the hornblende schist in association with the Loch Tay limestone is sometimes found below the limestone and at some distance from it, or it may be resting on the limestone, or it may be at some distance above it. It must be borne in mind that the hornblende schist represents an intrusive rock, and that the region was intensely folded after the intrusion. It is not surprising, therefore, to find that sections along the line of strike show variations of character in the relative numbers, sizes, and positions of the hornblende schist bands. In some sections the hornblende schist does not appear at all, while in the next burn, at a short distance, bands of various thicknesses may be found, either in the limestone above, or in the limestone below it, or in both of these positions. Along the steep west slope of Glendaruel, there are numerous burn sections, crossing the strike of the beds, which exhibit such differences; it is clear that the hornblende schists are not necessarily continuous, but that they represent a series of sills which have been injected along slightly different horizons and then subsequently folded; they vary considerably in width of outcrop, and this variation is certainly attributable in many cases to differences in the amount of folding. These different sills are all composed essentially of hornblende and felspar. Biotite is generally common. Garnets, epidote, magnetite, pyrites and leucoxene occur as accessories. Chlorite and calcite are rare. The hornblende may be green, dark green, or nearly black; the dark varieties are very common. In the finer grained specimens actinolite is often the common hornblende, disposing itself in needles along the foliation planes. In the coarser types the hornblende occurs in dark green stumpy prisms. In these types the foliation is less perfect than in the finer grained types, which are usually highly schistose throughout. The coarser varieties, especially when they occur in big masses, sometimes show very little foliation in their interior coarser portions, though they are finely schistose in their upper and lower portions. Even the coarsest specimens are finer in grain than some of the bands found to the north-west of them, near Stralachlan. J.B.H.

As elsewhere mentioned (p. 48) garnets sometimes occur in the hornblende schists within or near the garnet schist zone. Some are seen in a thin exposure within the limestone on the west side of Ardno burn $\frac{3}{4}$ mile south-east of Ardno, in a band under a limestone $\frac{1}{2}$ mile slightly north of west of the first "C" of "Cruach nan Capull" (133), and in a locality $\frac{5}{8}$ mile north-west of the same letter.

The hornblende schists that accompany the limestone which is a little south-east of the graphite schists, are of the same type as those with the Loch Tay limestone. The bands $\frac{3}{4}$ mile south-west of An Carr (152), and on the west side of An Carr, near the top, are also of this character.

The bands on the north-west side of those mentioned in the last paragraph are generally of a rather different character. Most of the hornblende is pale green in hand specimens and colourless in thin slices, belonging to the variety called tremolite. In the more sheared varieties chlorite is often abundant, and sometimes it prevails to the exclusion of hornblende. In some bands certain portions are unusually poor in felspar, and appear to be ultrabasic. We may call this type the St Catherine's type, as it occurs in two broad bands near this place. It is not probable that there is any sharp line between this and the Loch Tay type, and we may be in doubt as to which type a particular band should be referred, but in the extreme forms the difference is readily discerned.

The two broad bands at St Catherine's are of much the same character. The south-east band comes down to the sea-coast on the south-west side of Ard na Gailich, between 140 and 370 yards from the point, and shows itself in scars near the top of the wood between St Catherine's and Ardchyline. The north-west band is seen on the shore south-west of St Catherine's, from nearly $\frac{1}{6}$ mile to slightly over $\frac{5}{12}$ mile from that place, also at the back of the raised beach, and in the island-like knoll on which the chapel of St Catherine stands. There are old quarries on an extensive scale east of the knoll, and most of the stone used in building Inveraray Castle, and in rebuilding it after the fire, is said to have been procured from these. The stone is reputed an excellent building stone, extremely durable, and yet comparatively easy to cut. It has long been known under the name of "potstone."

A specimen of the chief variety obtained from the quarry 50 yards north of St Catherine's Chapel has been sliced (slide 5058), and is thus described by Mr Teall : "The main mass of the rock is composed of chlorite and white hornblende (tremolite). The tremolite forms irregular patches consisting of individuals confusedly aggregated. The chlorite also forms aggregates. It is faintly pleochroic in green and brown tints. . . . . One patch looks like unaltered hypersthene. With this doubtful exception, the whole rock must be looked upon as a pseudomorph. In the process the original minerals have not only been replaced by others of different composition, but their forms have also been destroyed. The rock in its present state may be termed tremolite-chlorite rock." In the hand specimen, the tremolite appears of a pale greenish tint. Some of it forms stout lumps nearly $\frac{1}{4}$ inch long.

Mr J. Wallace Young gives ("Miscellaneous Notes on Chemical Geology," *Trans. Geol. Soc. Glasgow*, vol. iii. p. 28) an analysis of foliated chlorite from St Catherine's, Loch Fine, but he does not state the exact locality from which he obtained the specimen. He describes the specimen as follows :—"Colour, blackish-green ; lustre, pearly.; consists of long, narrow foliæ cohering together, rendering the mineral almost fibrous in appearance. The thin leaves nearly transparent. When in a state of very fine sub-division, it is entirely decomposed by sulphuric acid. Sp. gr. 2·781 at 15·5° C.

Silicic Acid,	.	.	33·55
Alumina,	.	.	15·00
Ferrous Oxide,	.	.	10·78
Magnesia,	.	.	29·73
Water (by difference),		.	10·94
			100·00

"No chromium present. In one specimen the chlorite was associated with a ferriferous dolomite in rhomboidal crystals. Its composition is as follows :—

Sp. gr. 2·935 at 15·5° C.

Carbonate of Lime,	.	.	53·00
,, Iron,	.	.	8·16
,, Magnesia,	.	.	39·00
			100·16

Trace of Manganese."

Part of the rock on the shore 260 yards south-west of the pier, is very closely foliated and soft, and we could see no hornblende in it. But it did not seem possible to separate it by any sharp line from the more massive rock with clear hornblende on the north-west, and the soft rock itself contains small eye-like pieces that are massive. Twenty yards further southwest, the rock is very thinly foliated, and contains much black mica, and on the foliation planes there are closely felted dark green actinolite needles. The needles are as much as $\frac{1}{2}$ inch in length, and they lie in different directions on the foliation planes, but do not cross them. The main portion of the rock near this is massive, and weathers with a brownish-orange tint; in some parts the white elements are more prominent than others; black hornblende in "eyes" of the size of a pea is occasionally seen, as well as the tremolite. There are some veins, as much as 3 inches wide, which contain epidote prisms, dark green chlorite, and some calcite. The veins are often nearly vertical, and strike nearly east to west. The epidote prisms are sometimes seen penetrating into the calcite; they usually run across the vein from side to side, and sometimes attain a length of 2 inches or more. In some places they are distinctly bent; we see the bending most pronounced when the long axes of the prisms are at right angles to the foliation strike; but when the long axes are parallel to the strike, they may also be waved to some extent. Near them are also thin bands which consist of asbestus and chlorite, with here and there small eyes of hornblende encircled by the asbestus fibres.

On the shore $\frac{1}{12}$ mile south-west of the pier, the part of the band next to high-water mark is massive and hard and sharply jointed: blades of hornblende are prominent on the weathered face. Near low-water mark, and separable from the rock just described by a fairly definite line, is a much darker variety, weathering in big blocks with rounded outlines. This is much softer than the rock near high-water mark. In hand specimens it shows pale green "eyes" of hornblende, attaining the size of peas, embedded in a soft dark matrix in which brown mica is abundant. A specimen of the paler rock was examined by Mr Teall (slide 5057), and he sees in it "hornblende, chlorite, felspar, quartz, epidote, carbonates, and brown mica (scarce). Indications of micropegmatitic structure in places." The quartz specks are pale blue and opalescent in the hand specimen; the felspar is abundant and in moderately large pieces. The softer ultrabasic-looking rock from the same locality was also sliced (slide 5059), and examined by Mr Teall. He calls it a tremolite mica schist, and sees in it "tremolite, brown mica, a little green hornblende, and some aggregates of sphene granules." The mica flakes lie parallel to one another, and give rise to the foliated look of the rock; they are closely intermixed with the tremolite, not in separate layers.

Bluish opalescent quartz, like that in specimen 5057, is also seen in a massive portion of the south-east band of epidiorite at a point $\frac{1}{12}$ mile north-east of Ardchyline. Mr Teall thus describes this rock (slide 5056), "hornblende (irregular), quartz, epidote, felspar (scarce), white sphene with kernels of iron ore. Epidote-amphibolite or epidiorite. Almost certainly an altered dolerite." The sphene is in pieces quite large enough to be distinctly seen with the naked eye. In the exposure $\frac{1}{2}$ mile south-east of Ard na Gailich the sphene streaks are sometimes $\frac{1}{2}$ inch long. Many of the epidiorite blocks, whether distinctly sheared or not, in the sea-wall about 100 yards north-east of the pier, also show spots of blue opalescent quartz and sphene streaks with iron-ore kernels; the streaks often run parallel to one another and the other rock constituents.

Both bands of epidiorite are in places mixed with an unusually soft and thinly foliated rock in which no hornblende is to be seen by the

E

naked eye. Such schists are seen in the burn 260 yards south of, 150 yards
E.S.E. of, and ¼ mile south-east of the pier. We were long in doubt
whether these might not be chloritic green beds, rather than specially
sheared portions of the epidiorites. Green beds certainly do occur within
the Ardrishaig phyllites (see p. 37), and it was quite possible to suppose
they might be mixed with epidiorite sills most intimately. Similar
schists also occur alternating in thin bands with the phyllite and quartzite
schist next to the epidiorites. Sometimes the bands are only 6 to 8 inches
thick, but occur in a certain breadth of exposure in almost as great proportion as the phyllite, etc. We have, however, never seen any clear pebbles
in the thin schists referred to, and Mr Hill finds abundant evidence, on the
other side of Loch Fine, of the gradual passage of such schists into other
rocks of the igneous origin of which there can be no doubt, and this in
localities where no green beds are known. Mr Hill states that the more
finely foliated chloritic parts often occur near the outsides of the igneous
masses—parts which would inevitably be exposed to more intense shearing
than the interiors—and also that there are many thin sills which are composed throughout of soft chlorite schist. A specimen of one of these
schists from the burn 130 yards east of the pier, was sliced (slide 5060),
and examined by Mr Teall. It was found to be composed of chlorite,
carbonates, quartz, leucoxene, and a little white mica or talc. There was
nothing in the slide to indicate the original igneous character.

The two bands at Creggans Point are much like those at St Catherine's.
The north-west band is very schistose near the base, but its interior is
massive and shows abundant broad tremolite pieces. The north-east
continuation of this band is exposed at a place ½ mile south-west of the
"g" of "Creagan an Eich." It is very massive, and in some parts
displays hardly any indications of schistosity. The tremolite pieces are
sometimes nearly ½ inch in breadth : there is also a dark, almost black,
hornblende in some parts of the exposure. Still further north-east
the band seems to become thinner, and near the head of Eas Dubh there
is a series of outcrops on approximately the same horizon which are
generally so thin and so close together that in mapping they have to be
generalised. These thin bands are pale green chlorite schists, and do not
in themselves testify to their igneous origin.

On the shore 140 yards north of Stucreach (141) there is a very blocky
epidiorite in which the hornblende appears black in hand specimens. The
foliation is not pronounced, but the white and black constituents have their
greater lengths in rough parallelism. Near the base of the band on the
south-west side the grain of the rock is finer than in the interior, and
there are indications of porphyritic felspars. There are many veins of
chlorite and ferriferous carbonate, and thin streaks edged with yellow
mica, which cross the general direction of the rock constituents. Two to
three feet below the main outcrop, and separated from it by a quartzite
schist, is another thin band of epidiorite, also with indications of
porphyritic felspars.

The exposure on the coast ⅓ mile north-east of Leak is like that near
Stucreach, and probably is a continuation of the same band.

At Creag nam Faoileann (141), and several places on the coast as far as
¼ mile to the south-west, there is a close and evidently much folded series
of thin epidiorite schists. Some of them resemble the Loch Tay type
more than the St Catherine's.   C.T.C.

The epidiorites, etc., on the north-west side of a line drawn from Newton
(151) to a little east of Lephinchapel are rather allied in character to those
at St Catherine's, but are often more massive, and contain great eyes of

epidotised epidiorite or epidosite. We may speak of the type in this district as the Stralachlan type, as it is found in greatest strength between Loch Fine and the valley of Stralachlan, in the more siliceous part of the Ardrishaig phyllites.

In the area between Stralachlan and Loch Fine the epidiorites, etc. form huge irregular masses behaving on a large scale like sills with irregular protrusions, the longer axes of which coincide with the strike of the main mass, and of the quartzites in which they are intruded. One reason for the size of the masses found in the field compared to the exposures of the Loch Tay limestone type may be the different inclinations in the two areas. But apart from this, it is certain that the Stralachlan epidiorites far exceed in dimensions those of the Loch Tay limestone type.

The Stralachlan epidiorites differ amongst themselves in composition and texture. Not only so, but in the same intrusion various types may be observed. A few of these intrusions will now be described.

The strip of land between Newton Bay and Lachlan Bay is almost entirely made up of these intrusive hornblendic rocks. The ground is occupied mainly by two huge sill-like protrusions, each about $1\frac{1}{2}$ miles in length. The most easterly sill has an average width of 300 to 400 yards. This sill is separated from the sill to the west by a strip of sediment 30 to 100 yards in width. The westerly sill is more irregular, its eastern and western boundaries are roughly parallel with each other, and with the adjoining sill, but it is often sub-divided, and is nearly bisected by strips of sediment. In places the sedimentary schist appears to entirely isolate parts of the hornblende schist, but it will be convenient for the purposes of description to regard it as one mass, of which the widest part is $\frac{1}{2}$ mile across.

The most easterly or Bàrr nan Damh sill will be described first. The rock consists essentially of coarse hornblende and felspar in variable proportions. As a rule it is unsheared except near the outer edges, and may be regarded as a modified diorite. The mass as it is traced along presents every variety of structure. It is mostly a coarsely crystalline rock. Some zones are made up of little else but crystals of black hornblende packed together, in stumpy prisms of the size of marbles, but the hornblende most commonly occurs in green fibrous lath-shaped or tabular shapes. While in some places the hornblende occupies the greater part of the mass, in others the felspar predominates over the hornblende, and this perhaps at no great distance from the more basic zones. Again, in other parts there may be an intimate admixture of hornblende and felspar, the felspar usually occurring in lath-shaped forms, sometimes enclosed in fibrous hornblende. In many cases it is clear that the felspars crystallised out before the hornblende. The felspars are probably original, and some of the hornblende may be so also. Scattered crystals of green epidote and occasional grains of blue quartz are seen.

Mr Teall describes a specimen sent to him for slicing as follows :—

"4000 :—Coarsely crystalline massive rock. Pale green hornblende in blades often $\frac{1}{2}$ inch long in greyish matrix."

"Microscope :—Uralitic hornblende, a saussuritic aggregate of water clear felspar and granular epidote, irregular patches of sphene (leucoxene), and aggregates of chlorite. Epidiorite."

A common structure is the aggregation of felspar and hornblende crystals in shapes corresponding roughly to those of the hornblende prisms. These crystalline aggregates are very numerous in this area. Instead of porphyritic crystals of hornblende, the outlines of porphyritic hornblende crystals occur, but the mineral substance itself that builds up the crystal is an admixture of hornblende and felspar.

In some of the coarser epidiorites of the Stralachlan area, great lath-shaped crystals of dark fibrous hornblende attain a length of 2 or 3 inches, the crystals being free from any admixture of felspar. These rocks are, strictly speaking, amphibolites, but they pass insensibly into the ordinary epidiorites, and cannot be separated from them in the field.

At the south-west extremity of this Bàrr nan Damh mass, at Bàrr an Longairt, the rock assumes yet another character. It becomes so acid as to present quite a granitic appearance in the field. Porphyritic crystals of felspar are thickly scattered through the rock, some of them retaining a perfectly crystalline outline. Crystals of orthoclase can be made out with the Carlsbad type of twinning by the aid of a lens. These crystals of felspar are by far the most prominent of the minerals of which the rock is composed. Blue quartz grains without crystalline shape are scattered broadcast through the rock. Epidote is also common. Hornblende occurs in grains or as fibrous plates packed together. The hornblende alone in the rock gives an appearance of schistosity, the plates of hornblende usually lying along definite planes. Throughout the rock there are distinct planes of movement, the felspar crystals on approaching these lines of strain becoming pulled out, bent, and broken. Besides the fibrous hornblende a little mica occurs along the planes of schistosity. Between these undulating schistose planes the crystals appear to have been unaffected by the movement, and are arranged without any definite orientation; but as these planes are approached there is a tendency for the crystals to lie with their long axes parallel to these planes. In this granitic type the felspar crystals retain their crystalline form, and the surrounding schistose material is hornblende. In the more basic varieties described before, the hornblende stands out in large well-formed crystals, and the inclosing schistose material is felspar. In this latter case, however, the felspar is probably all secondary, and certainly some of the hornblende is, as sections under the microscope reveal uralite, and it seems likely that these basic rocks have been derived from gabbros. On the other hand, the granitic parts of the mass are distinctly more acid, and these have suffered the least change; the greater portion of the minerals we now see being original. The more basic portions have broken down earliest, and it is these masses that have been converted into hornblende schists, which will be referred to later in dealing with the westerly sill.

The following is Mr Teall's description of a specimen of the more acid type from this area.

"4005:—Massive crystalline rock, epidote, felspar, and blue quartz, easily recognised with a lens.

"Microscope:—Felspar (sometimes striated), in large and more or less idiomorphic individuals, and also as a constituent of a quartz felspar mosaic. Epidote in local granular aggregates, green hornblende in ragged aggregates, brown mica, iron ores, and apatite. Quartz in large grains, and also as a constituent of a mosaic."

Before dealing with other varieties of epidiorite in this area it will be convenient to describe a feature that is common to all of them, from the most acid to the most basic type, and from those of the coarsest crystalline structure to the finest-grained schistose varieties. In all these intrusive masses cores or eyes of a yellowish-green material are common. These eyes vary in size from a few feet to a few inches, and seldom show schistosity: where the rock is schistose the foliation planes encircle them: sometimes they are made up almost entirely of granular epidote. Often hornblende crystals are scattered through them, especially in the more crystalline masses. They may also contain blue quartz, and in some instances the cores differ little from the surrounding rock, except in the colour which is due to the presence of granular epidote. The colour of these eyes makes them very conspicuous objects in the field, and from their general massive habit and the tendency of the foliation planes to flow round them, it is evident that they belong to the earlier stages of the history of the mass. The hornblende has been the first mineral to crystallise out, occurring generally in the form of short stumpy prisms, but occasionally in long blades of fibrous hornblende or actinolite. In these coarser eyes a tendency to foliation can sometimes be seen in the arrangement of the hornblende crystals with their long axes in one direction. These eyes sometimes pass insensibly into the inclosing mass, or they may be seen sharply divided from it. In the mass at Creag Bhreac, just above the schoolhouse at Stralachlan, the hornblende schist is coarsely crystalline. The hornblende crystals are as large as peas.

In this mass one of these epidotic eyes attains a length of 5 to 6 feet, its texture being very fine-grained compared to the inclosing mass. The line between the two is very sharp. Quite close to this eye, however, a smaller eye is seen about 1 foot in length; in this case the hornblendic rock and the epidotic eye pass gradually into one another, hornblende crystals similar to those occurring in the main mass being scattered through the outer zones of the eyes, but in this instance the central portion of the eye is free from hornblende, and is of an ordinary fine-grained granitic texture. In some cases the hornblende crystals occupy not only the outer zone of the eye but are scattered through the central portions, and there is nothing to distinguish these portions from the main mass excepting the yellowish-green colour due to the presence of epidote, and the more acid character indicated by the less quantity of hornblende.

In the very fine-grained chlorite felspar schists and chloritic hornblende schists the eyes are much smaller, and are homogeneous throughout, consisting entirely of epidosite.

A specimen of one of these eyes from a fine-grained chlorite hornblende schist was examined by Mr Teall—his report is appended.

"4003 :—Eye in epidiorite. Kilbride Chapel, Stralachlan.

"Compact yellowish-green rock veined with quartz.

"Microscope :—Granular epidote with very little quartz, containing carbonates as well as quartz."

There occur in the Bàrr nan Damh sill two smaller sills of a different type, which appear to be distinct in character and of a different age to the rock in which they are found. One is seen on the western boundary of the sill about ½ mile south-west of the summit of Bàrr nan Damh, the other is quite inclosed by the sill, but occurs near its eastern boundary about ¾ mile south-west of the summit of Bàrr nan Damh. The former is about 400 yards in length and 30 yards in width, and the latter about 80 yards in length and 20 yards in width. In the field they can be readily distinguished by the rusty colour they assume on weathering. At a distance they look like decomposing basalts, and in their harder portions they exfoliate. The smaller sill is so decomposed that it is difficult to obtain a fresh specimen, but is in reality more fissile and schistose than the epidiorite surrounding it. It consists mostly of a very fibrous hornblende circling round crystals of felspar, many of which are somewhat large : it passes insensibly into the hornblende schist on either side.

The larger rusty mass, ½ mile south-west of Bàrr nan Damh, is much fresher in appearance. It weathers rusty brown, and shows a peculiar coarse flaggy structure parallel to the foliation dip. It is composed principally of fibrous hornblende, and it is almost impossible to obtain

a fresh specimen with the hammer,—in striking contrast to the fresh
epidiorite surrounding it which is very compact and fresh. . No clear
junction can be detected between this rusty hornblende schist and the
epidiorite. It rather seems as if the epidiorite was later than the rusty
sill, and had modified its outer boundary, which has been partially
recrystallised. It is not certain that the rusty sill is the earlier of the
two, but the evidence points in this direction. This type of sill is
exceedingly rare, and these are the only instances of their occurrence in
this area, but we have observed similar sills in the neighbourhood of
Lochgilphead. A specimen from one of these Lochgilphead sills was
examined by Dr Hatch. It is composed of colourless blades and prisms
of tremolite, embedded in a confused felted aggregate of fibres of hornblende and scales of chlorite, together with a little magnetite in dispersed
granules.

The big easterly or Bàrr nan Damh sill having now been described, the
great westerly mass will next be dealt with. As previously stated, this
mass lies close to the Bàrr nan Damh sill, being separated from it by a
strip of quartz schist. It is in the form of a compound mass giving off
numerous branches. Besides the coarse types of rock described in the
Bàrr nan Damh sill, its outer portions and some of its branches consist of
finer-grained hornblende schist just as foliated as the hornblende schists
of the Loch Tay limestone type. The epidotic eyes or cores are as
prevalent throughout this mass as in the easterly sill, and are as abundant
in the fine-grained schistose varieties as in those of coarser crystalline
texture.

The westerly boundary of the mass is very fine-grained and highly
sheared. The mass extends near the coast in a sill over $\frac{1}{2}$ mile long,
and between 60 and 70 yards wide. This sill meets the coast nearly
$1\frac{1}{4}$ miles south-west of Bàrr nan Damh. It is so sheared that in some
places hornblende crystals can no longer be detected, and the hornblende
has largely been converted into chlorite.

Mr Teall thus describes a specimen sent to him :—

" 4006 :—Green schist with pronounced lustre on the surface of
schistosity.

" Microscope :—Felspar, water clear and often showing traces of lath-
shaped forms, chlorite, and irregular patches of leucoxene. Quartz and
carbonates both present, but not abundant. Felspar chlorite schist."

This fine-grained felspar chlorite schist can be followed into the main
mass, which is a coarsely crystalline epidiorite. Close to the locality
where this fine-grained portion occurs, bands and patches of the coarsely
crystalline rocks are found, one of which appears to be later than, and to
have cut through, the felspar chlorite schist. The boundaries between the
two rocks are obscured by basalt dykes, and their relations are not clear.
It is possible that the basalt dykes may have come up along lines of fault,
which the dykes have succeeded in hiding. Against this supposition,
however, must be set the fact that the felspar chlorite schist does not seem
to have been thrown by faulting : the sill on both sides of the coarsely
crystalline epidiorite taking an even course as if there had been no
disturbance. It seems probable, therefore, that the basalt dykes have
seized on the intrusive junction of the epidiorite with the felspar chlorite
schist, and having made their course along the margin the original
junctions have now been destroyed. The intruding mass is a very
coarsely crystalline hornblendic rock with hornblende crystals of the size
of peas closely packed together in a dark matrix. The igneous rock which
it has cut through is a felspar chlorite schist.

About $\frac{2}{3}$ mile further south there is another instance of one hornblendic rock cutting another which has not been complicated by the presence of basalt dykes. Slightly to the south-west of Bàrr an Longairt and ½ mile north of Castle Lachlan a small mass of epidiorite occurs, about 130 yards long by 70 yards broad. It evidently belongs to the same set of intrusions as the other igneous masses in the vicinity which have been described. The rock is composed of dark green hornblende in crystals ¼ inch to ½ inch in size, intermixed with a good deal of felspar. In the centre of the mass there is a dyke-like band 4 or 5 feet wide, far more acid in composition and possessing no foliation, the structure being granitic. This band is composed largely of crystals of pinkish felspar, mixed with scattered grains of epidote, and a small amount of hornblende in little prisms. It varies considerably in texture, in parts resembling a very fine-grained diorite. It is sharply divided from the hornblende schist, with which it is associated, but it appears to belong to the same general set of intrusions, and probably represents one of the latest phases of irruption. It is significant that the general diffusion of epidotic material which is a prevailing characteristic of the epidiorites of the Stralachlan district is present here also, the yellowish-green tint due to the presence of epidotic materials being very marked. The fact that the acid part of the irruption is the latest, follows the usual rule and is what we should expect.

Mr Teall gives the following report on a specimen of this dyke-like intrusion, sent to him for examination.

" (4008) :—Medium-grained, massive crystalline rock.

" Microscope :—Plagioclase felspar in allotriomorphic grains forms the main mass of the rock. In the felspathic aggregates are seen more or less idiomorphic prisms of green hornblende, grains of epidote and iron ore (scarce). Diorite."

At Barnacarry, Loch Fine (151), very massive epidiorites occur along the coast, and form the wooded hills between Barnacarry and Stralachlan river. These massive beds are good examples of the epidiorites in which schistosity has not been carried far. On the whole they are more acid than the rocks to the north-westward already described. They are coarsely crystalline, and, as a rule, appear to be made up of a nearly equal admixture of green hornblende and plagioclase felspar. Here and there where the rock gets finer-grained, a foliated structure is developed, the hornblende crystals lying with their long axes in one direction. Many portions of the mass, however, which are fine-grained, show no schistosity. Epidotic eyes and cores like those already described are dispersed throughout the mass, and rarely show foliation. The hornblende may occur in short, stumpy prisms, or in long blades : the arrangement of the latter, with long axes parallel, gives the rock a foliated appearance. The mass is often traversed by veins 2 or 3 inches in width, different in texture from, and of more acid composition than, the main mass. At the edge of the mass on the north side the rock is very schistose for about 6 or 8 feet from its margin ; inside this, although coarsely crystalline and full of epidotic cores, the rock is markedly pulled out, but without destroying the coarse granitic character ; there has been merely a tendency for the crystals along certain lines to arrange themselves with their long axes in one plane.

At the point on the north side of Lachlan Bay close to the ruins of the old castle, mica slates and epidiorite sills occur associated together. The hill above is made up of massive quartzites, which have, to a large extent, resisted the earth movements. The argillaceous beds below have, however, completely succumbed and suffered the most intense folding and shearing, so that the original bedding planes have been destroyed, and the accompanying sill or sills of epidiorite have been so folded and sheared, that their original characters have been obliterated, and they are now represented by highly fissile green schists, in many parts of which no hornblende is visible, and the rocks are nothing more than highly folded and crumpled chlorite schists. They are so folded with the mica slates that in places the divisions are too small to be drawn on the 6-inch Maps. Some of the junctions, however, with the sedimentary rock show contact metamorphism, and the igneous nature of the rocks is clear. Scattered through some of the seams too are small eyes of epidosite so characteristic of the epidiorites of this area. In a few instances a coarser crystalline texture may be observed ; this is notably the case in a limb that is seen twisting round from north to south, and finally assuming an easterly course ; parts of this limb are moderately coarse in texture, and contain eyes of epidosite. The north part of the limb after bending round is highly sheared for some distance and converted into a fissile chlorite slate, and at the junction, is folded and mixed up with the argillaceous schists. At the southerly junction, however, the bed is more massive and crystalline, and only shows extreme fissility near the junction with the sedimentary schist. It is similar in composition to the felspar chlorite schist (4006) already described on p. 70, which has been traced into a coarsely crystalline epidiorite. In this instance, the greater shearing and alteration of the hornblende schist has probably been due to its intercalation in the phyllites, and their position relative to the quartzite overlying them to the north. The latter owing to its greater power of resisting the effects of dynamic movements, has tended to increase the deformation of the underlying, more yielding strata. Hence we find that the phyllites and the hornblende schist (now converted into felspar chlorite schist) are traversed by fissile planes having no clear relation to the bedding.

J.B.H.

CHAPTER VI.

SERPENTINE.

Inellan and Toward Serpentine.

Serpentine occurs in, or closely associated with, the schist boundary fault near Inellan and Toward, and is generally in an extremely crushed state, so that its original mode of occurrence is not in most places clear. It is seen in the coast sections a little south of Inellan Pier and by the east gates of Toward Castle policies (194 south-east), and in various places inland between these, e.g., at the back of the raised beach ¼ mile south-east of Inellan Pier, and 300 yards N.N.E. of Chapel Hall (194 south-east), and at the junction of the burns ¼ mile north of Toward Taynuill (194 south-east).

The Inellan section shows a crush breccia of green serpentinous rock in which the pieces, all of them serpentine, sometimes exceed the size of a man's head. The main band runs rudely parallel to the schist foliation, but in one place, about 180 yards from the pier, a portion runs off for a little way to the north at the back of some schist. Fresh fractured specimens show a mottled dark grey or green matrix without any porphyritic constituents. At times the pieces are so angular and so coated with fibrous serpentine and slickensided films of talc or white steatite that no one could doubt the crushed character of the rock. The pieces examined showed no traces of schistosity acquired before the advent of the fault movements, and we were for some time in doubt whether the rock should not be regarded as a crushed dyke which had been intruded along a pre-existing line of crush. The evidence at Toward to be shortly adduced makes it probable, however, that this is not so. A portion of one of the pieces was examined microscopically by Dr Hatch, and is described as follows : "A confused interlacing mass of lamellæ and blades of serpentine, the arrangement of which bears some resemblance to the 'Balken' structure of serpentines derived from augite-bearing rocks. No ferro-magnesian constituent could be detected."

Both at Inellan and Toward there is an alteration of the schists near to the serpentine. They are, for instance, harder, break less readily along the foliation planes, and weather whiter than usual for a few yards near the junction : also, they contain many thin streaks and little spots of chlorite and brown mica, which may be as large as a small pea. These spots occur both in phyllite and schistose grit. They sometimes have hexagonal outlines and contain grains of garnet, so that it may be considered certain that the constituents in them are pseudomorphous after that mineral. Garnets or pseudomorphs after it are unknown in this part of Cowal excepting close to the serpentine, but here the little spots referred to are so abundant that on some rock faces they compose nearly $\frac{1}{4}$ of the area. In slide 2835 the central portions of the spots are formed of chlorite or a granular aggregate of felspar, and the peripheral portions are of biotite arranged in short confusedly arranged lamellæ. The spots are often longer in one direction than another : this may be attributed to the same action which has elongated the pebbles in the schistose rocks, and be thus an independent indication that the serpentine belongs to the schist series.

The Toward section is better than that at Inellan, and in one place, a little west of the broad north-east running basalt dyke, the serpentine is distinctly schistose, independently of any effect produced by the fault movements. The exposure alluded to shows on one side a greenish pebbly schist, next a band, about 8 inches thick, of yellow or reddish schist without pebbles, then a 6-inch band of green serpentine, and lastly one of the fault breccias. The breccia runs rudely parallel to the junctions between the rocks mentioned, but the foliation in the schist is at a marked angle to the junctions, and continues through the serpentine. Hand specimens of serpentine from here show a fine even schistosity throughout. It is thus clear that at this place the edge of the serpentine is a schist, though parts of it elsewhere are quite massive. The edge of it must have been more affected by the movements which produced the schistosity than the interior parts were—a phenomenon, of course, constantly met with in the igneous masses of schistose areas. Several bands of serpentine occur at Toward besides the one just alluded to, but, in the main, they have been so much affected by the fault movements that they form now what we might call a new schist. A little west of the basalt dyke a red exposure of such crush rock occurs close by high-water mark. This rock contains large lenticles and eyes, some as large as a man's head, weathering mottled red and grey. The rock surrounding the eyes is very soft and unctuous, and microscopical examination (slide 4839) shows it to consist of carbonates and talc, with the greater lengths of the streaks and specks of the former roughly parallel. Some of these streaks are evidently of new formation, along lines of discordance and rupture, and sometimes cut off the edges of contorted talc layers. The lenticles and eyes consist of crystalline ferriferous dolomite, with pale grey patches of talc. In the hard red crush rock sticking up on the east side of the basalt there seems but little serpentine incorporated. At the sides of the basalt, before coming to either of the red rocks just mentioned, there is a crushed and contorted band of green and dark grey serpentinous rock : in places this contains small spots and streaks of carbonate of lime.

Near the east boundary of the crush area, and within 10 or 12 yards of high-water mark, we noticed a good many blocks, some as much as 3 feet in length, of a yellowish weathering ferriferous dolomite, speckled with irregular spots and streaks, often about $\frac{1}{4}$ inch long, of some serpentinous mineral. The dolomite in fresh fracture is deep red. These blocks are probably nearly *in situ*, and may perhaps be regarded as belonging to some vein which has been formed near the serpentine : they are of essentially the same character as the eyes in the red crush rock west of the basalt.

Both at Inellan and Toward the crushed serpentine is cut by dykes of uncrushed basalt of supposed Tertiary Age.

At the junction of the burns $\frac{2}{3}$ mile north of Toward Taynuill, a large portion of the crush area consists of a soft greenish-grey aggregate of minute crystalline scales, probably antigorite, mixed with redder portions composed of some carbonate (slide 4837). Mr Teall states that the rock is a serpentine now, but what it was originally is uncertain : it does not resemble the serpentines formed from olivine rocks. This rock is much crushed along rudely parallel lines of rupture and slickens, but there are still eye-like masses, sometimes 3 feet in length, comparatively unaffected. In parts of the crush there is an admixture of white steatite, and it is stated that about a dozen years ago this was mined, and samples taken for trial as French chalk. There is also dark grey and green serpentine, resembling that at Inellan, near the north margin of the crush area in both burns. In the west burn this lies on the north side of the most northerly crush of any prominence, but there are several thin crushes on the north side of it : the schist near it has the appearance of contact alteration, with spots of biotite, etc., already remarked on.

Three hundred yards N.N.E. of Chapel Hall part of the crush breccia contains fragments of steatite embedded in a soft unctuous red matrix, but most of it is a green or streaky green and dark blue rock, showing movement structures.

All the above occurrences of serpentine run rudely parallel to the general strike of the schists adjoining, and probably they may be regarded as belonging to an old intrusion of sill-like character, along the course of which the different movements of the schist boundary fault have subsequently occurred. In the district between Loch Lomond and Aberfoil there are frequent exposures of crushed serpentine and serpentinised gabbro in the fault crush that separates the Middle or Lower Old Red Sandstone from the older rocks. The character of the original rock appears to have varied considerably in different places,—in some places containing no felspar, in others a considerable proportion. This variability seems best to accord with the supposition that the rocks in the different exposures once belonged to a large mass during the consolidation of which considerable differentiation in the magma took place. The amount of alteration in the schists near the serpentine at Inellan and Toward is great, especially when one considers that these rocks are generally separated by crushes, which may have brought together rocks once lying at some distance from one another. The alteration can, therefore, be regarded as more probably due to a large intrusion than to a thin dyke.   C. T. C.

Glendaruel Serpentine.

Three-quarters of a mile north of Achanelid, a mass of serpentine extends on the western side of Glendaruel for rather over a mile along the strike. Its greatest width of outcrop is 500 yards. It has been cut across prominently in two places by north-west faults, and at these places it has narrowed itself, along one fault to 150 yards, and along the other to 100 yards. After the main mass ends, the serpentine is continued for another mile in a series of five smaller disconnected patches, the largest of these attaining a length of about 300 yards. For a large portion of its length the western boundary of the big mass coincides with a fault. This is also the case as regards the largest of the smaller patches to the north. The Loch Tay limestone has been mapped between the serpentine and the bottom of the glen, and affords a valuable guide for determining the stratigraphical horizon. The eastern boundary of the big serpentine mass, and of the smaller masses to the northward, keeps the same distance from the outcrop of the limestone ; as the outcrop of the limestone varies in direction, so does the outcrop of the serpentine. It is evident, therefore, that the serpentine is in the nature of a great sill or sills. The smaller patches are foliated throughout ; in the larger patches the serpentine becomes massive in the centre, and schistose on either side. In the big mass there is little or no foliation at its lower or eastern portion, but the foliation increases as we approach the upper portion or western extremity, where the rock is not only well foliated, but much more shattery than at its base. Fibrous serpentine occurs abundantly along joint planes.

In the largest of the small patches occurring to the north of the main mass, the interior of the rock has so much of the appearance of an epidiorite that a specimen was sent to Mr Teall for microscopic examination, who described it as follows :—

" Pale green, and colourless hornblende, and turbid decomposition products, after felspar " (4368).

Mr Teall thinks the hornblende is secondary, and the original rock was a gabbro. It appears, therefore, that a gabbro has been associated with peridotites, the latter of which have become serpentinised, so that there is a transition in space from altered gabbro to serpentine.

Mr Teall examined a specimen from the main serpentine mass, but the original minerals could not be determined.

The serpentine, as described on p. 153, has been pierced and veined by basalt dykes. Contact metamorphism, produced by the intrusion of the dykes, has sometimes been considerable, and its effects extend for some distance from the edge of the dykes. The dykes themselves within the serpentine mass seem to decompose readily, and have themselves become partially serpentinised. In addition to the dykes there are two or three small dyke-like veins which we cannot say for certain belong to the basalts, but they probably do. They occur as dusky black veins in the serpentine, are soft, and effervesce with hydrochloric acid.

A specimen is described thus by Mr Teall :—" (4370). Black massive compact rock. Microscope :—Extremely minute felspars. Microlites and much disseminated opaque matter between the felspars."

Mr Teall is doubtful as to the microlites being felspar, the rock being too fine to make anything out satisfactorily. J.B.H.

CHAPTER VII.

MINERALS IN THE SCHISTS. DIRECTIONS OF STRETCHING.

Among the minerals which may occur in the schists as part of them, or in veins which seem of the age of some part of their manufacture, are the following :—quartz, albite of the albite schists, an albite of earlier age than the former, biotite, chlorite, sericite and white mica, tourmaline, garnet, epidote, rutile, sphene, ilmenite, magnetite, hæmatite, calcite and a ferriferous carbonate, iron pyrites, copper pyrites, and an almost opaque yellow mineral forming thin flat tables.

The mode of occurrence, etc. of the albite of the albite schists has already been described. Though clearly of later origin than most of "the anticline" movements, it is not quite certain that it is always later than all of them : in some microscope sections there is a suggestion that it may have been slightly moved (see p. 43).

A felspar of earlier age than the albite of the albite schist often occurs with the quartz veins. Sometimes, as in the veins in the Dunoon phyllites, the colour is pink, but more commonly opaque white. The veins at Dunoon are described on p. 18. The felspar in these veins has been determined as albite. The growth in rods and streaks parallel to the direction of stretching, the bending and occasional cracking across their length, and the subsequent infilling of the cracks with new quartz, are features which readily distinguish them from the albite pegmatites of the albite schists, and indicate their earlier age, provided it is satisfactorily established, as we suppose, that the movements in the areas with rodded pegmatites are not later than the movements in the areas with pegmatites without rodding.

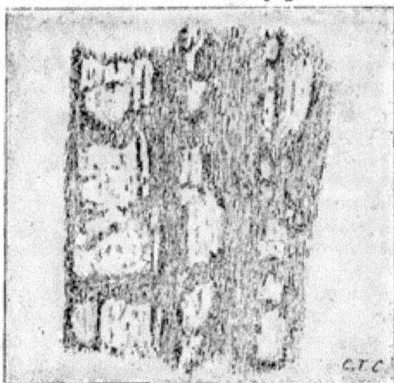

FIG. 27.— × 1. From a loose block ½ mile E.N.E of Cruach na Cioba (141). Pegmatite with interrupted streaks of opaque white felspar.

The folia of biotite, chlorite, and white mica, which occur so abundantly along the planes of "the anticline" rocks, have often been waved and crumpled. But in Inverchapel Burn, in the bend near the "n" of "Burn" (164 south-east), there are flakes of green mica, sometimes $\frac{1}{10}$ inch in length, which cross the foliation planes of the rock, and contain inclusions which show that the rock was a schist, already somewhat crumpled, before the formation of this mica : the inclusions consist partly of small granules of quartz or felspar, and partly of opaque yellow prisms of the doubtful mineral mentioned on p. 35. A white mica presumably of the same age shows a similar arrangement not unfrequently.

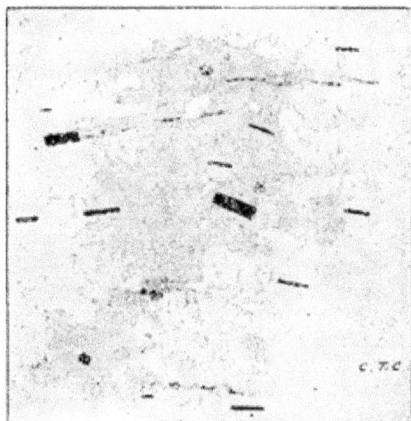

FIG. 28. — × 3½. From a thin slice of mica schist in Inverchapel Burn, cut at right angles to foliation. Seen with natural light. Thin ochreous veins near top and bottom of sketch. The black prisms are the doubtful opaque yellow mineral. The dark circular spots are surrounded by pleochroic halos.

The most striking instance of this is shown by some thin bands in the albite schists : these are nearly entirely made up of white micas lying across the foliation planes (see p. 43). Some of the flakes of white mica in the albite schist, from ½ mile E.N.E. of Ard a' Chapuill (182 northeast), have at times their flat surfaces lying nearly at right angles to the planes of schistosity (see Mr Teall's description in Appendix). Brown or black mica, too, has at times the same arrangement, e.g., in the burns ½ mile north-east of Sith an t-Sluain (182 north-east), and south-east of the hamlet of Glensluain, and in the albite schist ¾ mile slightly east of north of Binnein an Fhidleir (134). The chlorite pseudomorphs, after garnet and actinolite, are mentioned on p. 50.

FIG. 29. — × 4. Bent and cracked Tourmaline needles from the albite schist of A' Chruach (182).

The occurrence of tourmaline in the albite schists and green beds, and in veins in these schists, has already been mentioned, pp. 35, 40. It also occurs not unfrequently in the more micaceous varieties of schist which do not seem referable to either of these types of rock, e.g., in Tamhnich Burn, near the "m" of "Tamhnich" (172 north-east), and on the hillside a little more than ½ mile slightly east of north of Cruach nan Cuilean (172 north-east). In the last locality the needles are sometimes over 1 inch in length. Near the sharp bend of the stream ½ mile north-west of Ardacheranbeg (172 north), they occur in hard white thin slightly calcareous laminæ, which are lying in more micaceous schist. Far the best locality, however, that is known for tourmaline, in the south part of

Cowal, is A' Chruach (182 north-east). Here it occurs in the more micaceous schists whether these contain albite or not. Some of the needles are 1 inch in length and $\frac{1}{20}$ inch in breadth. They lie especially along the foliation planes, and in some cases have also a general direction along these planes, but in others there is no clear parallelism. These needles are often much bent, and even broken, by the numerous small "anticline" movements. This is seen particularly well at a point about 80 yards south of the top of the hill. Tourmaline has not been noticed more than about 2 miles south-east of the centre of "the anticline," and it may probably be regarded as one among the other signs of the greater metamorphism of "the anticline" rocks.

In the north part of Cowal small tourmaline needles are common in the albite schists and the more phyllitic mica schists, and also, in longer needles and sheaths, in the quartz veins in these schists. The isolated needles in the schist in the Garbh Allt Mor (126), near the " r " of " Mor," are some of them $\frac{1}{20}$ inch thick. On the south slopes of Binnein an Fhidleir (134) the quartz-tourmaline veins seem particularly abundant, and rich in tourmaline. The shores near Ardkinglas are thickly strewn with boulders of such vein stuff, which have probably in great part come down the Kinglas Water from this and neighbouring localities. In the burn $\frac{3}{4}$ mile north-west of the " B " of " Beinn an t-Seilich (134), the tourmaline needles in the quartz veins are sometimes 2 inches long, and coalesce into sheaves 2 or 3 inches wide. In some parts isolated needles are seen spearing through the white quartz of the vein, but there are also

smaller quartz strings which cross the tourmaline areas, almost at right angles to the lengths of the needles. The tourmalines are frequently waved: as far as noticed the waves are roughly parallel to those which affect the foliation of the schist. The quartz strings which cross the tourmalines have probably been formed subsequently having filled cracks formed during the bending. In these later quartz strings, ferriferous carbonates are also common. Again, in a loose bit of a quartz-tourmaline vein found $\frac{1}{3}$ mile above the foot of the Cab (141), isolated tourmaline needles seem to penetrate into presumably later quartz.

FIG. 30.— × 1. From a loose block $\frac{1}{3}$ mile above foot of the Cab (141). Sheafs of Tourmaline in a quartz vein are crossed by strings of quartz. The sheafs are shown in their natural position to one another. Near " a " Tourmaline needles seem to penetrate the cross strings of quartz.

In the quartzose schists tourmaline is rarer, but in a quartzite schist in the burn 1 mile N.N.W. of Ben Donich, it occurs in lines of closely congregated small needles: the direction of each needle is the same as that of the lines of occurrence.

Tourmaline and the albite of the albite schists are in this district close companions. The areas with prominent albites are also those in which tourmaline is most conspicuous. Tourmaline is often seen lying across the foliation of the schist (slide 3418, described in Appendix), but has frequently been subjected to bending and fracturing, as just described. It is occasionally included within albite (slide 5215, from a specimen at the head of Stuck Burn), and so must be of earlier date than it.

In the area east of Loch Riddon and Glendaruel garnets have only been observed near the Toward and Inellan serpentine. Here, as already mentioned, p. 73, they seem to be due to the contact metamorphism induced by the serpentine. On the north-west side of the Glendaruel limestone there is a zone in which garnets are common. For full description, see p. 48. The garnets along this zone are often deformed and stretched in one direction, and intersected by thin cracks filled with secondary quartz; or, still more frequently, they are represented by chlorite pseudomorphs elongated in two opposite directions. The direction of elongation is usually N.W. or N.N.W.

Epidote occurs in abundance in grains in the green beds, and the veins traversing these beds; also in small granules in the epidiorites and hornblende schists (slides 4740, 5056, 5061), and in long prisms in the veins in and at the sides of these beds (see p. 65). An occurrence in albite schist close to its junction with a lamprophyre is mentioned on p. 114. In an albite schist $\frac{7}{12}$ mile slightly east of north of the Ordnance Station 2658 on Binnein an Fhidleir, there is a thin quartz vein which contains needles, about 1 inch long, which probably belong to epidote.

Sphene occurs in slide 3795, obtained from a green bed at Tighnuilt, Inverchaolain (183 south-west). Mr Teall's description of this rock is given on p. 34. A quartzite schist (slide 3434), from $\frac{3}{4}$ mile S.S.E. of Aird Cottage (141), also contains it in small grains. In the epidiorites and hornblende schists it is common as a decomposition product (leucoxene) of ilmenite: see, for instance, p. 67.

Ilmenite occurs in thin plates, which are sometimes 1 inch in length, in the quartz veins of Glen Massan and other areas near "the anticline." The small opaque grains which occur along the foliation planes are, in some of the schists, ilmenite and not magnetite. In a garnetiferous mica schist near the top of Cruach na Capull (133 south-east), small black shining plates of ilmenite, about $\frac{1}{12}$ inch in length, are abundant throughout, and seem quite a part of the schist. In an albite schist $\frac{1}{2}$ mile east of the Ordnance Station 2658 on Binnein an Fhidleir, ilmenite plates were also discerned by the naked eye. In the albite schists north-east of Lochan Mill Bhig, ilmenite plates, partly decomposed to leucoxene, are very abundant: they are generally parallel to one another and to the schist foliation, and are often in close company, forming a considerable proportion of some hand specimens: they are sometimes an inch long, and show slight bends. Kernels of ilmenite, surrounded by sphene, are common in the epidiorites and hornblende schists.

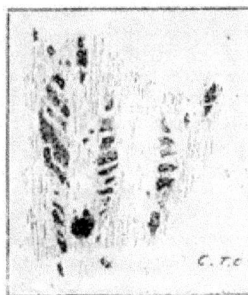

Fig. 31.—×1½. ½ mile S.W. of Strachur House. Trains of magnetite grains on a foliation plane of phyllitic mica schist. The shading shows the direction of the longer axes of the micaceous minerals.

That the small black opaque grains referred to in the above paragraph are sometimes magnetite was shown by Mr Teall in the course of his exhaustive examination of the albite schist from ¼ mile E.N.E. of Ard a' Chapuill (see description in Appendix of slide 3418). These small grains sometimes give the impression that they have been formed by the shearing, or dragging to pieces, of larger lumps. That this is so is rendered probable, too, by the fact that in various places such a shearing down of magnetite occurs on a scale large enough to be made out macroscopically, *e.g.*, on the shore of Loch Fine, 200 yards north-east of Tigh ua Criche (see p. 57). Magnetite crystals, about the size of a small pea, are common throughout "the anticline" area, and also in the green beds on the south-east side of this area. They are perhaps most common in the albite and more micaceous schists, but also occur frequently in massive greywackes. They are so abundant in some scars, *e.g.*, in those about ¼ mile north of A'Chruach (182 north-east), that a handful of loose crystals could readily be gathered from the rock ledges, on which they have tumbled and gradually accumulated during the process of rock weathering. The magnetite in the sands* of the Kyles of Bute has no doubt been derived, in part at all events, from the magnetite bearing schists of the neighbourhood.

In the green beds on the east side of Lephinkill (172), and in the albite schist, ½ mile E.N.E. of Lochgoilhead Established Church, it is not uncommon to find magnetite crystals with quartz streaks extending for a little space on two opposite sides. The length of each streak is often about ¼ inch. This is probably a parallel case to those in the phyllade aimantifère of Monthermé, described by Prof. Renard (*Bull. Mus. Roy. Hist. Belg.*, vol. ii. p. 134, and pl. vi., 1883), and mentioned more recently by Mr Harker in a paper "On 'Eyes' of Pyrites and other Minerals in Slate" (*Geol. Mag.*, vol. vi. No. ix. p. 396). To quote Mr Harker: "By its strong cleavage this rock gives evidence of considerable lateral compression. Prof. Renard has shown that prior to this compression, the magnetite crystals already existed in the rock, and were surrounded by a coating of chlorite. The crystals yielded to the pressure much less readily than their matrix; and the latter, having already a firm consistence, became separated from the crystals, carrying the chlorite with it, and was displaced along the planes which are now cleavage planes, that is, in a direction at right angles to that of the pressure." And again: "in yielding, it was able to leave vacant spaces, which were filled at a later date by crystalline quartz." Mr Harker also points out that similar phenomena often occur around crystals of pyrites in slate, *e.g.*, in North Wales and Ballachulish. As stated, pp. 50, 57, we have in Cowal also found the same phenomena around pyrites and garnet crystals.

Hæmatite, in the form of iron glance, occurs in a pegmatite which also contains felspar and chlorite, on the hillside ⅝ mile N.N.E. of Ardtarag (173 west).

The occurrence of iron pyrites in the Dunoon phyllites, in crystals and nodules, and thin films running along the foliation, has already, p. 9, been referred to. In the green and purple phyllites, on the shore between Dunoon and Kirn, cubical crystal shapes, reaching ½ inch in length, are not uncommon. The main portion of these now usually consists of limonite, but there are "hearts" of iron pyrites. The phyllite, for about ¼ inch from these shapes, is bleached to a cream colour.

A black phyllite a little above the upper end of the Dunoon reservoir (not shown in 1-inch Map, but lying about 1 mile west of the pier) has quite a rough surface from the abundance of hard projections, covered by black phyllite adhering firmly to them. These projections are longer in one direction than another, often from ⅛ to 1/18 inch long and less than 1/20 inch broad. They are now hollow in the middle, but have quartz at the sides. We suppose that the central hollow was once occupied by some mineral, perhaps marcasite, of which all trace has now disappeared,

---

* See "On a Magnetic Sand from East Bay, Rothesay," by D. C. Glen, *Trans. Geol. Soc. Glasgow*, vol. v. p. 158. Mr Glen states that only 9 or 10 per cent. of the heavy black grains is attracted by the magnet. He takes the rest to be ilmenite, and looks on the neighbouring trap dykes as their source of origin.

and that the quartz at the sides is analogous to that at the sides of the maguetite of Lephinkill. In several places in the Ardrishaig phyllites, and especially 200 yards north-east of Tigh na Criche (Strachur), iron pyrites is seen in a partially deformed state, with streaks of quartz at the sides (see p. 57). The occurrence of quartzose schist, unusually rich in pyrites, and with small proportions of galena, blende, silver and gold, is described on p. 58.

Marcasite occurs in the tourmaline bearing schists of A' Chruach (182) and in the epidiorite of Mid Hill (173 south-west).

Decomposed crystals of copper pyrites, or of an intergrowth of copper and iron pyrites, have been noticed in the schists, and as quite a part of them, without any apparent connection with any fault or vein, on the east side of the road, ½ mile north of the Clachan of Glendaruel. A little carbonate of copper has been formed from the crystals by weathering. They have only been seen in a band an inch or two thick, and this band was only traced some 3 or 4 feet. The containing rock is a somewhat pebbly schist. Crystals of some copper mineral also occur in the schists of the headland lying north of the Burnt Islands, Kyles of Bute, in a little cove about 60 yards N.N.W. of the Ordnance Station. These also are now largely converted into the green carbonate, but within this there is usually seen the red and black oxides. The band in which these occur runs ·roughly with the foliation : it is somewhat thicker than that in Glendaruel, but can only be traced a few yards.

The ferriferous carbonate of the veins seems in many cases to be an almost pure carbonate of iron, and has perhaps been used in old times for smelting purposes, in hillside bloomeries. See further, on p. 289.

The opaque yellow mineral mentioned at the end of the list given in the beginning of this chapter is that referred to by Mr Teall in his description of slide 3831, from a green bed about ⅔ mile slightly north of east of Lephinkill. For this description, refer to p. 35. The same mineral is of common occurrence in the more micaceous schists of "the anticline" area. It is particularly prominent in some of the bands in the crags about 1 mile south-west of Ardentinny.

Directions of Stretching.

The phenomena of stretching, or elongation of constituent parts, have been already described as they are seen in the coast section south of Dunoon (p. 17). They are also alluded to in the green beds (p. 35), the garnetiferous mica schists (p. 49), the Ardrishaig phyllites (pp. 56, 57), and some few other places (pp. 46, 100, 185, 196).

In the published maps the direction of stretching is indicated by a special sign placed on the locality of observation. In the section south of Dunoon the direction is liable to vary somewhat even in the same planes (see p. 11). Similar small variations are to be noticed commonly, but they do not prevent us seeing that in each locality there is usually one general direction. In the Dunoon section referred to, the general direction is 10° to 11° nearer east to west than the direction of the foliation dip.

On the south-east side of "the anticline," the general direction is much the same as at Dunoon, or else more nearly parallel to the foliation dip. Instances of such parallelism are seen, e.g., 200 yards east of Corran Lochan (153), in Blairmore burn near the "a" of "Blairmore," at Strone Point (174), in the burn ⅝ mile north-west of the "l" of "Kil- . mun," ¼ mile N.N.E. of Corlarach (194), the coast ½ mile north of Strone Point (194), and ⅛ mile north-east of Altgaltraig Point (182). 150 yards

F

north-west of Carrick Castle, on gently dipping planes, the direction is either with the dip or 12° or so nearer east to west, much as at Dunoon. In Inverneil burn (183), on planes dipping about 40°, on the other hand, the direction is about 30° nearer north to south than the dip.

On the north-west side of "the anticline" the usual direction is either with the dip or else nearer north to south than it. It is more north to south than it in the following, among other places, ¾ mile south-east of Cairndow ; ½ mile S.S.W. of Ardkinglas House (134), nearly north to south ; north-west side of Stob an Eas, sometimes nearly north to south; ¼ mile E.N.E. of Cruach nan Capull (133) ; ¼ mile north of Beinn Lochain (141) ; ⅝ mile west of the "C" of "Carnach Mor" ; ¼ mile W.S.W. and ½ mile N.N.E. of the "B" of "Carn Bàn" (163) ; by the first "r" of "Cruach Mhor" (162), nearly north to south ; ¼ mile S.S.W. of the "A" of "An t-Suil" ; and ¼ mile north-east of the "k" of "Dalleik" (172).

We have not noticed any clear case of alteration in direction in consequence of the reversal of the plane on which the indications occur. The foliation planes have not usually been reversed in this district by the action of "the anticline," but only made to incline in different directions. Consequently we cannot explain the fact that on the north-west side of this anticline the direction is often more north to south than the dip while on the south-east side it is more east to west, by supposing the stretching lines to be roughly continuous lines which have been given different directions by "the anticline" folding. But we suppose the planes on which these lines occur have been reversed before the anticline was produced through the agency of important synclines near the present centre of "the anticline" (see p. 86), and it is therefore allowable to consider that the stretching directions were once more uniform than they are now, the differences now distinctive of each side of "the anticline" being due to pre-anticline reversal of the planes on which they occur.   C.T.C.

CHAPTER VIII.

GENERAL PHYSICAL STRUCTURE OF THE SCHISTS.*
REMARKS ON THE METAMORPHISM.

The most conspicuous feature in the physical structure of the schist area is the great anticline of foliation which has already often been referred to as "the anticline" running in a direction about 35° west of south, through the heads of Loch Goil, Loch Striven, and Loch Riddon, and the hills 1 mile or so north-west of Tighnabruaich. In most parts of Cowal there is no particular line we can point to as the exact centre of the fold, for the schists on either side are excessively crumpled and with alternating dips for some distance. Illustrations of this point, and descriptions of various minute structures accompanying the anticline in the area north-east of Loch Riddon are given on pp.   .

About the heads of Loch Striven and Loch Riddon it is observable that the foliation dip on the south-east side of the anticline is generally steeper than that on the north-west side : averaging, perhaps, 40° or more instead of 20° to 30°. The steepness of dip on the south-east side increases

* Near the end of the volume a sketch map of the solid geology is inserted.

as we go north-east, and near the north margin of 1-inch Map 29, is at times vertical, or even, as on the hill 1¾ mile north of Ardentinny, slightly reversed. Further north-east still, out of Cowal, the prevailing dip is north-west, e.g., at Aberfoil and Callander. South-west from Loch Riddon, however, the dip on the south-east lessens in amount, while that on the north-west generally increases, so that we reach an area in which they are about equal on either side.

There is no doubt that this anticline is a true arch of an early foliation. Later foliations and other structures have been developed together with it, as already described ; but the most prominent foliation of the district, and an enormous amount of folding of the same age as this foliation, were already in existence before it, and were folded by it.

It is clear, then, that "the anticline" cannot be a simple or exact anticline of bedding. It should rather be looked on as an anticline of the limbs and axes of the early folds which affect the bedding. To what extent it departs from being an anticline of bedding must depend on the amount of this early folding. There is hardly a coast section which does not show how great this amount is, and we have shown, p. 24, that, even where it may not at first be noticed, this is not because it did not exist, but because it has been obscured by later movements.

We have not felt able to make any estimate of the amount of this pre-anticline folding, so as to say how much less than the breadth of present outcrop the original thickness of any bed is likely to have been, or how many times it may have been repeated by folding. In the areas near the anticline centre, rock exposures are more continuous than on the flanks, but in these there has been so much later movement that we do not regard them as promising localities for making such an estimate. Nor are the hillsides away from "the anticline" more favourable, for they are often extremely smooth and covered with drift or grass. The coast sections are more promising, but as they rarely attain a height of more than 20 or 30 feet, it is not possible to trace individual beds in them with confidence for any distance across the folding. We had at one time hoped to make an estimate by the help of a careful comparison of the best coast sections of a particular band or group of rocks—to compare, e.g., the section of the Dunoon phyllites near Cove and Barons Point with that at Dunoon, and these again with those at the foot of Loch Striven and in Bute. The least number of exposures of a bed of particular character in any of the sections would show that the number in excess of it in the other sections was due to repetition by folding. This least number might, of course, still be in excess of the original one, but it would be nearer to it than the others. To make such a comparison with the care and detail necessary would require the preliminary mapping to be done on a very large scale, but we believe that the labour involved in this task might be well repaid.

After the remarks we have made on the great amount of folding in the coast sections, it may surprise those who examine the Maps to see such comparatively small evidence of this structure in the run of the beds mapped out. This we suppose is largely due to the want of sections along the margins, and to the general equality of the amount of folding in different areas. Each line drawn to run evenly, no doubt, represents one which is repeatedly plicated by small folds, but these are of much the same frequency and importance at different parts of the line, and so the general direction remains uniform so long as the foliation dip and the slope of the ground do so. For the same reason the breadth of many of the bands mapped continues much the same, the original thickness being increased about equally in different areas. Our want of knowledge of the original succession of the rocks is, too, so great that we do not feel it safe to assume, in the case of thin outcrops lying a little off a thicker band of the same character, that these are either folded outliers of the thicker one among apparently lower beds, or inliers of it among higher beds. Doubtless they are so over and over again, but when we come to consider each particular instance we are not usually, for want of clear exposures, able to say whether they are or not.

The folding in the section south of Dunoon (already described, pp. 9–18), is so great that we should not, before mapping the adjacent hillside, have been surprised to find the beds practically flat. But they evidently cannot be so, from the way the north-west margin of the Bull Rock greywacke runs up this steep hill: for about 400 feet of vertical height this margin makes quite a good line, and must, on the 1-inch scale, be mapped as a simple unfolded one. So again the thin greywacke on the east side of Glendaruel, by Dalleik, can be traced to a height of about 600 feet above exposures near the glen bottom, and must be mapped as an even running band. These and the other straight running schist boundaries, where represented by uninterrupted lines, seem to show that, great as the amount of folding is, it is still possible to exaggerate it. The beds in the area generally cannot be considered as essentially flat, though they are so in particular parts, e.g., along the strike line between Loch Loskin (184 north-east) and Blar Buidhe, and again near Blairmore, where we regard the hornblende schist running almost at right angles to the foliation as an old igneous rock intrusive along the bedding, and on Loch Fine side near Leak and Ardchyline.

As a preliminary to the discussion of how far "the anticline" departs from being an anticline of bedding, it is necessary to compare the succession of rocks found on either side of it. On the south-east side the chief groups are, taking them in order as we advance north-west from the boundary fault :—

1. Phyllite and schistose pebbly grits, with pebbles chiefly of opaque white quartz, often getting up to the size of small beans. Thin limestones in places, but much rarer than in the Dunoon phyllite series. The proportion of phyllite also less.

2. Bull Rock greywacke schist. Average breadth of outcrop about ⅓ mile. A large proportion of pink felspar pebbles. Almost unmixed with phyllite, but some thin bands with a breadth of 4 or 5 feet.

3. Dunoon phyllite series. Phyllite, with many thin limestone and schistose grit bands. The grits near the south-east end are often coarsely pebbly with opaque white quartz.

4. Band of "green beds." Thickness on Blar Buidhe immensely exaggerated by folding. North-east of here it is probable that the original thickness gradually decreases, and it is doubtful whether there is any representative of it on the coast near Hunter's Quay.

  Alternation of schistose grits and greywackes and phyllites. Generally larger proportion of two former than in the Dunoon phyllites, and no limestones.

  Band of "green beds." Thickness south of Sandbank greatly due to folding. This again is not seen on the coast between Hunter's Quay and Sandbank, but it is possible that it may occur in an obscure area.

5. Alternations of schistose greywackes, phyllites, and albite schists, etc. Calcareous lenticles and stripes, rarely exceeding a few inches thick, common in the quartzose pebbly beds. Distinct albites do not generally occur until we get some 3 miles north-west

of group 4, but these are due to secondary change, and may have no stratigraphical value. Group keeps on to the centre of "the anticline," a distance across the foliation strike in the south part of Cowal of about 5 miles.

In the broad group 5 we have only been able to map out one band for any distance. This is the greywacke horizon, which is seen so well on the headland north-west of Fearna Bagh. It does not consist throughout of greywacke, and its margins are not always well-defined, but still it was worth while to follow it in default of anything better, and it has been a means of recognising various faults. So our ignorance of the physical structure of this broad area is greater than of that occupied by the groups 1 to 4, which altogether take up no more area than 5 alone.

In concluding that there is a succession much as indicated above we think we stand on fairly solid ground. Group 1 could not be group 3 repeated by folding : the characters of the groups are too different. The "green beds" are of such a peculiar character that they could not be confused with the Bull Rock greywacke. Between the north-west side of 4 and the centre of "the anticline" there is certainly nothing to match the limestones of the Dunoon phyllite series. •

Whether group 1 or group 5 is the oldest there is in this area nothing that we know of to tell. That 1 now lies over the others is, of course, no indication of their original relative positions. A very little greater strength of push from the north-west during the time of the production of "the anticline" would have made 5 the apparent top of the succession instead of the bottom. As already stated, p. 83, the dip of these beds near Aberfoil and Callander is north-west instead of south-east, as in Cowal, and group 5 is apparently the top of the series. This alteration in dip on the south-east side of "the anticline" was long ago called attention to by Professor Nicol (" On the Geological Structure of the Southern Grampians," *Q.J.G.S.*, vol. xix., 1863, p. 180).

Let us now turn to the beds on the north-west of "the anticline." Beginning at the north-west side of the district we have :—

1. Ardrishaig phyllites. Soft calcareous sericite schists, with many outcrops of quartzite schist and some limestones.
2. Graphite schists, dark graphitic limestones, and quartzose schists intermixed with phyllitic garnetiferous schist.
3. Alternations of thin banded mica schists, commonly garnetiferous, with some limestones and "green beds."
4. The Glendaruel or Loch Tay limestone. Coarsely crystalline marble, with some parts mottled black and white, and others of calcareous quartzite. Thickness in all perhaps 40 to 50 feet, but the lower part is often more calcareous quartzite than limestone.
5. "Green beds" group. "Green beds," mixed with perhaps an equal proportion of schistose greywacke and other mica schist.
6. Albite schists, schistose greywackes, and other mica schists. Extends in the south part of Cowal for a breadth of about $2\frac{1}{2}$ miles across the foliation strike up to centre of "anticline," and in the north part for a greater distance. Essentially of the same character as group 5 on the south-east side of "anticline," and with no satisfactory dividing line from it. In this group there is an horizon which consists mainly of massive greywacke, and the north-west margin of it has been mapped in some places : the south-east margin is much less readily traced.

We will, for distinction between the groups, add the letter "a" to the numbers of those on the south-east side of "the anticline," and the letter "b" to those on the north-west side: thus 1a, 2a, 3a, 4a, 5a, and 1b, 2b, 3b, 4b, 5b, 6b. There is little doubt that the "green beds" of groups 5b and 4a were of essentially the same character originally. The differences between them, the greater development of chlorite, black mica, and albite, in the former, compared to the latter, are only what we should expect from a study of the other schists associated with these green beds in the different areas. But the arrangement of the beds strictly entitled to the name of "green beds" in these groups is not very similar: there are many alternations of schistose greywackes and other mica schists in 5b, while in 4a there are two main bands, each comparatively unmixed.

Is it possible, for all this, that groups 5b and 4a are the same, folded over by "the anticline"? If it were possible to answer affirmatively, we should be afforded a most interesting indication of the amount of "pre-anticline" folding in the area between the two groups, for, in the south part of Cowal, 4a is about 5 miles distant from what we may roughly call the centre of "the anticline," while 5b is only $2\frac{1}{2}$ miles distant, in spite of the average foliation dip in the former case being not less than 40° and in the latter only 25°. If the groups had been the same, and if bedding and foliation had coincided, the thickness of the beds on the south-east side of "the anticline" between its centre and 4a would have been over 17,000 feet; and we could not have got this group on the north-west side at a less distance than $7\frac{1}{2}$ miles. Or, to look at the matter from another point of view, the apparent breadth of beds on the north-west side is about 6000 feet, while on the south-east side it is about 17,000. The beds on the south-east side would then have to be considered repeated by folding to nearly three times the extent they are on the north-west.

We are obliged, however, to regard these considerations as futile. In all probability the green beds on either side of "the anticline" are not the same folded over. If they were, the Glendaruel limestone, and the beds on the north-west of this, would have also to be the same as the Dunoon limestones and phyllites. The character of the former limestone, both in thickness and other respects, is, however, so different from any of those at Dunoon, and it keeps this different character for such a distance along the strike, far away north-east out of this Map, just as the Dunoon limestones also keep theirs along their strike, that we consider that they cannot represent one another. However greatly the original distance between the localities of occurrence of these beds may have been reduced by folding, it is not likely to have rivalled that for which the Glendaruel limestone is known to retain its usual character, and we have no reason for supposing the original character to have changed more rapidly in a direction across the present strike than along it.

So we regard the groups 1b, 2b, 3b, 4b, 5b as unrepresented on the south-east side of "the anticline." Groups 6b and 5a naturally go together to form one group, which we will call 6b 5a. It is possible, then, to divide the whole of the schist area into 10 zones, which are as follows, beginning at the north-west corner of the Map: 1b, 2b, 3b, 4b, 5b, 6b, 5a, 4a, 3a, 2a, 1a; and we may hold, with a fair amount of confidence, that this represents an original succession, though which group is top and which bottom we have so far seen nothing to determine. In the first five and part of the sixth of these the present superposition would indicate that 1b was the newest and 1a the oldest: in part of the sixth and the last four it would indicate that 1a was the newest and 1b the oldest.

We suppose that somewhere in group 6b 5a there is a great folding of "pre-anticline" age, which, roughly speaking, counterbalances the effect of "the anticline": that there has been here an important early syncline. The north-west margin of the greywacke horizon, a little on the north-west side of "the anticline," can hardly be the same line as the south-east margin of the greywacke mapped on the south-east side. We may temporarily look on the whole area between these lines as belonging to one greywacke group—of which the south-east margin of the south-east greywacke is one margin, and the north-west line of the north-west greywacke another margin : the former line we suppose never appears on the north-west side of "the anticline," owing to the lowness of the present ground surface : the latter line may either be folded back before it reaches the present ground surface, or, if it does reach it, it is soon folded back and comes out again.*

A large part of the area close to the centre of "the anticline" in the south part of Cowal is low and drift-covered, e.g., the area between Auchenbreck and the head of Loch Striven, and Glen Tarsan, and so the opportunities of examining structure are not good. We cannot point to any section in this area which confirms the supposition of the existence of an important syncline of "pre-anticline" age near "the anticline" centre. In the north part of Cowal, however, the hills west and south-west of Carrick Castle do afford some evidence in favour of it. The tops of Cnoc na Tricrich, Sgor Coinnich, and part of Cruach a' Bhuic, are formed of an apparently thick albite schist. This can be traced down from the hill top a little south of Cruach a' Bhuic both in a north-east and south-west direction, and in both of these it rapidly decreases in apparent thickness, the schistose greywackes on either side getting gradually nearer. We suppose these greywackes are in fact the same, the albite schist lying in a syncline between them. Further details of the structure in this locality are given on pp. 201–205.

We do not wish to speak with much confidence. There may be some other better explanation of the general structure of the district, but we do not think it possible that the general order of succession can be invalidated.

The average strike of the schist bands on the north-west side of the anticline is slightly more north of east than it is on the south-east. For instance, the lowest green beds on the north-west side, near Socach, are, in a direction measured at right angles to the strike, about 7 miles distant from the top of the schistose greywacke horizon at the junction of Loch Long and Loch Goil. But in the area between Glendaruel and Loch Riddon the corresponding horizons are within 4 miles of one another, i.e., they are above 3 miles nearer one another than they were. It is not possible to fix on the actual centre of the anticline, but it is clear that in the Loch Riddon district the horizons mentioned are both nearer the approximate centre than they are in the Loch Goil district.

It is impossible to follow any particular foliation plane any distance, but it seems as if the foliation planes also, as well as the bedding, have, on the north-west side of the anticline, a somewhat more northerly direction than on the south-east side. Near Lochgoilhead there is a greater width of ground, measured across the average strike of the foliation, where the foliation seems to be essentially flat, with alternating dips now north-west and now south-east, than there is at Loch Riddon. It is possible that in the former locality, too, the lowest foliation planes disclosed may be somewhat below the lowest at Loch Riddon. If this be so it follows that the crest or axis (using this word now in its strict sense—see footnote on p. 10) of the anticline must have a slight inclination to the south-west,† and it is probable that the numberless folds which together make up this anticline have usually an inclination of crest in the same direction. As a

* See Plate X.
† Or, to use the expression common among American geologists, has a slight "pitch" to the south-west.

matter of fact any inclination there may be is too slight to have been noticed in field sections.

Lines of actual rupture contemporaneous with the schist-making, comparable to the "thrusts" of the north-west of Scotland, probably do not occur anywhere in Cowal, except on the smallest scale, represented by the first foliation-throws seen in the Dunoon section, and the different strain slips already so often mentioned, and some few doubtful cases, which we will describe. I do not think I have ever observed the throw of a strain slip to exceed a few inches, and, more commonly, it is only half an inch or so. In some cases, e.g. on the coast ⅔ mile north-west of Ardbeg (183 north-west) bands, reaching to 4 or 5 inches in breadth, have in certain parts of limbs of very sharp folds, an appearance as if, for a few inches, they had been drawn out to nothing; but we do not feel sure this is always the true explanation of the appearance. It may be that the band was originally unequal in thickness, or even non-existent for a little distance. In other places there are discordances in the folding of two beds, and it is suggested that this may be due to the presence of a thrust line between them. In a scar on the hillside, ½ mile north-east of Glenmassan (163 south-west), there is a section showing, in its upper part, a massive pebbly schist lying almost flat, and, below this, are much folded white quartzite schist bands, which lie in phyllitic mica schist. Along the line of apparent discordance there are a number of quartz veins, mixed with broad streaks and knots of black mica. It seems at least necessary to suppose a certain amount of differential movement between the over and the underlying beds in cases like this, unless it is possible that they were originally of an extremely false-bedded or occasionally curve-bedded* character, such as we see not uncommonly in the unaltered Torridonian rocks. On the south side of Kinglas Water, ¾ mile above Ardkinglas House, is an interesting section in an alternating series of quartzose schists and phyllitic mica schists, the latter containing occasional small albites. The upper part of the section is mainly phyllitic, the lower, quartzose; for a length of 2 or 3 feet the line between the two parts seems a discordance, the foliation and bedding planes of the upper part striking markedly against those of the under. The apparent discordance line is in one place much crumpled; it is in places accompanied by short quartz veins; in other places there is nothing to mark it except the want of agreement in the beds on either side. A little above it, and running parallel, are short chlorite lenticles; these cross the foliation of the rock in which they occur, and may be supposed to mark the course of a subordinate discordance. In the same locality there are various other less marked discordance lines.

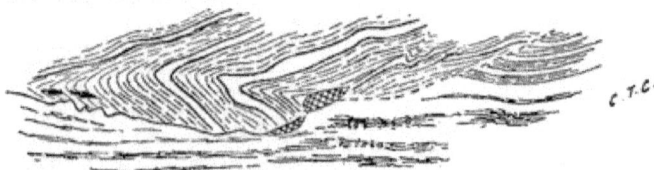

FIG. 32.— × ⅓. Vertical section S. side of Kinglas Water, ¾ mile above Ardkinglas Ho., in alternating albite and quartzose schists. The unshaded parts denote the latter. Direction of shading indicates direction of foliation. Black areas are chlorite. Cross-hatched areas are vein-quartz. The axis of fold in upper part of sketch hades N.W.

* We apply this term to beds of which the component laminæ are partly arranged in curves though the top and bottom surfaces are quite even.

The numerous faults which have been mapped and have effective throws are all later than the schist-making, and break up the minerals and planes of schistosity, instead of helping to form them. Not unfrequently we can see very thin, and perhaps almost horizontal, lines which make a prominent show in scars ; but these are all of comparatively late date, and occasionally can be proved to be later than the tertiary basalt dykes.·

Attention may be called to the wide range of the green beds. They occur every here and there all the way from the north-west side of the Dunoon phyllites up into the Ardrishaig phyllites on Loch Fine side. The character of these beds is so peculiar that it may with great probability be assumed that they all belong to one great formation. It does not seem likely that the different outcrops can be portion of one formation folded in unconformably with another, for transitions from the green beds proper into the common schistose greywackes can be observed over and over again. The green beds therefore help to connect together the schists of the different areas, and their distribution renders it probable that the great mass of these schists belong to one great formation.

No cherts have been observed like those in which radiolarian structures have lately been discovered near Aberfoil, Callander, etc. Nor are there any beds at the south-east frontier of the Highlands in this district which are so little altered as the black shales and greywackes associated with the cherts.*

The curious conglomerate on the east side of Loch Fad in Bute, by the " h " of " Loch " and the " F " of " Fad," contains, however, large pebbles and boulders of extremely little altered grit and grey-green felstone-like rock, with others of reddish limestone, which, in state of alteration, remind one very much of the beds accompanying the cherts. The fragments in this conglomerate seem entirely composed of such little altered beds : we could find no pieces that could be confidently matched with the schistose rocks at the opposite side of the loch. It may, therefore, well be that in old times before the formation of this conglomerate there was an adjacent area of rock, belonging to the same series as those near Aberfoil.

The fragments in the Upper Old Red conglomerates near the boundary of the schists are almost entirely of adjacent Highland schists, or of vein quartz, or of igneous rocks, probably in the main of Lower Old Red Age. It is not certain that they ever include fragments like those in the Loch Fad conglomerate : at all events, such fragments must be rare. We may, therefore, probably conclude that the exposures, from which the fragments in the Loch Fad conglomerate were derived, did not make at all so large a show during Upper Old Red times as they had done previously. This, perhaps, indicates that the boundary fault along the Highland frontier represents a line of very old and repeated movement, and thus confirms what is said, p. 95, as to the probability of a large part of the crushing at present seen along it being older than the Upper Old Red Sandstone. In the country between Aberfoil and Loch Lomond there is a thick crush zone running N.N.E. near the Highland frontier which is certainly of earlier date than the Upper Old Red Sandstone, and very probably there is another still earlier line of discordance of a thrust-like character.

The direction of the boundary fault is nearly parallel to that of the centre of " the anticline." There are some thin crush lines, seemingly continuations or connections of the Tyndrum set of north-east or N.N.E. faults, which are also approximately parallel to these : e.g., the one crossing Glendaruel near Lephinkill, one near Balliemore at the head of Loch

* These and the associated beds match very well with some of the Arenig and Llandeilo rocks of the South of Scotland.

Striven, one crossing the east shoulder of Ben Donich (134 and 142),
and others a mile or two off this last on the north-west side. These may
give rise to prominent features in the landscape, but they do not seem to
have large throws. Some of the same class, that run with the graphite
schists near Strachur and St Catherine's, seem to be more lateral thrusts
than vertical' movements, and are also of special interest as being of
earlier date than some of the lamprophyres : most of these lamprophyres
are, as stated on p. 108, clearly later than most of the faults of the district,
and are affected by their throws. The early faults referred to cannot be
classed with the thrusts of the north-west of Scotland, for they tend to
destroy the schistose character of the rocks, not to develop it.

In the area east of Loch Riddon there are many important faults which
run nearly north and south, or between this direction and that of the
schist boundary fault, so that they gradually approach the latter as one
proceeds in a south-west direction. These have nearly always downthrows
to the east, the beds on the east of them having their outcrops advanced
somewhat to the north. They can often be made out to be in part of
earlier date than the basalt dykes, but to have had the movement along
them renewed again after the intrusion of these : for illustrations of this,
refer to pp. 143, 186.

On the south-east side of Glen Fine the strike of the foliation is deflected into parallelism with the adjacent granite boundary. For a distance
of $\frac{1}{4}$ mile or more from the boundary the dip is inward towards the
granite. Further away the dip is, in some places, away from the granite
towards the south-west : at other times it is west, without any clear relation to the visible granite margin.

Some of the evidence for different faults and folds, together with an
account of some of the more interesting sections which have not yet been
described, is given in separate descriptions of the different areas.

Let us return for a little to a consideration of the coast sections which
have been described in chapter II. Anyone who examined cursorily the
section at Dunoon and that in the Kyles of Bute, could not help being
struck with the greater mineral change in the latter locality. He might
very likely put this down as an indication of the greater age of the
schists in it, and suppose that, somewhere or other in the intervening
area, an unconformity would be found, the less altered Dunoon rocks
lying unconformably on those of the Kyles of Bute. The rocks are now
so crumpled, and original structures so distorted and obscured, that
perhaps it would be quite possible for such an unconformable junction to
be overlooked.

If now the same person were also to examine the Ardentinny section he
would find a stage of metamorphism which seems in many respects intermediate between the two previously alluded to, and doubts might enter
his mind as to whether his former theory were correct. Still more would
he doubt, if he had the opportunity of also examining the different
sections that come between the Dunoon and Ardentinny horizons and the
Ardentinny and Kyles of Bute horizons respectively. He would then find
that there is an insensible gradation between the stages of metamorphism
found at Dunoon and the Kyles of Bute, and that it is at all events
necessary to conclude that the metamorphism increases quite gradually
as we proceed north-west from Dunoon toward "the anticline," whether
the rocks are the same or of different ages.

A more complete knowledge of the Kyles of Bute section would also
show him, as we hope we have made clear, that those structures, strain
slips, etc., which occur in it and not in the Dunoon section, affect earlier
structures which correspond to those at Dunoon. Hence, whatever may

be the relative geological age of the original rocks in the two localities, the age of those secondary structures in the Kyles which do not agree with those at Dunoon are not earlier than these last, but later.

We have shown, too, that the mineral changes, both in the Kyles section and elsewhere, are intimately connected with the mechanical secondary structures just referred to. They may not be the direct result of the heat, etc., liberated during the formation of those structures, but they have undeniably taken effect either during or after their production. Therefore we are forced to conclude that the mineral changes in the Kyles, in so far as they affect structures of later date than those at Dunoon, are also of later date.

To what agent or agents may we attribute the general regional metamorphism of the district? It is perhaps premature to say much in reply to this question. The whole of the metamorphism seen in any particular rock can hardly be considered the result of the mechanical strains occurring in that rock itself, for we have seen that there is a distinct increase of metamorphism in the section south of Ardentinny, as compared with that at Dunoon, though the mechanical movements are approximately the same at both places. The Ardentinny section is, however, much nearer the centre of the greatly crumpled "anticline" area, and we may perhaps suppose that the heat engendered by the physical movements there would travel outwards for some distance into rocks which are not themselves so crumpled.

There is also another important point to be considered. "The anticline" is certainly an anticline of foliation, whether it is of bedding or not, and the rocks at present exposed in the centre of it must, before the ridging up, and subsequent denudation, have been lying under a much greater thickness of rock, and therefore presumably exposed to higher temperatures, derived from the earth's internal heat, than those now occurring far away on the flanks. The distance of the schists at the boundary fault at Inellan from the centre of "the anticline" may be taken as about $9\frac{1}{2}$ miles, and the foliation dip in the intervening area may be averaged as 40°. These data give 32,000 feet as the approximate thickness of the beds measured across the early foliation planes, and the covering over "the anticline" rocks was once, perhaps, as much as this thickness greater than the covering over the rocks near the boundary fault.

There are no exposures of igneous rock in the Map to which it is possible to refer the metamorphism in question. The alteration effected by the Glen Fine granite does not extend much more than 1-mile from the granite, and gives rise to various minerals, andalusite, sillimanite, cordierite, etc., which are quite unknown elsewhere in Cowal. On the other hand, the albites of the albite schists are practically no more prominent near the granite than they are 20 miles away. The character of the alteration around the granite, and its relation to that of the albite schists, is described more fully on pp. 98–101.

The metamorphism shown by the garnetiferous mica schist gradually increases along the strike in a north-east direction, but we are ignorant of the cause of it. This zone lies some distance north-west of "the anticline," and the metamorphism special to it seems of earlier date than that which produced the albites (see p. 52). North-west of the garnetiferous schist the metamorphism decreases, and continues to do so on the north-west side of Upper Loch Fine. C.T.C.

CHAPTER IX.

UPPER OLD RED SANDSTONE.

The Upper Old Red Sandstone rocks of Inellan and Toward closely resemble those which occur in Bute below the andesitic lava flows, and are probably continuous with them under the sea. They are exposed along the shore most of the way between Inellan pier and the east gates of Toward Castle policies, and also in most of the old cliff of the raised beach between these places. Between Inellan and Chapelton they form such a narrow fringe to the land that its width does not usually exceed that of the raised beach, and consequently we get the schists on the west side of them exposed on the old cliff for a considerable distance. South-east of Chapelton they extend further inland, but with one small exception, about ⅓ mile north of Toward Cottages, there are no inland sections save in the burns, all the rest of the area being covered with drift or raised beach material.

They consist in the main of red breccias and sandstones, mixed with occasional blood-red shales and variegated purple and green-grey clayey shales. Calcareous sandstones or cornstones and magnesian limestones are also numerous on a certain horizon. No organic remains have been found in any of these beds.

The sandstones are false-bedded and rapidly changeable in character; in one place they may be half-full of schist fragments; in another, a few yards away, they may be quite free from these. The inclosed schist pieces are generally angular, and may reach a length of 2 or 3 feet, but the bands which contain the larger pieces seem very local, and a length of from 1 to 4 inches is more frequent: among the common pieces are purple and black phyllite, sheared grits with large quartz pebbles, and some tough green grits. The phyllites may well have come from the Dunoon series or from bands on the south-east side of it: the quartzose grits are like those common in the schist area immediately adjoining: the green grits are of the gritty "green beds" type found on the north-east side of the Dunoon phyllites. There is a general absence of the more highly altered and crumpled mica schists of "the anticline" area (see p. 23). We may conclude from these facts, and from the angularity of the pieces, that the coast line which furnished the pieces was near at hand and was composed of different schists in much the same proportion as the present schist margin. Some of the pieces are reddened, while others close at hand are not.

Of the pieces other than schist, those of vein quartz are the commonest, and after these hard reddish-yellow or purple quartzites; of more rare occurrence are magnesian limestone, soft red and yellow sandstones, red and pale yellow jasper-like bits, and porphyritic igneous rocks with prominent black mica.

The sandstone and limestone pieces perhaps point to contemporaneous denudation, and some of the jasper-like pieces may also have come from chert bands in limestones of approximately the same age. The sandstone pieces are generally angular, but the vein quartz, quartzite, and igneous rocks are remarkably well rounded. One of the porphyrite-like pebbles was examined by Dr Hatch (slide 2836), and proved to be a nepheline bearing rock. It is described by him as follows :—"In a dirty brown cryptocrystalline ground-mass, consisting of an admixture of turbid with clear crystalline matter, are embedded a number of rectangular and hexagonal sections, the greater part of which remains dark during a rotation under crossed Nicols, only a number of irregularly distributed specks transmitting light. These are probably pseudomorphs after nepheline. The rock strongly resembles the so-called Liebenerite porphyry, except that the secondary mineral, which has replaced the nepheline is isotropic, (analcime perhaps), instead of giving, under crossed Nicols, the brilliant mosaic characteristic of the Predazzo rock (Zirkel, *Neues Jahrb*, 1868, 7 and 9)."

Throughout the sandstones may be found dirty grey balls, averaging about 1 inch in diameter, which contain hard dark centres, apparently due to segregation of the iron oxides from the surrounding paler area.

In some of the beds the contained fragments consist entirely of schist; in others of schist and vein quartz mixed in not very unequal proportion; in others again schist is somewhat rarer. The last type occurs especially in association with the limestones, and, as is seen more clearly in Bute, we suppose that it and the limestones mark a definite horizon, and overlie a lower coarser set of rocks in which schist pieces are very numerous. This calcareous horizon, as far as known in this district, does not exceed 100 or 150 feet in thickness. Of the breccias and sandstones no base is seen.

The limestones are seen best at Toward Point, on the shore west and north of the Perch, and in the burn above Toward Taynuill. They are most usually white or cream colour, compact or finely crystalline in texture, and vary in thickness from the thinnest courses up to 20 feet at least. The limestone on the west side of Toward Point attains the thickness mentioned, and may be thicker, as no top is seen. Thin bands of limestone or calcareous sandstone are frequently seen to occur in a lenticular manner, their outcrops rapidly varying in thickness or dying out. Reddish nodules of chert occur in the Toward west limestone, and nodules or thin bands of it also occur in the Toward east limestone, in the limestone mapped by the lower part of the "C" of "Chapel," and in the most northerly exposures on the shore. In the last locality the chert streaks are repeatedly broken and slightly shifted by minute faults, which are confined to the bed in which they occur; below the main mass of limestone white calcareous bands, several inches thick, descend vertically or approximately so; at other times lumps of purer white limestone occur irregularly embedded in a less calcareous purple or reddish rock. Spots about the size and shape of peas, and whiter than the containing rock, are in places common, and project slightly on the weathered face. In the burn above Toward Taynuill the irregular manner of occurrence of the limestone is also marked—the separating lines between the other beds and it not keeping at all strictly to the bedding. Both the bed and sides of this burn, from the small north-east fault, which crosses the burn at a point about 80 yards south of the schist boundary, to a point 90 yards south of this fault, consist of limestone, mixed irregularly with more sandy portions. Other exposures are seen both above and below here, brought up partly by faults and partly by change of dip. In 1886 Messrs Merry & Cunningham made a trial of this limestone for lining their basic Bessemer steel converters, but the sample tried, amounting to 20 tons, showed 10·8 per cent. of siliceous matter, and so could not be used alone. The limestone quarried was white and crystalline, with irregular purplish and reddish parts, which were softer to the knife; it was taken out of the east bank of the burn at points 50 and 100 yards south of the fault mentioned. The following analyses of Toward dolomite are kindly furnished by Mr R. Main of Messrs Merry & Cunningham.

	No. 1.	No. 2.*	No. 3.	No. 4.
$CaOCO_2$,	53·87	54·83	54·03	57·98
M_gOCO_2,	41·52	36·87	35·73	38·37
Al_2O_3,	1·20	1·50	1·20	·90
SiO_2,	3·30	4·60	8·70	2·80
	99·89	99·80	99·66	100·05

On either side of the basalt dyke, 300 yards west of Toward Cottages, are exposures of thin hard white limestone bands, mixed with green and red marls and red sandstone posts. Softer red freestones of apparently some thickness come on above these to the south. These limestones are evenly bedded, and contrast strongly with the lenticular habit of most of the thinner limestones of the series.

In the fields at points 100 yards east and 240 yards W.N.W. of Toward Cottages, trials for coal are said to have been made. This cannot have been since 1816, and no information can now be obtained of them. There is still the appearance of an old hole in the former locality. We suppose they were in search, more especially, for the seam which occurs immediately below the andesites at Loch Ascog in Bute.

As the limestones are supposed to mark a stratigraphical horizon, it is clear, as we examine the ground, that it is necessary to call in the aid of certain faults to explain their present different positions.

South from Inellan pier, for the first 500 yards we seem to keep on much the same horizon, the beds consisting of greyish-red sandstones with pebbles, chiefly of vein quartz and quartzite, but subsequently we pass into lower rocks, blood-red clays, and sandstones with thin cornstones, until we reach thicker limestones. These are seen every here and there, with gently varying dips, for about $\frac{1}{2}$ mile; then we come to red sandstones and breccias, with an apparent dip to the north of sometimes as much as 20° or 25°. If this can be taken as true dip, we should, on getting to the shore east of the Perch, be some distance below the limestones. That this is the case seems likely from the abundance of schist pieces in the breccias. Concluding this to be the fact, then, we continue south, crossing three basalt dykes running W.N.W.; shortly after passing the last of these we come again to limestones still dipping north, at much the same angle as the breccias. The limestones would seem at first to underlie the breccias. We suppose them to be thrown down into this position by a fault, with downthrow to south, running alongside the last basalt dyke referred to. This dyke hades south, and the beds for a few yards on the south of it also dip south. It does not show any signs of crushing, but this is no objection to the existence of the supposed fault, as in this district we repeatedly find faults of early date along which the basalts have subsequently proceeded.

South from here the limestone series still continues dipping north, so that we gradually pass into lower beds. On the shore near Chapelhall a crush is also seen, running W.N.W. with hade to north, and this helps to the same result. Thus we soon lose the calcareous bands and reach a set of red sandstones and breccias, which keeps on all the way to Toward lighthouse, a distance of 1 mile. For the north half of this distance the dip is still to north at perhaps 10°-12°, but in the south half the beds seem nearly flat. These sandstones may be taken to be the same as those at and to the north of the Perch; and like them, as we have seen, they dip below a limestone series to the north. At Toward pier and lighthouse, we are doubtless some distance below the limestones, but on a point on the

---

* Analysis is given as received, but it does not add up correctly.

shore about 50 yards south of the lighthouse, where the coast changes direction from north and south to east and west, we suddenly find a dip to S.S.W. of 15° to 25°, and see several outcrops of limestone striking against the sandstones. This is no doubt due to a fault with downthrow to the south, running in a west or W.N.W. direction. The west limestone was sought by Messrs Merry & Cunningham in a series of shallow bores, immediately behind the wall of the field west of the last villa west of Toward lighthouse, but as these are on the upthrow side of the fault, they of course failed to find it. At a point 100 yards south-west of the lighthouse, a basalt dyke seems slightly shifted by the fault, the south side to the west, or else it had originally a slightly different position on either side of this line of weakness. The dyke on the north side contains a thin band of pebbly sandstone and also a calcareous band, with geodes of quartz crystals. We may perhaps estimate the throw of the fault as 200 to 250 feet. It probably runs under the raised beach at a point about 140 yards east of Toward Cottages ; on the west of this limestones and calcareous bands again appear, with a general absence of the coarse schist breccias.

Some 200 yards before reaching the schist boundary disturbances we again lose the calcareous beds, and meet with sandstones with some bands very full of schist pieces. We suggest another fault here running in a N.N.E. direction, and with downthrow to the east.

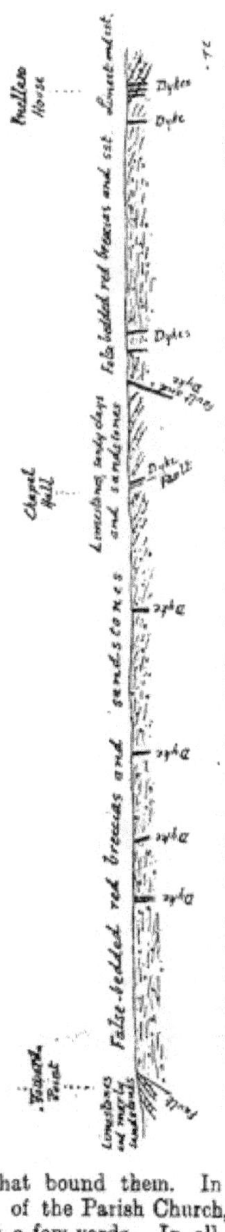

The sections that best illustrate the relations between the red rocks and the metamorphic schists occur on the shore close by the east gates of Toward policies, and at or just to the south of Inellan pier. Less satisfactory ones occur in the burn above Toward Taynuill, and in the burn and cliff near Chapelton. It is clear that the boundary of the schists is a line of strong crush, but it is not so clear that this crush is of later date than the red rocks. These rocks are nowhere seen distinctly crushed or disturbed near the boundary, and in the section at Toward they are lying nearly flat against the crush area, along a line which in one place changes sharply almost at right angles to its general direction, and still show no crushing. Neither could we satisfy ourselves that any fragments of them exist in the crush ribs that bound them. In the Inellan section, also, at a point about due east of the Parish Church, the boundary makes a sharp bend to the west, for a few yards. In all probability the boundary is an old line of weakness, along which movements have taken place at more than one geological epoch. It seems impossible

Fig 33.—Horizontal scale, 3 inch = 1 mile. Section showing supposed structure on shore between Toward Point and ¼ mile N. of the Perch.

to imagine that all the crushes in the Toward section—crushes in all 60 to 80 yards wide—could have been formed subsequently to the Upper Old Red Sandstone, and yet have left these rocks so little disturbed.

The absence of red staining in the schists seems at first a strong argument in favour of considering some of the movements of later date, but it is doubtful if this has the weight it at first seems to possess. The line of junction was probably steep, and there may not have been much tendency for the staining agents to percolate sideways through the earlier crush ribs. There is a wide basalt dyke in the Toward section, running in and parallel with the crush ribs, and this is quite unaffected by the crushing, so that we may be certain that no movement has taken place since its intrusion. Two similar uncrushed dykes occur in the Inellan section, running nearly at right angles to the direction of crush. The schists that bounded the red rocks at the time of consolidation of the dykes were then the same as those now doing so, and we know that no staining has taken place in the interval.

Still, it does seem necessary to suppose that some movements have taken place along this line subsequent to the time of the Upper Old Red Sandstone, and, though they may have done so without much crushing or contortion, the displacement effected may yet have been considerable. In the first place, we know that in the red rock area generally later displacements are common, and that rocks occupying different positions in the series now abut against the schists at the same level at different places—thus, in the cliff south of Inellan Free Church a magnesian limestone lies close by the schists, while at a point $\frac{1}{12}$ mile north-west of the Perch, red breccias do so. In the second place, the rocks now lying close to the schists are often much less full of schist fragments than we should expect if they had been formed at the immediate base of schist cliffs.

C.T.C.

CHAPTER X.

THE IGNEOUS ROCKS.—MARGIN OF THE GRANITE NEAR CRUACH TUIRC. HORNBLENDE-PORPHYRITE AND FELSITE. HYPERITE.

Margin of the Granite near Cruach Tuirc.

A small part of the igneous complex of Garabal Hill and Meall Breac, described by Messrs Dakyns and Teall (*Quart. Journ. Geol. Soc.*, vol. xlviii. p. 104), comes into the north-east corner of 1-inch Map 37. We make no attempt to describe the whole of this part, but give an account of the edge of it near Cruach Tuirc, and the metamorphism near this edge, chiefly with a view to contrast the metamorphism special to its neighbourhood with that found further away, near the anticline of Cowal, etc.

The edge of the complex between the River Fine and the east margin of 1-inch Map 37 consists, as far as the drift allows one to see, almost entirely of coarse granitic rocks, frequently hornblendic, and in some places so much so as to be worthy for small areas of the name of diorite. There seem all gradations from granite, or rather granitite (using this last term to imply a granite which contains much plagioclase and a black mica, but no original white mica) through hornblendic granitite, into a rock of much the same coarseness of grain in which the hornblende is in excess of the biotite. It would create a false impression if a sharp line were drawn to separate the different varieties, and we have in the Map massed them together as granite. There are, however, besides the coarse dioritic granitites which pass gradually into the normal granitites, one or two small exposures of a much darker fine-grained diorite which occurs close to the margins of the complex, and is probably sharply separated from the adjoining granitite. Unfortunately none of these exposures is clear, and they are of very limited extent,—too small to be satisfactorily shown on the 1-inch Map. But we shall shortly describe them in detail, as it is possible that they indicate the earlier existence of an extensive mass of diorite near the present edge of the complex, which mass has now in most places been intruded through by later granitic rock.

The granitic rocks are generally characterised by abundant porphyritic crystals of orthoclase felspar (*op. cit.*, pp. 105, 108), sometimes as much as 2 to 3 inches in length. These vary in prominence a good deal in different parts : in some they are quite rare : in others they form nearly half the area of certain rock exposures. They are usually, as shown most clearly in some of the clean-washed boulders on the shore of Loch Fine, distinctly zoned, the outer zones being of paler felspar substance than the inner parts, approximating in fact to the colour of the small felspars of the ground mass. Small inclusions of biotite frequently lie along the sides of the zones, and have their greater lengths parallel to the sides along which they lie. The biotite is, as already mentioned, usually accompanied with hornblende in varying quantity. Under the microscope it is seen to contain many inclusions of apatite, sometimes arranged in stellar groups. The interstitial quartz is not always conspicuous in hand specimens, and probably varies in quantity a good deal. It is sometimes of a slightly milky opalescent colour. Sphene is an abundant accessory mineral, occurring sometimes in large brown crystals. It is quite distinct to the naked eye in the granitite in the River Fine by the " F " of " Fine," and in some fine-grained dark patches included in the granitite. In these dark patches (slide 6115) biotite and sphene are more abundant than in the enclosing rock, but hornblende is absent. Slide 6113, from the rock in the River Fine, $\frac{1}{12}$ mile slightly south of east of Newton Hill, shows plagioclase in large porphyritic crystals. Mr Teall says of this :—" A coarse-grained grey hornblendic granite. Very large idiomorphic crystals of zoned plagioclase, idiomorphic green hornblende, biotite and sphene with interstitial orthoclase and quartz. The sphene occurs in very large crystals. . . . . Iron ores and apatite occur as accessories."

Dark basic inclusions, most of them probably referable to diorite, are common, and are sometimes 1 to 2 feet in diameter. They are rounded in outline, weather in depression, and are not conspicuously longer in one direction than another. Granitic strings are sometimes seen cutting through these inclusions : in some of the strings the full width may be taken up by a porphyritic felspar. No marked signs of fluxion structure have been observed in this area, but here and there a sub-parallel arrangement of the porphyritic felspars may be seen.

The granitic rock is as coarse in grain at its junction with the schists as in the interior. There is no definite line for the edge of it, but for a little space there are alternations of schist and granite, the latter gradually increasing north-east until it forms the whole of the rock. On the east side of Cruach Tuirc the area of veined schist extends perhaps 120 yards, or rather more, from the granite boundary drawn in the Map. In this area not only are there many thin approximately vertical granitic strings, sometimes several feet in width, but also, particularly near the main granite mass, numberless granitic strings which run with the foliation of the

G

schist. The thickest of these last are often about 1 foot thick, but there are others much thinner and very short, which sometimes give rise to an appearance of almost isolated crystals of felspar along the foliation planes of the schist. Neither the vertical strings nor those running with the foliation are finer at their edges : the felspar crystals within the thinnest of them are often about the size of peas, and occasionally take up the full breadth of a string. In this locality we did not observe the large porphyritic felspars, until the main granite area was reached, and they are not always apparent there. For 30 yards or so west of the line taken as the generalised granite boundary, the alternations of schist and granite are almost equally divided, in bands 10 to 30 feet thick, which run roughly parallel with the schist foliation. The granite bands here have often thin pegmatites, 2 to 3 inches thick in them : in these pegmatites the patches of quartz are occasionally 1 to 2 inches thick, and are somewhat opalescent: the biotite folia sometimes attain a length of $\frac{3}{4}$ inch.

The schists into which the above strings intrude are an alternating series of micaceous albite schists and more quartzose schists. The general dip of the foliation is inwards towards the granite mass. Andalusites in brown prisms, sometimes $\frac{1}{2}$ inch long, are extremely abundant in the more micaceous and albite bearing schists. They form a close network which projects slightly on the weathered face, and so are readily discerned. In fresh fractures it is much more difficult to detect them. The prisms cut across the foliation in any direction, and show no bends or fractures. The albites in the albite schists are not uncommonly changed into a pink tint. This change seems due to the oxidation of the magnetite specks inclosed within them (see p. 101), and is of the same nature as that seen in the neighbourhood of other intrusions in Cowal. Streaks of iron pyrites and nests of black mica are common in the more micaceous schists. In the quartzose schists the alteration near the granite is not so evident ; nests of black mica occur here and there, but not commonly : the black mica along the foliation planes seems also to be in unusually large flakes—lengths of $\frac{1}{8}$ inch being common. In addition to the andalusite, two other well-known contact minerals have been detected in the microscopic examination of slides from the altered andalusite rock or "hornfels." These are sillimanite and cordierite. The former occurs in slide 6098, in bundles of very thin colourless fibres, with transverse jointing. It has also been noticed preserved in quartz in a slide of a boulder found $\frac{7}{8}$ mile north-east of Achadunan. This boulder, of andalusite hornfels, is traversed by thin strings which are composed at their sides of a coarse-grained pale acid rock with abundant quartz, and no mica or hornblende : in the central parts of the strings there is no macroscopic quartz, but hornblende is abundant. In slide 6099 (andalusite-hornfels) cordierite is recognised by the occurrence of yellow pleochroic halos in a mineral otherwise resembling untwinned felspar.

In the river $\frac{1}{4}$ mile W.S.W. of the "F" of "Fine," the schists also consist of an alternating series of micaceous albite schists and more quartzose schists. The general foliation dip is east. The albites are reddened : the matrix of the containing schist weathers smooth and black, and the red albites scattered through are prominent. Many fine granitic strings intersect the schists and run with a general E.N.E. direction. There are also coarser bands of hornblendic granitite which occur partly as dykes and partly as irregular sheets along the foliation : these are as coarse at the edges as in the middle : when they occur as dykes they have the same direction as the finer dykes. Near the south-west and west sides of the south-west island, there are good sections showing the permeation of the schists by thin streaks and irregular patches of the granitite. If it were not for occasional clear intrusive sections some parts of the granitite might be mistaken for an altered albite schist of coarser character than usual. Sixteen yards north of the north end of the south-west island, a vein of dark diorite, 3 inches thick, is seen intruding in the schist. Sixty yards N.N.E. of the north end of the same island another exposure of fine-grained black diorite occurs. On examining vertical sections, the constituents of the rock seem to run parallel in a nearly vertical direction. The diorite contains scattered grains of white felspar attaining the size of a pea. Some parts of the exposure are composed of a coarse paler diorite, which occasionally runs in straight strings through the finer rock and appear to be intrusive in it. East from this place we come on an area of schist 4 to 5 yards broad, and then the fine black diorite is seen a third time, and again with a parallelism in the constituents. In the north-east channel of the river, 16 yards north-east of the south-west end of the big island, the fine black diorite and a fine-grained granite or porphyrite are in contact. The diorite has a foliated appearance in vertical sections. It seems underlain by the granitic rock, and to have the course of its constituents cut across by it. On the east side of the exposure there is more schist before we reach the main granite area. The granitic rock does not get finer at its junction with the diorite : it looks very like the porphyrite which occurs in sheets within the schists, etc.

In the burn ½ mile E.N.E. of the first "a" of " Meall Beag," there is a dark fine-grained diorite, with porphyritic felspars sometimes ⅛ inch long, close to a porphyrite, but the actual contact of the two is not seen. In the part of the same burn extending for ½ mile below this, and on the E.N.E. side of Lochan Mill Bhig, there are good exposures of the altered albite and quartzose schists. The rocks are all harder than usual, and weather with a brown crust. Thin pyrites strings often occur along the foliation. The matrix of the albite schists is darker than usual and has a rather conchoidal fracture : the albites themselves often appear red, but not always : the same bands in which they occur also contain andalusites, and small nests of black mica : we are inclined to say, too, that both in the albite schists and in the quartzose schists there is a greater abundance of black or purplish-brown mica on the foliation planes than usual. Epidote, generally in short irregular strings and sometimes mixed with quartz in veins, is in some places common. These strings are not absolutely confined to the schists : they occur also in granitic veins. Garnets and ilmenite also occur abundantly in the albite schists N.N.E. of Lochan Mill Bhig, but it is not probable that these have any genetic connection with the andalusites. The garnets are like those in the garnetiferous mica schists described on pp. 49–52. In the burn ⅛ mile N.N.E. of the "M" of "Meall," strings of fine granite or porphyrite are common, and sometimes about 1 foot wide. These have a general direction almost at right angles to the foliation strike and the adjacent edge of the granite mass. In one place an 8-inch string is edged on either side by 1 inch of opaque white quartz, and thin streaks of the same mineral occur here and there in the middle also. Some of the strings are accompanied by throws at least a few inches in amount. In this locality none of them seems to contain porphyritic felspars, but ⅛ mile further up the burn there is a 3½ to 5 foot vein in which the porphyritic felspars are sometimes 1½ inches long : 10 yards north of the burn these crystals are so abundant and large that occasionally in an exposure of several square feet they compose about half the area.

In the River Fine, 70 yards above the bridge by the "G" of "Glen," there is, on the south side of the fault in the river course, a band of schist about 1 or 2 feet thick, which contains dark green hornblendes. It is pale in tint and evidently too quartzose to be classed as a hornblende schist derived from an epidiorite. The band also contains much black or brown mica, and breaks with a marked conchoidal fracture. The lengths of the hornblendes are in a common direction—14° west of north. Two or three feet below the band, on the east of it, the albites of the albite schist have often a red altered look, and there are some layers almost entirely composed of andalusite mixed with purplish-brown mica. In other bands andalusites are scattered sparingly through a rock mainly composed of white mica. The andalusites interlace together to form a network, yet, when looked at collectively, they seem to have one direction more frequently than any other, and the more frequent direction is much the same as that of the adjacent hornblendes. Fourteen yards further east, the uppermost part of a massive quartzose band also contains dark hornblendes for a thickness of about a foot. The distance of this exposure from the nearest part of the main granite edge is rather more than $\frac{3}{4}$ mile The dip of the schists is nearly due east. We suppose the hornblende to be probably due to contact metamorphism, and the state of the rock in other respects speaks in favour of this. But only one other instance is known where hornblende appears to have been specially formed near the edge of the granite. This other case occurs in the foliated lamprophyre sheet at the first " u " of "Cruach Tuirc" (see p. 125). The common orientation of the hornblendes, and the tendency, though but slight, towards a parallelism in some of the andalusites, are matters of importance. They seem to suggest that these minerals may possibly be due to the influence of some intrusion earlier than the granite. Some of the nests of black mica found near the granite edge suggest the same idea. In the burn $\frac{1}{2}$ mile E.N.E. of the outlet of Lochan Mill Bhig, $\frac{1}{3}$ mile south-east of Cruach Tuirc, and just above the Eagles Fall, these nests are longer in the direction of foliation than across it. The strike of the foliation of the schists is deflected into parallelism with the adjacent granite boundary. For a distance of $\frac{1}{4}$ mile or more from this boundary, the dip is inwards towards the granite. Further away the dip is in some places away from the granite, towards the south-west : at other times it is west, as in the exposure of hornblendic schist just described, without any clear relation to the visible granite margin. This alteration of strike and dip show that there has been a certain amount of local movement near the granite during or since its intrusion, or since the intrusion of some other igneous rock which had approximately similar boundaries. Perhaps this movement may have been strong enough to bring about the orientation of the hornblendes, and the other phenomena mentioned in connection with it. Or perhaps it is not necessary to have recourse to such an explanation. Earlier structures in the schists may have had influence enough to guide later formed minerals along their own direction to some extent.* The indications of fluxion structure in the granite itself, in this locality, are so slight that it does not seem likely that this rock was subjected to severe stresses during or after consolidation. The obscure exposures of dark diorite near the edge of the granite show more parallelism of structure. If it is safe to suppose that these exposures are remains of a once much larger mass which existed before the granite came up, and which possessed an outer edge not

---

* Since the above was written, we have noticed in the altered Cambrian limestones of Ben Suardal (Skye) that the contact minerals, tremolite, etc., developed near the tertiary granophyres, have in some cases a distinct tendency to occur with their greater lengths along the lines of "stretching" developed in the rock during the great "thrust" movements. These thrusts are certainly of earlier date than the Old Red Sandstone.

far removed from that of the granite, we might attribute the metamorphism partly to the pre-existing diorite. Most of the dark inclusions within the granite could be referred to the same mass. The parallelism of structure in the diorite might be taken as an indication that earth movements were going on during its consolidation on a greater scale than during the consolidation of the granite. We may mention in this connection that in a boulder, no doubt derived from the Garabal Hill and Meall Breac complex, on the shore of Loch Fine, ½ mile south-west of Rudha Bathaich Bhain, a thick streak of some black mineral, apparently magnetite, continued for some distance. In some boulders in Glencroe (134 south-east) there are streaks of opalescent quartz which run parallel to one another for about a foot. These boulders have probably come from the diorite mass between Ben Arthur and Ben Ime (1-inch Map 38), a mass which we regard as probably of the same age as the Garabal Hill and Meall Breac complex.

It seems certain that the metamorphism which produced the andalusites, the alteration of the appearance of the albites, the epidote veins, the black mica nests, etc., is due either to the granite or to some earlier intrusion belonging to the same great phase of igneous activity, the margin of which was not far off that of the granite. For all these phenomena are confined to the near neighbourhood of the granite. We have noticed no andalusites more than 1 mile from the granite edge. The reddening of the albites extends a little, but only a little, further than this: in the River Fine it is first met with ½ mile below the bridge near the "G" of "Glen." The universally unbent character of the andalusites shows that there can have been no close folding or shearing since their production. The reddening of the albites is like that caused by the broader basalt dykes (see p. 149), and these dykes are certainly of later age than the albites. But this alteration of colour depends on an alteration produced in the inclusions in the albites more than in the albite substance itself, and possibly it is not quite safe to assume that the alteration is of later date than the formation of the albites. It might be argued that the oxidation of the enclosed magnetite grains had taken place before they were surrounded by the albite substance. But apart from this, the continuous extension of the albite schists for many miles away from the granite—at least more than 20 miles—while the andalusites, etc., are confined to the close neighbourhood of the granite, is itself enough to indicate that these minerals are due to different causes and are probably of different ages. The albites are of regional occurrence, are found independently of any visible igneous rock, and may be regarded as one of the last minerals formed in connection with, but slightly subsequent to, the completion of the mass of the movements going on during the schist manufacture. The andalusites are due to the influence of a local intrusion of later date than the schist manufacture, and are in all probability of later date too than the albites.

The relations of the porphyrite sheets and the foliated lamprophyre intrusions to the granitic rocks are referred to on pp. 102, 125 respectively. The sheared character of the camptonites near the granite, which character is not, however, universal, is itself a sign of their greater age. It may perhaps be connected with the foliated aspect of the dark diorite exposures near the edge of the granite. In the area near the granite there seem to have been earth movements of importance at a later date than elsewhere in the district, and after the mass of the schists were essentially as they are now.

We know of no mica trap within or near the edge of the granite.

Hornblende-porphyrite and Felsite.

The felsite and hornblende-porphyrite are classed together, because it is not always possible, in the field, to decide to which of these groups a particular band should be referred. The same porphyritic constituents, felspar, black mica, or chlorite, sometimes hornblende, and scattered quartz blebs, occur in both, and the general colour of the ground mass is also the same—pink or orange. The state of crystallisation of the ground mass it is often impossible to decide with the aid of the hand lens alone. There is also a probability that the same band may be felsite in one area and porphyrite in another.

These rocks behave in the field in the same way as the lamprophyres, generally forming sheets which run roughly with the foliation of the schists, and more rarely dykes. As stated elsewhere, p. 125, there is some evidence for regarding the felsites as later than the basic micaceous lamprophyres. On the other hand, the hornblende-porphyrites are later than the foliated lamprophyres, and later also than the granite. In the district of Cowal the range of the felsites is of much less extent than that of the lamprophyres. It is almost confined to the part of the water-shed of Loch Fine above Strachur, and to the neighbourhood of Acharosson (181 and 182). The hornblende-porphyrites are of still more limited range : none are certainly known south-west of a line between Achadunan and Butterbridge, but felsites are abundant south-west of this line : on the other hand, no certain felsites have been observed within a mile of the edge of the granite of Glen Fine. Putting these facts together it almost seems as if the hornblende-porphyrites took the place of the felsites in areas near the granite.

In a little burn ½ mile south-east of the first "e" of "Glen Fine" a thin band of hornblende-porphyrite occurs in the coarse granite, and presents chilled margins to it : the hornblende-porphyrite weathers in more sharply jointed blocks than the granite, and is not so decomposed. On the hill-side, ¾ mile south-east of Cruach Tuirc, an exposure of hornblende-porphyrite within the granite has been traced for a little more than 200 yards in a N.N.E. direction. The breadth occasionally exceeds 50 yards. At the sides the ground-mass is in some places distinctly more fine-grained than in the interior, but we could not make out that this was always so. A thin band on the west side of the broad band shows a chilled margin against the granite, and the colour of the base is changed from pink to dark grey. We could see no difference between the hornblende-porphyrite intrusive in the granite, and that intrusive in the schists. We have not traced any band from the one area into the other, but the length of granite boundary examined by us does not much exceed 1½ mile, and half of this is hidden under drift.

The hornblende-porphyrites within the schist area occur in several more or less irregular sheets, as much as 20 feet thick, on the south and east slopes of Cruach Tuirc. The general colour of the matrix is pink or buff, but for an inch or two near the margins it may become dark grey or chocolate-brown. The red porphyritic felspars and black micas continue to the very edge of the intrusions, and form a striking contrast in colour with the matrix. This change of colour is seen at points ½ mile W.N.W., and ½ mile slightly north of west of Cruach Tuirc, and ½ mile slightly north of west of the "E" of "Eagles Fall." Together with the change of colour the grain grows finer. The quartz that is seen with the hand lens is always in the form of subrounded pyramids and blebs : it is not abundant and cannot be noticed at all in some hand specimens.

Slides 6109 and 6110 are respectively from the middle, and 2 inches from the side of the sheet, ½ mile slightly north of west of Cruach Tuirc. Of 6109 Mr Teall says :—" Phenocrysts of felspar, biotite, and hornblende in a light grey compact matrix. The large crystals of plagioclase are idiomorphic and zonal—allied to oligoclase. The hornblende occurs in long prisms, and the biotite in hexagonal tablets partly changed to chlorite. The ground-mass is mainly composed of small idiomorphic felspars which often give rectangular sections and interstitial quartz, together with some fine microcrystalline material. Magnetite occurs as an accessory in small idiomorphic crystals. This rock represents the porphyritic phase of a hornblende-granitite or quartz-mica-diorite. The ferro-magnesian constituents are by no means abundant.

The hornblende-porphyrites near the fault in the burn south and south-west of Cruach Tuirc are decomposed, and the forms of the porphyritic felspars are filled by a soft yellowish-green substance, which reminds us of the agalmatolite formed from the felspars of the pegmatites of the Lewisian gneiss in Sutherland, etc. Slide 6117 was prepared for the further examination of this substance, but threw no light on its nature. Mr Teall noticed, however, one interesting point in the slide. It contains two subangular grains of quartz, each of which is surrounded for a considerable distance by a zone of irregular breadth which attains its maximum darkness when the grain itself is extinguished. Mr Teall supposes there must have been a period of quartz building followed by one of corrosion, and again one of quartz building.

In the River Fine, ½ mile south-west of the " G " of " Glen Fine," there is a dyke of felsite or hornblende-porphyrite which runs slightly north of east. It is about 10 yards wide, and crossed by flaggy joints. The porphyritic felspars are sometimes ½ inch in length, are in stout shapes, and zoned. A similar dyke, or perhaps the continuation of the same, comes into the river again ⅜ mile below this place.

A sheet running by the foot of Lochan Mill Bhig expands in places into an outcrop 100 yards wide. This has not been sliced, but probably it should be classed with the hornblende-porphyrites. It is seen in Eas Riachain by the last "a " of " Riachain." One hundred and sixty yards further up the burn a similar band is thrown into the stream by a fault hading east. This may be the same band repeated by the fault. On the north-east side of the burn, between the two exposures in the stream, there are two other small outcrops surrounded by drift.

The sheet in the burn near the second " I " of " KILMORICH " does not seem connected at the surface with the one near Lochan Mill Bhig, but when the one dies out the other appears and runs on in much the same direction. The sheet by this " I " is traceable for about 1½ miles, and passes over the hilltop a little east of Binnein Fhidleir into the watershed of the Kinglas. About 200 yards south of the Ordnance Station 2658 it is 30 feet thick. Then it gradually thins. In the burn ¼ mile south-west of the above station it is 12 feet thick, and soon after that it ceases to be traceable. Another outcrop of similar rock appears, however, on a lower horizon, and can be traced ½ mile south-west : in this course it is thrown by at least three faults, each of them running in burns.

On a lower horizon than either of these is another sheet which has been traced 1½ miles. This crosses the hill-top ¾ mile east of the last " e " of " Binnein an Fhidleir." In the burn a little more than ⅓ mile south-east of Binnein an Fhidleir it is 30 feet thick, and has a lamprophyre sheet close above. This lamprophyre is seen in close company every here and there for nearly a mile north-east, sometimes without any separation of schist and sometimes with only a very thin one, but their relations in age are not clear. In the Kinglas, just below the " g " of " Glen Kinglas," the two rocks are again seen in close contact. The evidence favours the later age of the lamprophyre, but it is not decisive.

There is a big sheet of felsite not far off the Loch Tay limestone, in the area between Ardno and Ardkinglas. At Ardkinglas it is above the limestone, but near Ardno below it. The main change of horizon is at the north-west fault, ¾ mile E.N.E. of Ardno, and is not visible at the surface. The chief joints are parallel to the surfaces of the sheet: this is noticeable at a point ½ mile E.N.E. of Ardno. Near the same place there is a quarry in the felsite, and it has been extensively worked for road metal. In Coire No, there are bands which probably represent the south-east continuation of this sheet: one band occurs some distance above the limestone, and is seen in the stream ¼ mile south-west of the "N" of "No"; the other occurs below the limestone, and is seen at the burnside by the "o" of the same word. A band of felsite, probably representing one of the last mentioned, is seen in all the burns between the Coire No watershed and the "E" of "ARGYLLSHIRE," in the Cur by the "E," and ½ mile above the junction of the Cur and the Cab. Further S.S.W., felsite is seen in places near the course of the disturbances that run from N.N.E. to S.S.W. towards Bridgend. These disturbances traverse an area in which there are no other exposures of similar rock. The crush lines have afforded ready openings for the intrusion of the rock, and it has come up along them for some distance outside its usual area of occurrence. In Leamhanin, not quite ½ mile slightly east of south of Socach, a crush is seen at the side of the felsite, but the latter rock itself is not appreciably crushed. Between the "C" and "H" of "STRACHUR" an east to west fault shifts the crush with which the felsite runs. The felsite runs along the fault for 20 yards, and it is not clear that it is crushed at all in this distance: it would seem that the east to west fault, as well as the N.N.E. crush, is of earlier date than the intrusion. The most S.S.W. point at which we have noticed felsite along this line is about ¾ mile N.N.E. of Bridgend. Here there are two bands, the east band being accompanied by a crush 6 feet wide.

There is sometimes more than one sheet of felsite on the south-east side of Beinn an t-Seilich (134). As stated elsewhere, p. 125, one of these is clearly cut by a later lamprophyre sheet which contains stellar groups of hornblende needles. In the burn ⅔ mile S.S.W. of Stob an Eas there is a good section of one of these sheets about 30 feet thick. The felsitic matrix is of a deep cream colour. The chief joints are parallel to the surfaces of the sheet, and split it up into coarse flags. Porphyritic felspars are as large and abundant at the sides as in the middle of the sheet. Slide 4741 was prepared at the surfaces from a specimen near the side of the sheet. Mr Teall describes it as follows:—" Pale coloured massive rock composed of phenocrysts of soda-felspar and pseudomorphs after hornblende (?) and biotite in a felsitic matrix. Under the microscope one or two small more or less rounded grains of quartz may be recognised, but these are scarcely present in sufficient abundance to make the rock a quartz porphyry. The groundmass shows a tendency to a microlitic structure in the development of the felspars." Further south-west the same sheet is cut by a dark micaceous lamprophyre (see p. 125).

In the three burns, ⅓ mile slightly east of south, ½ mile south, and ¾ mile slightly west of south, of the "n" of "St Catherine's," there is a 3 to 6 foot sheet with a dark grey matrix, which contains distinct quartz blebs and biotite flakes.   C.T.C.

Some fine-grained red felsitic intrusions, behaving as very irregular sills, occur ½ mile south-east of Kilfinan (181). The largest of these is well exposed in the Acharosson Burn. It varies considerably in width of outcrop: in one part it attains a breadth of 100 yards and then suddenly narrows down to about 12 yards. On the south of the burn it is 200 yards wide. A specimen was sent to Mr Teall and described by him as follows:—"4013. Red felsitic sill. Pink rock consisting of numerous small porphyritic crystals of felspar (orthoclase and plagioclase) in a felsitic matrix." Several other masses that occur to the south of Acharosson Burn, as far south as Loch na Melldalloch, probably belong to the same set of intrusions.   W.G., J.B.H.

There are two sheets of hornblende porphyrite near St Catherine's. One of them occurs on the shore 100 yards east of Ard na Slaite, and can be traced in a south-west direction for nearly 2 miles : it is distinctly shifted here and there by faults. The other occurs within a band of epidiorite 230 yards south-west of Ard na Gailich : there is also a good section of it in the burn ⅛ mile south-east of St Catherine's pier. The first mentioned band is the soundest, and has been quarried a little below the road near the top of the first "h" of "Tighe Claddich," and on the hillside ⅛ mile south-east of the small "s" of "St Catherine's." The matrix is pale greenish-grey : in this are small grey porphyritic crystals of felspar, and larger dark green chlorite pseudomorphs after hornblende. Some of the pseudomorphs are ½ inch long and $\frac{1}{10}$ inch broad : they are sometimes collected into stellar shapes. A specimen from the first mentioned quarry is thus described by Mr Teall (slide 5063):—"Greenish-grey porphyritic rock. Matrix compact. Pseudomorphs in chlorite, and a carbonate after idiomorphic hornblende. Also more or less altered plagioclase. Both occurring as porphyritic constituents. Matrix of felspar microlites and chlorite. . ." The rock shows beautiful fluxion structure of the felspar microlites round the larger porphyritic crystals of felspar and the pseudomorphs after hornblende.

The sheet on the shore 230 yards south-west of Ard na Gailich is of a cream or orange-brown colour. It breaks up into flags roughly parallel to the original surfaces, and shows many small rusty-weathering, lath-shaped, decomposed hornblende crystals. Near the surfaces these small hornblendes are parallel to one another, and to the surface adjoining, but they change direction near the porphyritic felspars and the larger hornblendes, so as to bend round their edges. In certain places amygdules are abundant and very large, sometimes more than 2 inches long and nearly 1 inch broad. The outer parts of the amygdules are composed of a ferriferous carbonate and barytes : the insides are sometimes hollow. In the shore section the flat sides of the amygdules are parallel to one another and the flaggy joints of the rock.

Hyperite.

A specimen from a rock exposure in the burn 1¼ mile E.N.E. of Donich Lodge is described by Mr Teall as follows (slide 4743):—"Medium to coarse-grained massive rock. Hypersthene, augite (diallagic in part), biotite, plagioclase (more or less lath-shaped), iron ores and interstitial quartz. Quartz-biotite-hyperite." The augite and hypersthene are sometimes surrounded, or partially edged, by hornblende : it is more probable that this is due to variations during the original growth of the rock than to subsequent dynamic changes. The felspar has sometimes a pinkish tint as in the mica dolerites. The biotite is hardly recognisable in the hand specimen. The exposure to which this specimen belongs is largely drift covered, and its margins are obscure. Enough is seen, however, partly in the main burn and partly in the streams on the north-west side of the burn, to show that it is of large extent—in the map we have represented it as $\frac{1}{2}$ mile in length and $\frac{1}{8}$ mile in breadth—and thicker than any lamprophyre in the district. It seems probable that it belongs to an intrusive boss, the alteration in the adjoining schists being greater than any elsewhere observed in the district excepting near the granite of Glen Fine, and, perhaps, near Sith an t-Sluain (152). The albites of the albite schists become pink (see p. 149 for the explanation of the similar change caused by basalt dykes), and the matrix in which they are imbedded is considerably darkened, and speckled with strings and spots of iron pyrites. In two of the side streams, the west one of those shown in the Map, and another one, not shown in the Map, near the south-west end of the outcrop, we see a peculiar rock, apparently a hardened and altered breccia made of the different schists next to the intrusion. It seems a case of brecciation in situ. We can see the darker more micaceous bands of altered schist becoming more and more broken by small, and often coalescing, fault lines, until at last the connection between their different patches is lost, and they appear as fragments isolated in a grey rather quartzose matrix. There are no pieces of the igneous rock to be observed in the breccia. No other breccia like this is known in the neighbourhood, and we suppose it may have been formed either at the same time with, or but little later than, the intrusive rock, and as an accompaniment of it. The intrusion itself is also much slickened and veined with calcite, or broken up into thin breccia courses, but we cannot say that these have any connection with the breccia outside it. Their direction is very various, and often they are nearly horizontal.

About $\frac{3}{4}$ mile north of the top of Ben Donich (134 south-east) there is a small irregular intrusion, sometimes sheet-like and sometimes with vertical walls, which closely resembles the one in Donich Burn. Under the microscope (slide 4744) hypersthene is seen to be less conspicuous than in 4743, and there are also one or two (?) pseudomorphs after olivine. Some of the felspar is clearly idiomorphic with respect to some of the augite, but at other times the converse relation prevails.

A small exposure in a burn $\frac{3}{4}$ mile S.S.W. of the head of Loch Restil is perhaps also referable to hyperite.   C.T.C.

CHAPTER XI.

THE IGNEOUS ROCKS (continued).—LAMPROPHYRES AND MICA TRAPS.

So many different varieties of rock are included under the terms lamprophyre and mica trap that it is hardly possible to describe them in general terms. They are generally much decomposed, and their minerals are largely replaced by carbonates. Some bands are subject to rapid variations in character, or they enclose fragments and strips of different colour and grain: in one part they may contain little or no microscopic mica, in another part, only a few yards off, this may be exceedingly abundant. Instances of this occur in the small burn $\frac{1}{2}$ mile south-west of Ballochyle (173 south-east), and on the west shore of Loch Striven, $\frac{1}{8}$ mile E.N.E. of Dun Mor. In the last locality the middle of the band is quite breccia-like, the general mass of smooth weathering rock enclosing numerous rough pieces which weather in depression. These pieces contain more prominent mica than the rest of the rock, and are coarser in grain. The lower portion of the band consists almost entirely of the more micaceous rock, and so also does part of the upper portion.

Where amygdules are present, these are in the great majority of cases prominently drawn out, in planes parallel to the surfaces of the sheets, and they may thus be a help in distinguishing these sheets from basalt, in cases where unsatisfactory exposures might otherwise leave one in doubt. In the basalts, as stated, p. 129, the amygdules are almost without exception roughly spherical, and the flow motion must either have ceased before their production, or have carried them along with it without any distortion of shape: in the lamprophyres they were generally produced before the cessation of the motion, and have been somewhat dragged out by it. In some few cases, however, they have not been deformed, or only very slightly, e.g., in one of the zones of the band ⅛ mile E.N.E. of Springfield House (182 north-east).

The amygdules are generally composed of calcite, and weather hollow on hill exposures. The rock rim immediately enclosing them often projects slightly beyond the rest of the band, as if less readily weathered than it, and it has also a more pronounced red tint. In decomposed rocks the last character has often been found a help in distinguishing them from the basalts. The prevailing red tint of so many of the rocks in this group depends on a reddish cloudiness in the felspars.

It is only when the mica is macroscopically prominent, that the term mica trap seems applicable, and in many of the outcrops it is so inconspicuous that it was necessary to adopt some other term to include the different rocks of the series, whether rich in mica or not, which had the same habit of occurrence in the field, and which were, as far as could be judged, of approximately the same age. The term lamprophyre, introduced by Gümbel, has been adopted in this sense. We use it to include all the rocks which fulfil the conditions mentioned, except some of the more evidently acid rocks, and some few hornblende-porphyrites. The fact relating to the comparative age of these excepted rocks are mentioned on p. 125. The chief varieties of rock included in this unsatisfactorily defined lamprophyre group are noticed as we proceed. But we shall, in the first place, describe the general mode of occurrence, and the characters which are common to different varieties.

In the north and east parts of Cowal the lamprophyres and mica traps are exceedingly numerous, being in some of the schist areas quite as prominent a feature as the basalts, but they are not known within a distance of 4 or 5 miles of the Upper Old Red Sandstone boundary. They are represented on the Maps by a dark shade of Rubens' madder: in the broader bands this is sufficiently distinct from the basalt colour, but in thin lines they may be sometimes confounded. In such cases it will be a help to notice the mode of occurrence of the bands where the colour seems doubtful, for the general habit of the lamprophyres is different from that of the basalts. In most of the district they rarely form vertical dykes, and when they do so these dykes, more often than not, run in a different direction to the basalts,—e.g., a vertical band on the hillside ½ mile east of the foot of Loch Eck runs north-east, and is crossed by four or five basalts almost at right angles to it. A north-east direction is the commonest for the lamprophyre dykes in the south part of Cowal, but in the north part there are many east to west dykes of a camptonite variety. Sometimes an intrusion which is generally sheet-like, its surface making only gentle angles with the horizon, changes and acts as a vertical dyke for a short distance. This irregular mode of occurrence seems more frequent with the lamprophyres than the basalts : a good example of it may be seen in the burn ⅓ mile south-west of Mid Hill (173 south-west). But most frequently they occur as sheets with varying and not very steep hades to the horizon. They do not keep to the bedding or foliation of the schists, but may be constantly seen cutting straight across all their crumplings : it is clear that all the movements which were taking place at the times of the production of the different foliations of the schist had ceased before their intrusion.

It is rare for these bands to exceed 20 or 25 feet in thickness. The dark bands on the Toman Dubh (182), on the hillside 1 mile to the west, and crossing the lower part of the Tamhnich Burn (172), must be thicker than this in most of their course, but appear in places to dwindle almost to nothing, e.g., the Allt Glac na maill (182) is crossed in three places by one of the above bands, and in the middle place it is only 3 feet thick. A thickness of from 3 to 10 feet is perhaps most commonly met with, and from this down to the thinnest strings which cannot be mapped.

The paths of the sheets have in some cases, perhaps in many, been determined by the pre-existing joints, and we can see them changing horizon from one joint plane to another at a slightly different level, the change being made sharply along an almost vertical line, but the new horizon, once acquired, is kept to for some distance.

A section on the hillside ½ mile south-east of Uig (above the Holy Loch) well illustrates their inconstancy. Here are two sheets, the upper one 10 feet thick, and the lower from 3 to 6. The lower is itself in one place split up into two by a band of schist, from 1 to 6 inches thick, which runs on for a distance of several yards, rudely parallel to the surfaces of the sheet. In one place the sheets come within 3 feet of one another, or rather less, for the upper one sends out a short arm, as if to try to reach the other : in another place it sends down a 2-inch string in the same direction. On the west side of the section the sheets are seen to unite, and on the east also they are inferred to do so.

In another section in a scar ½ mile slightly west of south of Cnoc Madaidh (west side of Loch Striven) a reddish fine-grained sheet, reaching 12 feet thick, is seen to die out for a distance of 3 feet, and then it makes its appearance again.

In spite, however, of their thinness and inconstancy, they are of considerable interest in working out the geological structure of the district, and a greater help than the basalts. This is because the majority of the faults are later than them, and throw them, so that they afford a means of finding out the different faults, and of estimating their throws independently of the schists. The hillsides between Glenmassan and Sgarach Mor, and 1½ miles W.N.W. of Ardentinny, Clach Beinn (163 south-east), the west slopes of Beinn Mhòr, the south-east side of Stob Liath (142 north-east), all afford excellent examples of this faulting. But, as might be gathered from what has been already said of the inconstancy of the bands, it is not safe to assume that all the interruptions in them are due to faulting : in most cases we can with certainty infer that this is so from the evidences of crushing to be seen, but occasionally there are bands, e.g., on Toman Dubh (182 south-east) at a point about 100 yards S.S.W. of Ordnance Station 963, and again about ¼ mile south-west of the same station, in which it is probable the intrusion came up originally along lines which were not connected at the present surface, or only by such thin strings that they escape observation in ordinary field sections.

It is certain that some of the faulting in the district was anterior to some of them. In the north part of the area some of the reddish-yellow fine grained bands with occasional mica must be later than some of the north-west running crushes, as they cut through the contorted and crushed rocks without being crushed themselves. And in other cases where north-east running dyke-like bands occur, even though these bands may now be crushed, it is not unlikely that there were previous movements along the same line of which they took advantage at the time of their intrusion.

On the north-west side of Eas Dubb (141), at a point about ⅜ mile north-east of the "h" of "Dubb," there is a section of contorted graphite schist underlying a crush line which hades north-west. The beds above the crush are quartzose schists. These are not contorted as the graphite schist is, but their foliation is roughly parallel to the crush plane. The contortion in the graphite schist is connected with the crush movement: it seems due to a motion of the rock mass overlying the crush plane in a north-east direction relatively to the lower rock mass. There are two decomposed ochreous-weathering lamprophyre sheets running with the crush, each of them from 1 to 2 feet thick. The lower one, quite close to the crush rock, has an even surface which is not affected by the contortion shown in the graphite schist. In other sections near, the lamprophyre is crossed with slickens to some extent, but is never so contorted as the graphite schist. It seems clear, therefore, that most of these crush and contortion movements must be of earlier date than the lamprophyres in the section.

Two lamprophyre dykes occur running with the strong N.N.E. crushes in the burn 1 mile N.N.W. of the "S" of "Strachur" (141), but we cannot say that the dykes themselves are crushed: one of the dykes is of a dark basic character. At a point about ½ mile south-east of the "n" of "St Catherine's" (133) a lamprophyre dyke, running north-west for a little way, is seen to turn north-east and run as a sheet along a crush line: the sheet is quite uncrushed.

We are inclined to suppose that the common north-east and south-west direction of lamprophyre dykes (not the sheets) is generally due to the existence of pre-lamprophyre faults running in this direction, along which the lamprophyres have subsequently been intruded, and that an east to west direction is in other circumstances their most general direction. In the Sleat district of Skye the east to west direction is certainly the prevalent one for these dykes, and in the north part of Cowal this is also so, though there are many instances in the latter locality where a north-east direction also occurs. In the south part of Cowal a north-east direction is the most prevalent. When the dykes run east and west, there may or may not be indications of crushes running in the same line with them, but when they run north-east there are almost invariably indications of crushing at some point or other along their course. The crushing in some places has affected the dyke so strongly that we cannot now determine whether there were also crushes of earlier date than the dyke; but in other places, as in those just cited, it is clear that there must have been. As instances of the former class, we may mention the dykes ½ mile south-east and ⅜ mile east of Socach (141 south-east). The crush stuff in the last locality is several yards wide, and consists in part of a stiff clay, from which the burn is locally called the Clay Burn. Formerly, the clay had a great local reputation for pottery purposes: we suppose it to be formed from the decomposed crushed lamprophyre.

If the Maps be examined, it will be evident how many of the north-east or N.N.E. dykes occur with, or along continuations of, lines marked as fault or crush lines. Even in the instances where little or no crushing has been observed there may still have been at the time of intrusion some thin break running in these directions. There are but few of the longer N.N.E. faults which are not accompanied, in some part of their course, with lamprophyre dykes; but individual dykes rarely continue far in this direction. More commonly we can trace one for only a little distance, and then see no sign of any for some way as we walk along the fault: then another dyke-like exposure occurs, and so on. In some cases, as in Gleann Dubh (152 north-west) there are several inconstant dykes, all running in the general direction of the fault, and partly overlapping one another.

As the basalt dykes decrease in thickness (p. 143) when they run out of their normal directions to take advantage of some pre-existing fault line, so apparently do the lamprophyre dykes when they change from an east to west into a N.N.E. direction. In the Kinglas Water (134) a yellow-weathering fine-grained dyke can be traced westwards in and near the sides of the burn for about ½ mile from the "G" of "Glen," the general direction being slightly north of west. One-eighth of a mile south-west of the "G," the dyke leaves its normal direction, and runs N.N.E. along a thin crush line, and diminishes at the same time in thickness, from 1½ to ¾ feet, or possibly to nothing in certain places. The dyke is not crushed when running N.N.E. It takes its old direction after following the changed path for the width of the burn, about 12 yards. One-eighth of a mile slightly west of south, and ¼ mile slightly south of west, of the "G," similar deviations of the dyke into a N.N.E. direction appear to occur again. In the last locality the thickness when running east and west is about 1½ feet; when running N.N.E. the dyke either ceases to exist at the surface, or is represented by a sheet varying from 6 to 10 inches in thickness.

A few instances of apparently north-west running lamprophyre dykes occur on the west side of Beinn an Lochain (134 south-east), but this face of the hill is very steep, and, the hade of the dykes being to south, we may get north-west running outcrops even when the direction on a flattish surface would be approximately east to west.

It has already been stated that the lamprophyre bands may change locally in their mode of occurrence, and after acting as sheets for a long distance may for a short space run as dykes. Conversely when dykes meet with low-angled strike faults, as in the locality ½ mile south-east of the "n" of "St Catherine's," they may run along these faults and act as sills.

Not unfrequently dykes and sheets of much the same characters occur in the same localities, and the relation between them may not be easy to determine. About ⅞ mile south-east of the "r" of "Ardkinglas House" (134 north-west) a 4 or 5 feet fine-grained yellow-weathering sheet with long needles of decomposing hornblende is intersected by a gulley or deep part in the burn. This gulley probably represents a dyke of somewhat similar rock, as such a one is seen in the line of the gulley at a little distance off. In this case the sheet seems earlier than the dyke. The general direction of the dyke is east to west: in the burn ¾ mile further west it is accompanied with crushes. On the hill ⅘ mile E.N.E. of the "C" of "Cairndow" (134) a sheet with decomposing hornblende arranged in prominent stellar groups, is shifted and crushed along an east to west fault, and in the course of the fault a fine-grained yellow-weathering dyke, without prominent hornblende, occurs.

One quarter of a mile slightly south of west of the "G" of "Glen Kinglas" the east to west dyke already referred to is seen close to a fine-grained yellow-weathering sheet. This sheet is compounded of two parallel bands, there being a chilled surface near its centre. In lithological character the two sheets seem much alike, but the lower half is well cross-jointed while the upper is not. The middle portion of the lower band shows many small brownish mica flakes. A yard or two from the water, on the north bank of the river, the upper band separates from the lower, which continues to run evenly while the upper ends suddenly at a steeply inclined line, and then appears again at the other side of the line at a slightly higher horizon. The east to west dyke is seen within 6 inches of the base of the lower band, and it certainly does not cross the band in the continuation of its old line. So either the sheet must be the later of the two, or the dyke must shift its position a little to the south-west when it comes to the base of the sheet.

The great amount of faulting to which the bands have been subjected is evidence that they are of greater age than the basalts. But we are by no means left to depend on this evidence alone. In many places north-west running basalts can be seen cutting through them in clear section, e.g., on the shore 100 yards W.S.W. of Mid Letter (141 south-west), on the south-west side of the River Massan at a point ¼ mile south-west of Glenmassan, ⅛ mile N.N.E. of Glenmassan, in the burn 1 mile slightly south-west of Glenmassan, in the burn ⅝ mile north of Balliemore (173), on the west shore of Loch Striven, ⅝ mile E.N.E. of Dun Mor, and again in two sections ¾ mile north-east of Dun Mor, in the lower part of Tamhnich burn (172 south-east), on the east side of Loch Riddon ¼ mile north-east of Eilean Dearg, and in the burn 140 yards west of Braingortan (west side of Loch Striven). In the second locality mentioned a close jointing is set up in the lamprophyre parallel to the sides of the basalt, for a distance of several yards from it.

The earlier east to west basalts are also later than the lamprophyres. In a little burn ⅛ mile S.S.E. of the top of Larach Hill (164 north-east), a red fine-grained amygdaloidal band, 2 or 3 feet thick, occurs close on the south side of the broad east to west basalt dyke. As far as seen, which, however, is only for a few yards, the lamprophyre runs parallel to the basalt, but the latter is chilled at the junction line, while the former is not. In the River Cur, ¼ mile east of Strachurmore, a broad east to west basalt cuts a lamprophyre sheet with prominent black mica; the lamprophyre is seen on both sides of the basalt. One-third of a mile W.N.W. of the "B" of "Beinn an t-Seilich" (134) the east to west basalt is exposed in, and on the west side of, the middle burn shown in the Map; two lamprophyres, one a coarse rock with much black mica, are seen on either side of it at only a little distance off; they do not cross the basalt, and are therefore inferred to be crossed by it. Half a mile west of Loch Restil outlet (134 north-east) a lamprophyre dyke is seen a little from the sides of the east to west basalt; the latter is certainly not intersected by the former. The east to west broad basalt dyke on the east side of Loch Riddon makes a prominent slack near the ruined chapel, where it approaches the broad lamprophyre, but no basalt is actually seen between the lamprophyre outcrops on either side of the slack, and the presence of the slack is not by itself decisive of its presence, as the lamprophyre also sometimes forms slacks.

On the south-east side of Glen Kinglas, ½ mile south-west of Butterbridge, there is a dolerite-like sheet with porphyritic augite, which includes some parts much finer in grain and smoother weathering than the rest. The fine parts seem sometimes to be running in veins through the coarser. In an exposure nearly ⅓ mile north-west of this locality a sheet that is probably a continuation of the last mentioned occurs again, but now no fine-grained parts are observable. An exposure in Kinglas Water, nearly ¾ mile below Butterbridge, probably belongs to the same sheet, but it does not show such prominent augites. The coarse dark grey parts are frequently intersected by thin strings of a fine-grained reddish-yellow rock, and also often enclosed as fragments in it ; the fragments have sharp margins, and are sometimes several inches long.

There are difficulties in ascertaining to what extent variations in character along the course of one continuous band may occur, because bands of different characters frequently run in close contact, having each taken advantage of the same line of weakness. There are two ways in which it is possible that variations along the outcrop may have arisen. They may be due to differences originated before or at the time of cooling of any particular band, or they may be characteristic of bands which are different in their age and origin, though occurring along the same line of outcrop. It is rare to find exposures, between the points where the variations are observed, which are so continuously bare that we cannot be sure that the rock at one point may not at some place have suddenly ceased to come up to the surface, and been replaced by another of different character running in the same direction.

On the south side of the east to west fault ¾ mile S.S.E. of Sgor Coinnich (153) there is a band composed in its upper part of 12 feet of dark grey rock, weathering in big blocks, with stout porphyritic crystals of somewhat decomposed augite and small thin black needles of sounder hornblende. This overlies a 5-foot sheet of sharply-jointed reddish rock with stellar clusters of needle-shaped decomposing hornblende. The lower portion seems the later of the two, for it becomes finer grained near the junction. On the hill 100 yards south-east of the above section, the band seems to consist of the upper dark part alone, nothing being seen of the redder portion, either in contact with or near the other. To the north-west the redder portion appears to separate from the blacker, for in the burn a little over ½ mile south-east of Sgor Coinnich we find an upper sheet of the same character as the upper one near the fault, except that it is mixed near the top with some redder coarse bands. This is underlain by a strip of schist about 1 foot thick, and then we see a 5-foot reddish sheet with decomposing hornblende crystals in radiate clusters. Below this is a strip of schist 5 feet thick, and then a second band of reddish lamprophyre about 5 feet thick. One hundred and fifty yards E.N.E. of the burn section, the lower part of the upper sheet weathers in a breccia-like manner, owing to the inclusion of many bits of coarser grain within the finer rock. The band which runs round the south side of Cruach a' Bhuic (153 south-west) is entirely of a reddish fine-grained type, but it must be on the same line of intrusion as the upper dark grey sheet just described.

About 100 yards north of the "r" of "Meall Breac" (152) the upper lamprophyre band is compounded of two sheets in close contact. The upper one is a 4 or 5 foot reddish flaggy-weathering fine-grained rock, and the lower a 10 foot black blocky exfoliating band with porphyritic augite. Both rocks become finer near the line of contact with one another, and so we must suppose that one was first intruded and chilled against schists, and then the other came up along the same line and was chilled against the earlier one. There is nothing to show which portion is the earlier. To the west they are in contact for 10 yards, and then there is an obscure interval of 100 yards, after which the augite rock is seen again, but now it is unaccompanied with the red rock. A few yards to the east, a red flaggy sheet appears below the augite rock, though it does not appear as if such can have existed further west, and at the same time we lose the overlying red sheet, so that the presumption is that the augite sheet is the later of the two, and has altered its horizon with respect to the red sheet.

The sheet at the base of the crag, ½ mile N.N.W. of the "P" of

"Dornoch Point" (152), is a reddish fine-grained flaggy rock : the flaggy planes are generally parallel to the surfaces of the sheet but not always. The microscopic characters of the rock are described elsewhere, p. 118. It contains radiate clusters of sound black hornblende needles. The sheet just above this in the same scar is a black rock with porphyritic augite, and it weathers in very massive exfoliating blocks. The black sheet is hading more steeply north-west than the red sheet, and gradually approaches it in this direction until it is within 12 or 15 feet. Then it ends bluntly, but, without any surface connection, it somehow finds its way into the lower sheet and cuts through it. Then for 50 or 60 yards further west the two rocks keep together, the red above and the black below. After this, for 100 yards or more northwest, the black band alone appears, but it is possible that the red one also exists underneath it. Further north-west still there is an obscure area for 200 yards, and after this down to the alluvium on Loch Eck side only the red rock is seen.

Spheroidal weathering is common. It occurs in bands which are quite red in tint, as well as those which are dark grey or black. Where the rock is not spheroidal the prominent joints are generally parallel to the surfaces of the intrusions, and may be so close that the rock splits into rough flags. This is noticeable in the two little outliers on the hill ¾ mile south of Glenmassan, on the hill ¾ mile slightly north of east of the foot of Loch Eck, and at the north-west end of Creag Capull (173 north-east). Occasionally there are also indications of joints at right angles to the surfaces, but we have never seen them nearly as marked as in the basalts. Outcrops in scars are generally recognisable at a considerable distance, as they form more or less straight, and usually gently sloping bands, which weather in retreat from the harder schists above and below.

Intrusions always grow finer in grain as their surfaces are approached— in the micaceous varieties the mica flakes become gradually smaller until they cease to be visible macroscopically. Some of the exposures which are dull red in the interior change colour near the margins, passing through reddish-brown into dark grey. Both these changes are to be noticed in the burn 1 mile slightly west of south of Glenmassan. In some of the basic rocks, with porphyritic augites, the augites continue right up to the original surfaces of the intrusion, as large in size as they are in the interior.

Inclusions of schist, etc., are not uncommon. The band near the old chapel east of Fearna Bagh is nearly half full of hardened schist and vein-quartz pieces, which reach a length of 4 inches. Similar inclusions are seen again in the north-east continuation of this band where it crosses Allt Glac na maill. In the south exposure ¾ mile north-east of Dun Mor, an inclusion of vein quartz at least 1½ foot long is seen, and this has the cracked and splintery look so often found in the near neighbourhood of intrusions.

The sheet going round the north and west sides of Beinn Roithe (142 and 153), at a point 60 yards E.N.E. of the top of the hill, is quite full of schist and vein-quartz pieces ; the thickness of the sheet is 4 feet, and the pieces are sometimes 6 inches long. Nearly ¼ mile S.S.W. of the hill-top the sheet is 5 feet thick, and the middle for a thickness of 2 feet is full of such pieces, while the rest is almost free from them. One-seventh of a mile east of the hill-top the sheet is 6 feet thick, and the middle is again full of included pieces.

A dark blocky augitic sheet near the burn ⅝ mile south of the "e" of "Coire Ealt" (152 north-east) contains many pieces of vein quartz and

some of schist. The greater lengths of the pieces are parallel to the
surfaces of the sheet. In some bands, for a breadth of a foot or more,
the inclusions make up $\frac{1}{4}$ of the rock mass. The vein quartz has the
cracked and splintered look already referred to ; the main cracks in it run
almost at right angles to its surfaces. A quarter of a mile N.N.W. of Cruach
a' Bhuic, a dolerite-like sheet contains near its centre pieces of much altered
quartzose schist, as much as 6 inches long, and vein quartz, sometimes 9
inches long and 2 inches broad. These pieces also have their longer axes
parallel to the surfaces of the sheet.

No great amount of contact alteration is usually caused by the lamprophyres, but in some cases the alteration is much more pronounced. On
the west side of Loch Striven, close to the mouth of the little burn 100
yards east of Braingortan, a great hardening of the schists is effected by
three thin sheets with dark matrix and large black mica flakes. The
albite schists near the dark sheet in Kinglas Water, not quite $\frac{3}{8}$ mile
below Butterbridge, are distinctly hardened and speckled with iron
pyrites ; the albites, too, acquire a reddish tint. One-quarter of a mile
north-west of Beinn Reithe, epidote occurs in an albite schist within a few
inches from the side of a lamprophyre. No epidote was observed anywhere else in the neighbourhood, and perhaps it is due to the proximity
of the lamprophyre. The epidotes are much the same in size and shape
as the albites of the albite schist further off the intrusion.

As examples of rocks rich in mica we may mention the exposures
on the south-west side of the River Massan $\frac{1}{4}$ mile south-west of Glenmassan, the repeatedly faulted band between Glenmassan and Sgarach
Mor, the middle band on the west side of Glen Tarsan, the rock in the
burn 140 yards west of Braingortan, the bands on the Toman Dubh, the
bands on the east side of Cruach nam Mult (134 south-west), and the
lower band $\frac{2}{3}$ mile N.N.E. of Stob an Eas (134).

The bands on the Toman Dubh are of the same type as those at the old
chapel east of Fearna Bagh, in the lower part of Tamhnich Burn, on the
hillsides $\frac{4}{5}$ mile west of Glenmassan and east of Cruach nam Mult, the
lower band $\frac{1}{3}$ mile N.N.E. of Stob an Eas, and the east to west dyke just
north of Laglingarten (133 south-east). These may all be classed as
mica-dolerites. The amount of mica varies considerably in different exposures ; the Toman Dubh bands are generally rich in mica ; the band $\frac{1}{4}$
mile south-west of Glenmassan poor. The mica-dolerites have generally
a blackish aspect macroscopically, owing to the great abundance of stout
augite prisms. At times they weather in unusually massive blocks,
which exfoliate into a loose sparkling sand, giving rise to green grassy
banks or slacks. Between the augite prisms can sometimes be seen
macroscopically small pinkish specks, the felspar having very much the
same tint as in the redder lamprophyres. The red tint is sometimes,
as in the lamprophyres generally, especially pronounced just at the edges
of the amygdules, e.g., in the band $\frac{3}{4}$ mile south-west of Cnoc Breamanach
(182 north-east). At the old chapel east of Fearna Bagh, and in the
band $\frac{3}{4}$ mile E.N.E. of Cruach nam Mult (134), the dark rock is crossed
by redder finer grained strings, about $\frac{1}{2}$ inch thick, which project slightly
on the weathered surfaces.

An average specimen of the chapel rock was examined microscopically
(slide 3417) by Mr Teall, and is thus described by him : " Large and
more or less idiomorphic crystals of augite in a ground-mass of lath-shaped
plagioclase, brown mica (allied to that of the mica traps) iron ores and
ophitic calcite. The rock may be termed a mica-dolerite. It appears
to have affinities with certain rocks which have been called kersan-
tites."   C.T.C.

Along the east side of Stralachlan Glen various exposures of micaceous lamprophyre occur, not as sills, but as oval patches—this form being due to their occurrence on a dip slope. A specimen from one of these was examined by Mr Teall, and is described by him as follows:—"4012. Barnacarry, Loch Fine. Massive dark grey medium-grained rock, containing pink patches. Felspar often striated and more or less lath-shaped. External zones extinguish at low angles, internal zones at high angles gradually changing through a range of 20°. Brown mica, magnetite, carbonates, and green or grey pseudomorphs after indeterminable minerals. Pink patches are mainly composed of felspar giving rectangular sections, probably orthoclase. Kersantite."   J.B.H.

In slide 4755, obtained from the lower band $\frac{3}{4}$ mile N.N.E. of Stob an Eas (134), irregular grains and crystals of olivine occur, but they are much smaller than the porphyritic augites. They are sometimes quite surrounded by biotite. The porphyritic augites appear almost colourless in the slide : they are far larger than any other constituent of the rock, but there is also a set of smaller colourless augites without any approach in size to the large augites. The small augites are clearly idiomorphic with respect to the felspar. In the west of the two dykes $\frac{1}{2}$ mile E.N.E. of Cruach nam Mult (134) the biotite is very conspicuous, occurring in large ragged plates, which show under the microscope (slide 4752) inclusions of the later developed rock constituents. The augites are also large and abundant, but have so generally decomposed into a dull greenish tint that they do not strike the eye readily.

By a gradual decrease in the amount of mica, the mica-dolerites may pass into rocks of more normal dolerite aspect. In all these early dolerites which have been examined, the augite differs from that of the tertiary basalts in belonging to the pale form, malacolite, and in generally occurring in the porphyritic form. It is usually idiomorphic with respect to the felspar, but in some cases, e.g., in slide 4744, from an intrusion $\frac{3}{4}$ mile north of Ben Donich (134), some of the felspar is idiomorphic with respect to some of the augite, while other portions are not. Well characterised ophitic augite has never been observed. The augites are generally sufficiently large to be readily seen by the naked eye : they frequently attain the size of a large pea. In some bands they are remarkably abundant. In some thin sheets in the coast section north of the " L " of " Leak " (151) they seem to compose about $\frac{1}{2}$ the bulk of the rock. In the sheet $\frac{1}{2}$ mile N.N.W. of the " P " of " Dornoch Point " (152 south-east), the porphyritic augites continue right up to the original surface, and are as big there as they are further within the sheet.

When these rocks are examined under the microscope it is often found that the augite is associated with hornblende. As the latter gradually increases in amount the rock passes into a dioritic type, that of the so-called camptonites. Even with the naked eye we can see in certain bands that both augite and hornblende are associated together, e.g., in the upper sheet $\frac{4}{8}$ and $\frac{7}{8}$ mile south-east of Sgor Coinnich (153). The augite is generally in a decomposed state, while the hornblende remains fresh and with a brilliant black lustre in hand specimens. When the augites decompose so as to form pale greenish-grey almost opaque shapes, distinct indications of zonal growth may often be observed in them with a hand lens, the different concentric bands differing slightly in the depth of their tints. In cases of more thorough decomposition these pale-green shapes become rusty-yellow and cavernous, or are represented merely by hollows with a rusty lining. These different ways of decomposition are observed in the exposures of the sheet in Garrachra burn (163 north-west) near the "rr" of "Garrachra," and by the side of the little burn ¼ mile north-east of this place.

The east to west dyke in the burn a little over ½ mile south of Bathaich ban Cottage (134 north-west) contains conspicuous crystals of augite embedded in a medium-grained dark-greenish matrix. Under the microscope (slide 4746) brown hornblende is seen also to occur, so that the rock may be called hornblende dolerite, and it is to be considered one of the passage forms between dolerite and diorite. The porphyritic augites are pale in colour, and more or less idiomorphic. The groundmass contains smaller augites of the same characters, brown hornblende (sometimes surrounding the augite), plagioclase (lath-shaped in section), and iron ores.

In another dark-green east to west dyke, seen in the wood 100 yards south-east of Bathaich ban Cottage, the hornblende is more abundant than the augite, so that the rock may be called an augite-bearing diorite. Mr Teall describes the microscopic character (slide 4747) as follows:— "Medium-grained, dark-greenish massive crystalline rock. Idiomorphic brownish-green hornblende (Forms: {110}, {010}, {100}, {I11}?, and {001}), colourless augite (malacolite), and iron ores changed to leucoxene. The above minerals are embedded in a much altered felspathic ground-mass containing felspar, quartz, and carbonates. As the hornblende is considerably more abundant than the augite, this rock may be termed an augite-bearing diorite." The hornblende never becomes porphyritic, or so conspicuous as the augite in 4746, but still it is distinct to the naked eye, and is often about the size of a large mustard seed. The malacolite is of about the same size as the hornblende.

Another dyke, of much the same macroscopic aspect as the last, occurs ½ mile W.N.W. of the outlet of Loch Restil (134). This is described by Mr Teall (slide 4748) as follows:—"Similar to 4747, but somewhat finer in grain. Hornblende, similar to that occurring in 4747, one or two pseudomorphs after a mineral which was probably olivine. Malacolite is not now recognisable, but is probably represented by certain chloritic pseudomorphs. The felspars of the ground-mass show in places a strong tendency to a feathery mode of aggregation. Both this rock and 4747 are very similar to the basic rocks found as sills in the limestone of Assynt near Inchnadampf (see Geol. Mag., 1886, p. 346)."

The two dark-grey or black sheets occurring from ½ to ¼ mile south-east of Loch Restil Head (134) weather in parts in very massive blocks with rough faces and show in places lustre-mottled cleavages of some porphyritic mineral. In other parts they are much finer in grain. The microscopic character of a specimen from the upper sheet (slide 4753) is thus described by Mr Teall:—"The original rock appears to have consisted essentially of malacolite, hornblende and felspar. Carbonates and chlorite now abundant. The hornblende is sometimes ophitic with respect to felspar. This rock must evidently be classed with the diorites of camptonite type." The lustre-mottling is due to the inclusion of bright shining black hornblende portions within the more decomposed dull-looking augite. Similar hornblende can also be seen edging the augite. The augite occurs in crystals reaching ½ inch in length, and is much the most prominent constituent to the naked eye. The hornblende crystals bear evidence to a zonal growth, giving lighter tints, on rotating with crossed Nicols, near their margins than at their centres. As the hornblende may occur both as inclusions in the augite and also surrounding it, without there being any sign of a passage from the one into the other, it would seem that the growth of these two minerals was proceeding *pari passu*. The hornblende crystals within the augite may be zoned by less pleochroic margins just as the larger hornblendes are. The felspar is pink : it is only small in amount and is hardly noticed on the rough weathered face.

The dykes in the burns,—$\frac{1}{2}$ mile east of Monovechadan (124 north-west, slide 4749), nearly $\frac{1}{2}$ mile S.S.E. of Bathaich ban Cottage (slide 4750), and on the hill $\frac{1}{4}$ mile west of Loch Restil outlet (slide 4751), may be taken as examples of the camptonite dykes in which the hornblende is generally distinctly in excess of the augite. In these slides the hornblende is in more elongated needly shapes than it is in 4747, 4748, and 4753, and is either quite sound with black horny lustre, or decomposed into opaque grey or greenish-grey shapes. The shapes are not usually more than $\frac{1}{4}$ inch in length, and can frequently be made out, with the aid of a hand lens, to be idiomorphic with respect to the felspar. In 4750, the felspathic matrix is moderately fresh and pink in colour, but the hornblende is decomposed into a white substance which is almost totally opaque. Sometimes a margin of such opaque matter may be seen surrounding the porphyritic crystals of malacolite.

The dark-grey diorites or camptonites are specially abundant in the area between Lochgoilhead and the head of Loch Fine, and occur generally in the form of east to west dykes.

More common than the grey or black lamprophyres are those of red, reddish-brown, or yellow colour. The colour of the felspar in all the lamprophyres has a tendency to be red or reddish-brown, and the difference in colour between the redder and the blacker bands is supposed to be due, either to a greater proportion of felspar in the former, or to a deeper tint of red in this felspar. Judging from the rocks that have been sliced, the redder lamprophyres do not differ in any conspicuous way from those of darker aspect. A specimen of a reddish lamprophyre in Invervegain Burn (183 north-west), was taken from a place near the "r" of "Invervegain," and was chemically examined in the late Prof. Dittmar's laboratory at Anderson's College, Glasgow. A sample, after being dried at 100° C., gave a mean silica percentage of 47·98.

As examples of the red lamprophyres without prominent mica, specimens from the bands $\frac{1}{2}$ mile south-west of Cnoc Madaidh (east side of Glen Striven), and 300 yards north of the Craig (183 north-west), were sliced and examined microscopically. The former (slide 2838) is thus described by Dr Hatch.

"Diorite. Medium-grained rock with red felspar. Striped felspar in lath-shaped crystals of a reddish-brown colour due to kaolinisation. Hornblende in isolated brownish-green crystalline grains, sometimes bounded by $\propto$ P. Pleochroism as follows :—

$$\left\{ \begin{array}{l} \alpha \text{ pale yellow,} \\ \beta \text{ brown,} \\ \gamma \text{ greenish-brown,} \end{array} \right.$$

Maximum extinction angle measured to cleavage in vertical sections = 17°. Scattered grains of opaque iron ore (ilmenite). Secondary minerals are—chlorite, and patches of flakey, colourless mineral with high double refraction and iridescence (moiré) under crossed Nicols (talc ?). Both these minerals are alteration products of some vanished ferromagnesian constituent, not of the hornblende, for that is quite fresh, perhaps malacolite, if the rock bears any genetic relations to 2839."

The Craig rock (slide 2839) is described as follows :—
" Diorite (Augite Diorite). Mineral constituents (1) Dark mica in
long, laminated, and slightly contorted lathes, showing undulatory
coloured polarisation under crossed Nicols. Strong pleochroism :—

$$\begin{cases} a \text{ pale yellow,} \\ \beta, \gamma \text{ dark brown,} \end{cases}$$

(2) Hornblende in isolated crystals, giving long, lath-shaped, vertical
sections and lozenge-shaped cross-sections ($\propto$ P, $\propto$ $\maltese$ $\propto$, $\propto$ $\maltese$ ∞). Pleochroism
$$\begin{cases} a \text{ pale yellow,} \\ \gamma \text{ dark brown,} \end{cases}$$

(3) Numerous short prisms and irregular grains of a colourless mineral
with moderately high index of refraction and high double refraction.
Maximum extinction, measured to vertical axis = 40°. Optic axes
situated as in augite. These are the properties of a colourless augite
malacolite (salite). Forming a kind of ground-mass in which the ferro-
magnesian minerals are embedded, is a reddish felspar, the crystallo-
graphic nature of which could not be determined, owing partly to a
turbidity resulting from kaolinisation, partly to the indistinct and
undulatory manner of extinction under crossed Nicols. Iron ore in
largish opaque grains."

In some of the reddish lamprophyres, chiefly, we think, where mica
is not prominent, stellate groups of acicular crystal forms, attaining a
length of $\frac{1}{2}$ inch, are conspicuous, e.g., in the burn 1 mile north-west of
Glenlean, and at the top of a band in the burn $\frac{1}{2}$ mile south-west of
Glenmassan. These forms are nearly always much decomposed, and
are so largely replaced by calcite that it is not always certain what mineral
they represent. We presume it is hornblende rather than felspar, partly
because of the deep rusty colour they assume near weathered surfaces,
and partly because in a sheet $\frac{1}{2}$ mile N.N.W. of the " P " of " Dornoch
Point " (152 south-east) crystals of black hornblende are seen arranged
in similar clusters.

This last rock is described by Mr Teall as follows (slide 3453):—" This
rock is composed of long hornblende prisms in a pinkish matrix. The
porphyritic hornblende is idiomorphic and shows the forms (110), (100)
and (010). Colours: a very pale brown or yellowish brown, β and γ
brown. The matrix is composed of felspar stained with ferrite, small
prisms of hornblende and iron ores. A little calcite is also present. .
The felspar is too much altered to admit of determination." The black
hornblende needles continue to the original surface of the rock. They
are not always in clusters, and then they run parallel to the surface.

The rock with the stellate shapes in the burn $\frac{1}{2}$ mile south-west of
Glenmassan is thus described (slide 2958) by Dr Hatch:—" Under a low
power and in ordinary light, this rock appears to consist of a brown
glassy base, containing granules of felspar, clusters of magnetite granules
and scales of some greenish minerals. Embedded in this ground-mass
are large porphyritic crystals, apparently of felspar."

"But under a higher power and between crossed Nicols the brown
'base' is resolved into a mass of lath-shaped crystals of felspar—the
brown pigment (particles of oxide of iron) being accumulated along the
peripheral portions of the lathes. The green patches which are largely
distributed through the section consist of some chloritic mineral, which
does not, however, show any pleochroism and appears almost completely
isotropic. This chloritic substance is perhaps pseudomorphous after
augite, no trace of which can however be discovered. The rock is con-

siderably altered, patches of secondary calcite occurring in every part of the slide.

" The colourless portions with more or less crystalline contours, which appear in ordinary light to be crystals of felspar, are seen under crossed Nicols to consist of calcite—evidently an alteration product."

A reddish-brown north-east dyke occurs in the point between Loch Goil and Loch Long, and comes down to the raised beach on the south side of Mark. It has been traced for about $\frac{2}{3}$ mile, and is in places 10 yards wide. There is no mica discernible with a hand lens, but many prominent porphyritic crystals of hornblende occur, often quite fresh, which are at times as much as 1 inch long, and $\frac{1}{4}$ inch broad ; usually, however, their breadth is less in proportion to their length than this. They may have their long axes distinctly parallel to one another and to the sides of the dyke : no radiate clusters were observed. Occasional porphyritic crystals of felspar of a reddish colour were observed in the same rock.

A striking feature in some of the lamprophyres, particularly the redder ones, is the frequency of spherulitic shapes near their surfaces. These are seen in several sections on the west shore of Loch Striven, e.g., $\frac{2}{3}$ mile E.N.E. of Dun Mor, in the 2 exposures $\frac{3}{4}$ mile north-east of this hill, $\frac{1}{2}$ mile N.N.W. of Braingortan : on the east shore of Loch Riddon $\frac{1}{2}$ mile north-east of Eilein Dearg : at the sides of a reddish brown mica trap on the coast $\frac{1}{4}$ mile south-west of St Catherine's pier (133), and at the surface of a dark micaceous sheet $\frac{1}{2}$ mile south-west of the same pier : in a vertical dyke $1\frac{1}{4}$ mile south-west of the last " n " of " Coilessan Glen " (142 south-east) : in a band a little less than $\frac{1}{2}$ mile slightly south of east of Monovechadan (144 south-west) : in an east to west dyke 1 mile south-west of the top of Ben Donich : and in other places we need not specify. It should be observed that some of these localities have been already mentioned as showing intersections by basalts, &c.

In the north of the two localities $\frac{3}{4}$ mile north-east of Dun Mor there is a dyke like band, with a width of 4 or 5 feet. The colour is reddish, and there are many small flakes of black mica, especially in the interior. In many places near the sides, for a breadth of about 2 inches, there are aggregations of small spherical shapes, sometimes as large as a pea. The shapes are themselves red like the mass of the rock, but the interspaces, which weather in depression, and thus give prominent relief to the spherical shapes, are of a grey or greenish-grey colour. Occasionally two or more of the spheres are seen united. The most distinct forms are often arranged in rows which run rudely parallel to the sides of the dyke, at a distance from one another of $\frac{1}{4}$ to $\frac{1}{8}$ inch. The size of the spheres becomes less and less as the side is approached, until in the last $\frac{1}{2}$ inch the eye cannot distinguish them.

A lamprophyre 100 yards south of this section, in the form of a sheet about 1 foot wide, is in parts quite full of similar forms. The spheres are most noticeable on the weathered surfaces. A specimen close to the surface of the sheet was sliced (slide 2959) and is described by Dr Hatch as follows:—

"Dark-coloured glassy rock, through which minute needle-shaped doubly refractive microlites are seen under crossed Nicols to be plentifully distributed.

"The colouring matter consists probably of minute particles of oxide of iron finely disseminated through the base. Darker patches are due to aggregates of irregular masses or patches of oxide of iron with a few granules of magnetite."

"Crystals of one of the bisilicates or of olivine appear to have floated in, and been carroded by the molten magma ; for crystalline forms with opaque borders of Fe_2O_3 are most frequently met with. The interior portions of such crystals have been dissolved out during the weathering of the rock and replaced partly by calcite, partly by other alteration products. Granules of calcite are abundantly distributed through the whole rock."

In the section nearly ½ mile N.N.W. of Troustan (183), a sheet from 4 to 8 inches thick is crowded with pea-like forms throughout.

On the hillside ⅝ mile E.N.E. of Glenmassan similar forms are extremely abundant in the upper 4 inches of the sheet; just below this comes a band, about 4 inches thick, which is very amygdaloidal, the amygdules being prominently drawn out in a plane parallel to the surfaces. The amygdules weather hollow, and their insides are occasionally coated with small pea-like forms. Similar forms are not observed on the weathered surface of other portions of this zone.

In the band nearly ¼ mile slightly south of east of Monovechadan (134) the shapes, about the size of small shot, are red in colour and lie in a fine-grained almost black matrix. They often coalesce and form bands, sometimes as much as ⅛ inch in thickness, parallel to one another. On examining shapes with a lens a distinct radiate arrangement can be recognised, so that they may be called spherulites.

The dyke 1 mile south-west of Ben Donich (142) is exposed in a section in the burn, just above some steep scars. It is only 1 foot thick, but there is another parallel dyke close to the north side. The shapes reach ⅛ inch in diameter, and are as usual, of a redder tint than the matrix in which they lie. A hand lens suffices to show that they are built up of concentric layers which differ somewhat in colour. Three zones of colour can be made out in most of them, the central area being generally the deepest red. When two shapes join one another their outer zones coalesce without any visible line of separation, while the central areas remain isolated. The matrix in which the shapes are embedded is a dark greenish-grey rock, in which is seen a network of very thin hair-like needles of pale greenish-grey colour. The same needles penetrate the pellet shapes : it is not clear that their directions within these have any reference to the margins of the shapes.

Radiate spherulitic structures have been observed microscopically in slide 4197, obtained from a sheet ¼ mile north-west of Beinn Reithe (142), where it is in close contact with epidote schists (see p. 114). Mr Teall describes this slide as follows : " Junction of epidote schists with lamprophyre. The lamprophyre is composed of ill-defined felspar microlites, small scales of chlorite, grains and specks of iron ore. There are a few pseudomorphs (carbonates) after porphyritic constituents. Near the junction the felspathic matter is beautifully spherulitic : further off the microlites show a tendency to assume feathery forms. Brownish granular matter tends to obscure the characters of the rocks."

In the locality ⅝ mile E.N.E. of Dun Mor (172) a greyish-red fine-grained sheet 4 to 8 feet thick occurs, with a low hade to the north. This occasionally shows the spherulitic forms on either side : in one place the surface bulges outward and the breadth of the distinctly spherulitic area widens correspondingly, its outer boundary bulging out with the surface while its inner one keeps straight—a thickness of 1 inch is thus attained. Inward from the spherulitic zone we come to a band, about 2 inches thick, which shows beautiful fluxion structure parallel to the surface, the different flow bands being indicated on the weathered face by slight differences both of colour and hardness ; then comes an amygdaloidal zone, about 2 inches thick, in which the amygdules are not prominently elongated, and, after this, an area with rarer amygdules which show elongation parallel to the side.

An indication of flow, shown by a fine streakiness parallel to the surface, is seen in the weathered face of a 3-foot felstone-like sheet on the hillside $\frac{1}{4}$ mile north-west of A' Chruach (182 north-east).

One-quarter of a mile north-west of the top of Beinn Reithe (142 and 153) a $2\frac{1}{2}$-foot sheet shows, for a breadth of 8 inches from either surface, close parallel lines on the weathered face, at intervals of about $\frac{1}{4}$ inch. These lines run parallel to the adjacent surfaces of the sheet.

The upper dark sheet nearly $\frac{3}{8}$ mile E.S.E. of Sgor Coinnich (153) is in its upper part divided into bands of coarse and fine grain, which are parallel to its surfaces.

In the fine-grained band $\frac{1}{4}$ mile slightly west of south of Beinn Bhreac (152 south-west) there are various bands, about $\frac{1}{2}$ inch thick, which are rudely parallel to its sides, and display on their weathered faces a delicate lineation in the direction of the containing bands. These bands are finer in grain than the rest of the rock, and darker in colour. In places they have an appearance of slight contortion, due probably to folding while they were still in the liquid condition. Sometimes, too, there is a breccia-like appearance within the sheet, owing to irregular mixing of different varieties of rock—one variety is dark grey in colour and fine in grain, like the thin bands just mentioned, while the other is redder and coarser. In some places there are sharp lines between these different rocks, at others there seems quite a gradual change.

Foliated Lamprophyres.

In the extreme north part of the area being described, between the head of Loch Fine and the adjacent margins of the Glen Fine granite, we meet with a set of intrusive rocks which resemble the camptonites, or other lamprophyres, in general aspect, excepting that they have in places a cleaved or even slightly foliated structure. They generally occur as dykes, and run east to west, like the unfoliated camptonites. If they are rightly included with the lamprophyres, they must certainly be regarded as the oldest member of this group, for in this district none of the other members show any sign of cleavage or foliation. These foliated camptonite-like rocks are sometimes cut by the hornblende-porphyrite intrusions, and these intrusions are themselves probably earlier than some of the varieties of lamprophyre (see pp. 124, 125). Their foliation is extremely slight when compared to that of the schists into which they intrude, and no cleavage or foliation in the schists has ever been observed to correspond to that in the dykes.[*] The cleavage and foliation in the dykes seem quite confined to themselves, and do not exist in the schists at their sides : they are not, however, strictly parallel to the side of the dyke, but often diagonal to it. The foliation in the schist is always sharply cut by the sides of the dykes, and it is clear that at the time of the intrusion of the dykes these rocks were schists, and essentially in their present condition, with the exception of the change exerted on them by the metamorphic influences near the Glen Fine granite. So that we need not modify the statement made (p. 108) in speaking of the lamprophyres generally, that all the foliation of the schists had been produced before their intrusion.

[*] In this respect the phenomena observed differ distinctly from those usually seen at the sides of the Pre-Torridon basic dykes in the North-West Highlands (see "Recent Work of the Geological Survey in the North-West Highlands of Scotland," *Quart. Journ. Geol. Soc.*, vol. xliv. p. 378).

The relationship of the unfoliated camptonites of Cowal to the camptonites of the Assynt district of Sutherland has already been incidentally indicated by Mr Teall (p. 116). It is known that the Assynt camptonites have suffered from the thrusting and metamorphic changes which accompanied the great Post-Cambrian movements of their district, while the mica traps have not, so that it is clear they are of earlier age than the latter rocks. The finding in Cowal a set of foliated camptonites which are also earlier than other lamprophyres and mica traps, enables us to draw a parallel between the two districts, and renders it probable that the two classes of rock in the one area roughly correspond in age with those in the other. But if this be so, if the camptonites of Cowal are the same age as those of Assynt, at what time was the foliation of the Cowal schists produced? In Assynt it is not as yet certainly recognised that there were any great thrust movements of pre-camptonite age. In Cowal it is certain that the process of schist manufacture was practically completed before the intrusion of the camptonites. Can it be that the schist manufacture began in Cowal before it did in Assynt? Did the great movements begin at first far to the east of the zone of distinctly recognisable, thrust Durness and Torridon rocks, and then travel outwards from the mountain area—from the south-east towards the north-west—in subsequent ages?

Owing to the similarity in character and identity of direction between the camptonites that may be foliated and the unfoliated camptonites, and the fact that where the former occur they are common, while those of the other class are unknown or rare, we are inclined to suppose that the two represent one another in different districts, and are approximately of the same age. We cannot indeed say that all the camptonite dykes north of the most southerly foliated camptonite show foliation in one part or other of their course, but the greater number of them do so, and it is probable that if they could be traced more continuously we should find that all of them do. As the foliated camptonites are in some places foliated, and in others not, we need not be surprised that in some of the dykes, particularly in those that can only be traced short distances, no foliation can be seen. Supposing the two sets of rocks are really the same, there must have been some special causes in the district of the foliated set for the production of foliation in them, which did not occur in the region of the unfoliated set. What these causes were is not clear. The absence of foliation in the Glen Fine granite shows that the causes, whatever they were, had ceased before the solidification of the granite.

The most southerly of these foliated dykes occurs in the burn ¼ mile north of the "d" of "Achadunan" (126). It is a greenish rock, markedly cleaved, and with a slight lustre on the cleavage planes, but there is no distinct mica visible to the naked eye. It is so much less foliated than the schists at the sides, which consist of alternations of albite and more quartzose schists full of large flakes of mica or chlorite, that the observer concludes at once that it must belong to a rock later than they, though the sides of the exposure are not seen in this burn. A total thickness of 6 or 8 feet is seen. The direction of foliation is much the same as the apparent direction of the band, and the hade is to north at about 70°. There are closely aggregated white specks, about the size of grains of rice, which effervesce briskly with dilute hydrochloric acid. These shapes rather remind us of amygdules: in most places they have their greater lengths parallel to the foliation of the band, but in one place, near the south side of the band, they are much less elongated, and at the same time the foliation is less marked, but there is still a cleavage. Hand specimens from this place show a distinct plexus of minute thin needles, apparently of hornblende: there is no orientation of them in one direction. The same band appears again, in a direction slightly north of east from this section, in the south branch of Eas Riachain. Here it is divided into 2 or 3 bands, a yard or less from one another, and the intrusive character is perfectly clear. They are often about a yard thick, and are generally dyke-like, but sometimes they act as sills for a short distance. The most southerly part of the rock in this branch of the burn is nearly vertical, and is cleaved roughly parallel to its sides: amygdule-like spots elongate themselves parallel to this cleavage. In some places the cleavage is very poorly developed, and the rock weathers in a massive way with slightly projecting knobs, and is at the same time essentially free from the supposed amygdules, which in this locality characterise the more cleaved varieties. The cleavage, or weak foliation, is usually roughly parallel to the sides of the intrusion, but not always strictly so: it often varies 10° to 20° from it, sometimes in a more north-east and sometimes in a more north-west direction; in one place it was noticed to cross the band at about 45°—in this case the direction was more north-west. We could see no planes in the schists corresponding to those in the dykes: the planes in the schists are generally at a marked angle to those in the dykes, and we saw no tendency to approximation near the dyke sides. The dyke seems thrown along a feature running N.N.E. and hading east. On the east side of this it seems at first all together in one band perhaps 6 or 8 feet thick, and is generally massive and weathering with knobs. But soon it splits into two parts, each 2 or 3 feet thick, and often distinctly cleaved. After this, another feature also running N.N.E. and hading east, is met with, and we lose the dykes for some distance. A dyke of similar character and in the line of these is seen at the head of a landslip ¼ mile E.N.E. of the "E" of "Eas Riachain"; it is 18 inches thick. Further east still, at the east side of a north to south slack, which marks the line of a fault, two bands of greyish-green cleaved dyke rock are seen: they are each from 2 to 3 feet thick, and must be shifted by the fault in the slack.

The next exposure of this kind of rock to the north occurs in the burn ½ mile north-west of the "n" of "Lochan Mill Bhig" (126). It runs east and west, and is from 1 to 2 feet wide; the cleavage in it runs nearly parallel with the side. This dyke could not be traced more than about 100 yards. Nearly ½ mile due west of it, a 3 feet greyish-green lamprophyre dyke is seen, but this shows no sign of cleavage.   C.T.C.

An instance of a well foliated dyke occurs in the burn ½ mile W.S.W. of the "G" of Glen Fine" (126), about 300 yards before its junction with the River Fine. Just below the sheep fold, a dark green fine-grained dyke about 3 feet wide is bounded by a fault trending north-west. The dyke is schistose throughout, the planes of foliation striking obliquely at the fault which bounds the dyke. The neighbouring rock is a gnarled and contorted mica schist containing albites. The foliation planes of the dyke are nearly at right angles to the foliation planes of the adjoining rock, and show no tendency to penetrate the latter, the junction between them being quite sharp. The dyke contains inclusions of the albite schists, the inclusions being similar to the schist at the side. The foliation of the dyke passes round the inclusions which retain their older foliation, lying oblique to the foliation of the dyke. The included fragments are entirely unaffected by the foliation of the dyke. Close to the dyke two smaller dykes occur, one on each side of the central dyke and running parallel with it. They are only about one foot in breadth. The nearest of them is within a yard of the sheared dyke, but they possess no foliation whatever. The sheared dyke was examined by Mr Teall, and found to consist in the main of pale brown mica and carbonates, some colourless quartz or felspar, chlorite, and iron ores. The interior is somewhat coarser than the exterior portion of the dyke. According to Mr Teall the unsheared dykes consist of fibrous hornblende, biotite, iron ores, carbonates, and colourless felspars. The felspar forms a kind of matrix, and shows under crossed Nicols a tendency to a feathery mode of aggregation. It is evident, therefore, that the essential difference between the foliated and non-foliated dykes is the presence of hornblende in the latter, and chlorite in the former. As the breaking down and shearing of a rock containing hornblende frequently results in the replacement of hornblende by chlorite it is probable that the composition of the dykes before foliation was the same. The unsheared dykes are not seen bounded by any crush movement. The sheared dyke is at one side bounded by such a movement. It is suggested that some early movements along the side of the sheared dyke may have induced the foliation in it, though the crushing now seen at its side may be due to later renewed movement.   J.B.H.

The two east to west bands in the burn $\frac{3}{8}$ mile west of the " E " of " Eagles Fall " (126) seem both to be earlier than the porphyrite sheet on the west side of the burn, for there is no sign of their passing through it: the evidence is not, however, decisive, for they are not seen on the west side of the sheet. They are both greenish-grey in colour. The south band is cleaved at one side for a breadth of several inches. A powerful north to south fault occurs a little on the east side of them, and they are probably shifted considerably by it; on the east side of this fault we suppose them to be represented by the two W.N.W. dykes seen on either side of another north and south fault, about $\frac{1}{4}$ mile west of the " E " of " Eagles." The dykes show no distinct cleavage. One of them is intersected by a porphyrite sheet in the west bank of the burn. One-eighth of a mile south-west of the " E " of " Eagles " a 4-foot slightly foliated basic dyke makes a gulley in the schist scars. It is in the exact line of the dyke which is intersected by the porphyrite sheet, and we suppose represents either it or the companion dyke.

In the burn $\frac{1}{4}$ mile south-west of the " C " of " Cruach Tuirc " (126) a basic dyke with stout crystals, apparently of hornblende, can be traced in or close to the burn for a distance of 200 yards. The breadth is 7 feet, and the direction slightly north of west. No cleavage was discerned in it. One quarter of a mile S.S.W. of the same letter, a little on the north side of the burn, a basic dyke again appears, in the same line as that in the lower part of the burn, and this is distinctly cleaved. Further up the burn, either in or close to it, a similar cleaved basic dyke is seen in several places. On the west side of the north-west fault that crosses the burn, such a dyke is seen just above the porphyrite sheet on the north side of the burn. On the east side of the same fault a somewhat similar exposure occurs; the thickness is 2 feet, and the splitting planes of the dyke show small but distinct flakes of mica. The exposure lies immediately over the top of the porphyrite sheet, only a few feet above the water level of the burn; the porphyrite is, in one place, distinctly chilled against it, and acquires a felsitic matrix for a breadth of about an inch from it. A little further west a similar dyke is seen below the porphyrite sheet. It is possible that the two dyke-like exposures belong to one band now separated by the later porphyrite intrusion. The lower exposure is cut across distinctly by the porphyrite. The upper exposure shows a foliation which makes a marked angle to the direction of the band; the dip of it is about E.N.E. at about 50°. The dyke can be traced up the burn distinctly for

about 60 yards after the porphyrite sheet is lost against the fault running
up the burn. It is accompanied with parallel crushes and strings of carbonate of iron mixed with rarer strings of barytes, and does not, in this course,
always show distinct cleavage. A little over ¼ mile south-west of the
" C," a camptonite-like dyke, sometimes massive, and sometimes cleaved,
particularly near the side, is seen again. In one place it is in contact
with a porphyrite sheet, but the section is too much affected by faulting
to be readily interpreted. On the same line of intrusion two exposures of
cleaved dyke are seen ¼ mile east of the " E " of " Eagles Fall." In the
bed of the Fine, near the north side of it, about 180 yards above the cart-
bridge, there is a 5-foot dyke which may be the W.N.W. continuation of
the same line of intrusion. It contains distinct hornblende, and some
basic inclusions of coarse grain. No cleavage was noticed in it except for
a breadth of 1 inch at the south side.

One quarter of a mile west of the " C " of " Cruach," another line of
intrusions of the same class is seen. There are at least three different dykes
running roughly parallel, and close to one another, in a direction slightly
north of west, but they are inconstant in occurrence, when one dies out
another coming on at a little lateral distance, and they are only 2 or 3 feet
thick. Further E.S.E., along the continuation of their direction, a
similar dyke is seen crossing the crags on the south side of Cruach Tuirc.
In some places it is almost free from cleavage.

By the " u " of " Cruach " a foliated basic sheet, 4 or 5 feet thick, makes
a conspicuous ledge among the schist crags. This differs from the sheared
dykes described in showing micaceous spangles much more distinctly, and
also large radiate shapes composed of black mica. The arms of these
shapes are in places nearly 1 inch long. The black mica composing them
may be pseudomorphous after actinolite. The exposure is less than ¼
mile south-west of the great granite mass which has effected such a
striking contact alteration in the schists of the neighbourhood, and we
suppose that the exceptional character of the sheet may also be
attributed to its action.

Summary of the Relations of the Older Igneous Rocks in Regard to Age.

Of the different varieties of rock now described, not much can be said
of their relative ages. As already pointed out on p. 124, the foliated
camptonites are certainly earlier than the hornblende-porphyrite sheets
with which they come in contact. We suppose it probable that the
camptonites not seen to be foliated are of the same age as the foliated
ones, and that the hornblende-porphyrite sheets, etc., alluded to, are of
the same age as the felsites.

A section on the north-east side of Hell's Glen (134 south-west) between
the " a " and " g " of " Beag " shows a dark micaceous dyke cutting
through a pale felsite sheet. This shows that the basic dykes are not
all of the same age as the foliated camptonites—some are later than
the felsites, and, therefore, still later than the foliated camptonites.

Three-eighths of a mile slightly east of south of Beinn an t-Seilich
(134), a sheet with stellar groups of decomposing hornblende needles cuts
through a felsite sheet. Just south of the " G " of " Glen Kinglas "
(134) similar intrusions are also in contact. The evidence is not
decisive, but is in favour of the felsite being the earlier of the two.

In the section (p. 113) ½ mile N.N.W. of the " P " of " Dornoch
Point " (152), a red band with radiate hornblende seems earlier than an
augitic dark grey band. The dark rock, however, does not seem quite so

fine-grained at its junction with the red as it is at its junction with the schists, and it is not certain that the flag-forming joints of the latter rock are cut through. When the dark rock is in contact with the schist it is usually, but not always, marked by a set of joints at right angles to the junction, at intervals of 6 or 8 inches, which gradually die out in a distance of about a foot. Where it is in contact with the red rock we could see no such joints. Perhaps the cooling of the red rock was not thoroughly completed at the time the dark rock came up.　　　C.T.C.

CHAPTER XII.

THE IGNEOUS ROCKS (*continued*).—BASALTIC DYKES.
GENERAL DESCRIPTION.

In this chapter the examples given of particular structures, modes of occurrence, etc., have been chiefly collected from that part of Cowal which lies within 1-inch Map 29 on the east side of Glendaruel and Loch Riddon. The details observed along the course of the basaltic dykes in other parts of the area are mentioned subsequently.

The same colour, deep carmine, is used to express basalt, dolerite, augite andesite, and tachylite : and the same letter B is applied to all these, with the addition of a small T (thus B^T) in the case of tachylite, where this may occur in sufficient thickness to form a special feature in the exposure.

Basalts, dolerites, and tachylites, are so closely connected, forming often different parts of the same dyke, that it is impossible to separate them with satisfaction. Augite andesites have also much the same macroscopic appearance, and such a similar behaviour in the field, that they seem to form in this area a natural group with the others. All these rocks are distinguished by their dark grey or black colour in fresh hand specimens, their rusty and exfoliating weathering, and their habit in this area of forming approximately vertical dykes, which, with the exception of the tachylite, have prominent cross or longitudinal joints, or frequently both of these.

A glance at the Maps will show what a very important feature in the geology of the district these dykes form. Even the number shown is far from complete, for in areas covered by peat or drift, their outcrops have not been continued far, except in the case of some of the thicker dykes, the probable courses of which we desired to indicate, in order to show the connection of their different outcrops.

In rocky areas the dykes are generally easy to trace, this tracing being only a question of time and labour. They make features—rock-scars, green grassy hummocks, slacks, or flat terraces—which clearly cut across the strike of the bedding or foliation planes adjacent, and which attract attention even at some distance. When forming rock exposures their intrusive nature is at once apparent from the difference between the joints in them, and the rock at their sides ; when they form slacks or terraces we are not always able to tell whether these are due to a crush breccia or a dyke, but if we follow such a line for a little distance we are sure before long to meet with some evidence that settles the question, either some exposure of dyke-rock in the slack, or indications of contact alteration in the rock at the sides.

It is not certain why some dykes form rock-scars, resisting weathering more than the adjacent rocks, and others slacks and terraces, resisting weathering less than these rocks; but probably the reason is connected with the grain of the dyke-rocks, the coarser grained being most readily penetrated by water, and the other decomposing agents introduced by water. It is, at all events, a fact that the coarser grained rocks generally form either hollows, or smooth green hummocks and ridges, with but few rock exposures. This is very well illustrated along the course of the broad Dunoon Castle dyke, which, when free from drift, universally forms either well-marked hollows, or dry smooth banks and mounds clothed with short sweet grass, which contrast strongly with the dark peaty or "benty" ground on the sides. From the high beach near Glen Morag (184 south-west) for a distance of rather more than 1½ miles west, it forms such banks and mounds, not continuously but so near together that when you are standing on one you can usually pick out the next on either side. Similar green ground is formed by it in the areas ½ mile south-west of Bodach Bochd (183 east), and on the south and south-west sides of Coraddie (182 east). On the side of the hill west of Troustan (183 south-west), and between Allt Glac na maill and the high beach at Fearn'ach (182 south-east) it forms a conspicuous trench or slack.

Other dykes which we have especially noticed to form smooth green banks occur 1 mile south-west of Craigendaive (172), ⅝ mile E.S.E. of Balliemore, from ¼ to ½ mile south-west of Bishop's Seat (183), and ½ mile E.S.E. of Tom Odhar (184).

Where the more readily decomposed dykes are running not across but along the level contour lines of the hill they form flat terrace features rather than slacks, as any hollow they might otherwise form has a tendency to be filled up by movements of soil from the higher side. Such terraces are excellently shown in the rocky area just east of the foot of Loch Eck, on the east side of Stronchullin Hill (164 south-east), and on the hillface 1 mile south of Glen Lean (173 east). When their run steeply across contours they form slacks or lines of ready weathering, which we repeatedly find to have been taken advantage of by the streams. In some districts, e.g., on the sides of Glendaruel and Glen Tarsan, we can hardly find a burn which is not running along either a dyke or a crush breccia, or both of these combined, and some of the streams have worn such deep channels that it is extremely difficult to walk continuously along them, their sides being for considerable distances quite vertical, perhaps sixty or seventy feet high, and separated here and there by deep pools. The burn on the hillside 1 mile S.S.W. of Glen-massan is such an one, and it has consequently received the name of "Garbh Allt" (the rough burn).

Among the more prominent crags formed by dykes we may mention the following :—one ½ mile N.N.E. of Dun Mor (172 south-east) made by a somewhat fine-grained basalt; one in the Creag Mhòr 2½ mile north-west of Blairmore, which is a massive black scar continuing with but little interruption for a distance of nearly a mile ; one ¼ mile north of Garrow-choran Hill (184 south-west), and continuing thence for ½ mile north-west; and one at Cluniter (195 north-west), which is very prominent from the Clyde tourist steamers.

The joints in the dykes have always an evident relation to their sides or surfaces of cooling. It is perhaps rare to find one that does not show some joints both at right angles and parallel to the sides ; in some these different sets are nearly equally prominent, but in others they are so unequally developed that, in brief descriptions, we may call them simply either longitudinally-jointed or cross-jointed. Good examples of the longitudinal planes occur in the north to south dyke in the burn ¾ mile south-east of Corlarach Hill (183 south-east), and on the shore of the Clyde at places ⅛ mile south-east of Glen Morag (183 south-west), ¼ mile S.S.E. of Glen Morag, and ¼ mile N.N.E. of the Bull Rock (184 south-west).

Prominently cross-jointed dykes are perhaps as abundant as the above, but are themselves divisible into two classes, which, in their extreme forms, are markedly different in appearance. In one class the joints divide the dyke into rude five or six-sided columns, the cross diameters of which in one direction are not widely different from those in another. In the other class the dyke is divided into slabs or flags which, though only one or two inches broad, may cross most of the breadth of the dyke, and extend also for several feet vertically or diagonally. The columnar cross joints are well seen in the following dykes,—the Dunoon Castle dyke on the shore at Dunoon and on the east side of Loch Riddon ; in the dyke nearly ¼ mile slightly north of east of Cruach nan Capull (183 north-west), combined with longitudinal joints ; in the dyke just below the bridge over the burn that flows into the West Bay of Dunoon; in the dyke 150 yards above this bridge; in the Toward Church dyke, both at the church itself and on the shore (combined with longitudinal joints near its south side). The flaggy cross joints occur,—in the dykes ¾ mile N.N.W. of Cnocan Sgeir (172); ⅞ mile S.S.W. of Glen Lean (173 south-west), with dip of joints varying rapidly ; near the top of Leacann nan Gall (183 north-east), with flags dipping either north-west or south-east, or forming curves ; the top of Bishop's Seat (183 north-east), with planes vertical or between this and a dip of 50° north-west ; ¼ mile south of the top of Blar Buidhe (183 south-east), with dip to south-east ; in an exposure 90 yards north-west of this last locality with a steep dip to north-west ; ¼ mile north of Garrowchoran Hill (184 south-west); and from this locality for a distance of ½ mile north-west. In the dyke near the top of Leacann nan Gall we see both the flaggy joints and cross prisms of short diameter, which in places cut the dyke into columns about 3 feet long and 3 inches broad. It is clear from the particulars in the above list that there is no general direction of dip for the joint flags. This fact is particularly well seen in the dyke with central tachylite (to be subsequently mentioned) rather more than ½ mile slightly north of east of the Clachan of Glendaruel (172 north-west), where it is seen that the flags are not at all perfect or parallel, if closely examined : they may perhaps be regarded as prismatic columns which have had their cross diameter in one direction much increased at the expense of those in others ; the general direction of the flags may curve not only in the direction of their dip but also in that of their strike, as they are traced across the breadth of the dyke, though the dyke side continues straight.

The dyke by the north side of the "n" of "Glen Tarsan" shows good cross prisms in the south-east half of its course, but in the north-west half it is flaggy across the length: an obscure area separates the differently-jointing portions.

A north-west dyke in the burn ¾ mile south-west of Cruach na Caorach (173 south-west) has cross flags for a distance of one foot from its edge, and also in a part near the middle, but not so elsewhere.

The interiors of dykes are least affected by cross joints, and contain also the most abundant, and the largest vesicles. The margins of the most vesicular areas are rudely parallel to the dyke sides, and in those dykes which contain alternating layers of more or less vesicular portions the alternations are parallel with the sides. Dykes on the shore ¼ mile

N.N.E. of the Bull Rock (184 south-west), and 200 yards south-west of this rock, show the special occurrence of vesicles in the interior portions; in the last-mentioned dyke, which is the most northerly but one of a set of four, the amygdules are more common in certain bands than in others, and are sometimes of irregular shape ; some thin strings of basalt lie on the north of it, from one of which, only 4 inches wide, hand specimens can be obtained showing its full thickness, and many of the common characters of dykes, including the vesicular nature of their centres.

An 8-foot dyke on the shore ¼ mile S.S.E. of Chapel Hall (194 south-east) also shows vesicular bands in rows parallel to the side : the vesicles are filled with a soft green earth, or with this and calcite mixed. In the burn 100 yards east of Glenlean (173 east) the vesicular portions, whether near the centres or not, are clearly disposed parallel to the side. The most easterly of the north and south dykes near the head of the burn ¾ mile south-east of Corlarach Hill (183 south-east) has, a little within it, several bands, a few inches thick, of slaggy basalt, mixed with other bands of fine grain which are quite free from vesicles.

It has just been stated that the vesicles in the dyke 200 yards south-west of the Bull Rock are sometimes of irregular shape. In all the other cases alluded to they are roughly spherical, and are often about the size of a small pea. The spherical shape is in the basalt dykes of such constant occurrence, that we have not noticed more than 4 or 5 exceptions in addition to that just mentioned. In the dyke ¼ mile south of the top of Blar Buidhe (183 south-east) the amygdules are at times distinctly pulled out with their long axes parallel to the dyke side. In the north-west dyke in the burn bed ¼ mile north-east of the " K " of " KILMODAN " (162 south-west) the central portion is unusually vesicular, and the vesicles, which are usually filled with calcite, are sometimes as much as ½ or ¾ inch long : they are often distinctly longer than broad, and have their greatest lengths, as seen in a flat surface, parallel to the dyke side. On the shore of Loch Fine, ¼ mile south-east of the " C " of " McPhun's Cairn," (141 north-west) a dyke shows calcite amygdules, especially near the centre, and these are partly elongated parallel to the dyke's direction. In the burn ¾ mile north-west of Sgarach Mor (173 north-west) the vesicles in a north-west dyke are abundant and most irregular in shape, joining together by numerous processes so that they may, for a small area, equal in bulk the rock in which they occur.

In the middle of a 5-foot dyke 130 yards south-east of the East Gates of Toward Policies there are in places short bands of quartz and calcite which attain a width of 4 inches, and sometimes end bluntly : they run parallel with the dyke side, and continue for some distance vertically ; there is no appearance of crushing in the dyke near them. The dyke on the north side of the fault 100 yards south-west of Toward Lighthouse has also a calcareous band in it with geodes of quartz crystals ; these also seem independent of any crushing.

The special occurrence of amygdules in the centres of dykes is an indication that these were the last portions to consolidate, and that the previous partial cooling of the outer parts was accompanied by a contraction in bulk, which led to a diminution of pressure in the interior. From the spherical shape of the vesicles we conclude that the motion in the fluid rock must usually have ceased after the vesicles were formed, or else that they were carried on without any distortion. The setting free of the gas and steam, or their sudden expansion so as to form visible cavities, would lead to an absorption of heat from the surrounding mass, and this would tend to cool and solidify it and stop the motion.

Calcite is much the most common mineral found in the vesicles, and,

I

as this is readily removed by weathering, they generally appear empty
in natural exposures. A white radiating mineral which projects on
weathered surfaces, and is probably some zeolite, is not uncommon:
examples occur in the bend of the broad dyke $\frac{1}{2}$ mile north-east of Cruach
na Caorach, in the dyke by the big bend in Inverchapel Burn, in a thin
dyke on the west shore of Loch Striven 200 yards north-west of the Craig,
and in the most easterly dyke of the close set which occurs $\frac{2}{3}$ mile slightly
south of east of Uig (173 and 174): the amygdules in the last mentioned
attain a length of $\frac{1}{2}$ inch: those by the big bend in Inverchapel Burn
are supposed by Dr Hatch to be natrolite (slide 2842). A dyke,
about 4 feet thick, 1 mile N.N.E. of the Badd (183 north-east) contains
some amygdules of white agate, mixed with others of calcite. In the
dyke already mentioned $\frac{1}{4}$ mile slightly east of south of Chapel Hall
" green earth " occurs, sometimes filling the entire vesicle. On the shore
of Loch Fine 100 yards W.S.W. of Mid Letter (141 south-west), the soft
earthy substance in the amygdules in a freshly broken face is pale green
in colour, but after exposure to light it changes to dark green in less than
twelve hours, and ultimately becomes quite black. This substance seems
to be the Chlorophæite of Macculloch (*Western Isles*, i. p. 504). See
also Heddle, *Trans. Roy. Soc. Edin.*, vol. xxix. p. 84.

On examining with a hand lens specimens of the coarser dyke rocks,
the lath-shaped forms of the felspars of the ground-mass are often quite
evident, and the darker augite patches may appear to be intersected by
them : but the last character can, of course, be made out much more
satisfactorily when the rocks are examined in thin slices under the microscope. The distinctly crystalline rocks seem always to show this "ophitic"
character, *i.e.*, the augite areas are penetrated and cut up into small
portions by lath-shaped crystals of felspar, the felspar having been the
first of the two to crystallize out on the solidification of the rock.

The felspars and other minerals of the ground-mass become less and less
as the edges of the dyke are approached, so that the general grain of the
rock grows finer and finer. Usually they attain the greatest dimensions
only in the interiors of the broader dykes.

But besides the felspars of the ground-mass, larger and more stoutly-
formed felspar crystals are in some dykes very prominent as porphyritic
constituents, and these may occur as abundantly and in as large forms
close to the edges of the dykes as elsewhere. They would seem consequently to have been in existence in the fluid dyke rock at the time of
its intrusion.

On the south-west side of a little burn about $\frac{1}{2}$ mile N.N.W. of the
shepherd's house in Glen Tarsan (173 north-west), for an inch or so from
a dyke side, porphyritic felspars, giving nearly square sections, sometimes
about $\frac{1}{4}$ inch in length are common, so that they nearly equal in area the
ground-mass ; for the next $1\frac{1}{2}$ inch they are decidedly rarer, occupying
hardly $\frac{1}{8}$ of the area exposed ; then for a breadth of 3 inches they come
in again nearly as abundantly as at the side ; and after that, for 6 inches
or more, they become rarer again, occupying perhaps from $\frac{1}{8}$ to $\frac{1}{12}$ of
the whole mass. Again, in a north and south dyke, several yards wide,
in Gleann Laoigh burn, nearly due east of An Socach (163 south-west) porphyritic felspars sometimes 1 inch in length are common about 2 feet from
the side, but are practically absent in the rest of the dyke. A porphyritic
dyke on the shore 300 yards west of Toward Cottages also shows marked
variations in the abundance of the large felspars in different parts ; spaces
of a square foot or so can be selected which show hardly any of them,
while in other parts not far away they are common ; in this dyke it is
also evident that often the porphyritic crystals do not consist of isolated

individuals, but of groups of crystals connected together, so that the rock may be called "glomero-porphyritic." Probably this structure is common in the porphyritic dykes. A 2-foot dyke, on the shore $\frac{1}{8}$ mile N.N.E. of Inellan Pier, has on its south side a thin string which is, in certain parts, very rich in porphyritic felspar, but these parts are only small and scattered.

In a dyke in the burn $\frac{1}{8}$ mile south-west of the "G" of "Gleann Laoigh," we have noticed the porphyritic felspars to attain an unusual size. In one small exposure a crystal with a length of $3\frac{3}{4}$ inch was seen, another was 3 inches by 2 inches, and others 2 inches long; the edges of the crystals are, for $\frac{1}{8}$ inch or so, of a yellower tint than the insides, as if owing to the action of the magma on them.

Among other porphyritic dykes are,—the two on the west shore of Loch Striven about 1 mile N.N.E. of Dun Mor (172 south-east), the two on the shore a little over $\frac{1}{2}$ mile south of Craigendaive, the continuation of of one of these in the hill exposures north-east of A' Chruach, and $\frac{1}{4}$ mile, and $\frac{3}{4}$ mile north-west of A' Chruach, the dyke $\frac{1}{2}$ mile N.N.W. of Cnocan Sgeir (172 north-east), the north-south dyke $\frac{1}{2}$ mile S.S.E. of Cruach na Caorach (173 south-west), the dyke in the burn $\frac{4}{8}$ mile W.N.W. of Sgarach Mor (173 north-west), a thin dyke on the shore a little over 1 mile south of Fearn'ach (182 south-east), the north-east portion of a compound basalt band which runs $\frac{1}{2}$ mile south-west of Cruach nan Capull (183 north-west), the dyke in the burn $\frac{2}{3}$ mile south-west of Kilbride Hill (183 and 184), some of the north and south bands in the burn $\frac{1}{2}$ mile south-east of Corlarach (194 north-west) and a thin dyke $\frac{1}{8}$ mile east of Cluniter, just at the edge of the high beach.

It is suggested in several places that the occurrence of porphyritic felspars is liable to vary not only in isolated dyke exposures, but also, and to a larger extent, in the direction of their lengths. We cannot, however, mention any case that is quite clear; intrusions in this area are so abundant and so variable in occurrence that we cannot safely conclude that one dyke that appears in the same line as another at a little distance is really the same, unless there is a continuous exposure between the two. An apparent replacement of a porphyritic dyke by one that is not porphyritic occurs in the east and west burn $\frac{3}{4}$ mile N.N.W. of Bodach Bochd (183 north-east). The most easterly north and south dyke in the burn $\frac{3}{4}$ mile south-east of Corlarach Hill (183 south-east) is in one place very porphyritic, but is not at all clearly so elsewhere.

Olivine in small greenish grains, and pseudomorphs after it, have been observed by the naked eye in,—a porphyritic basalt, which occurs in the shore 200 yards E.S.E. of Brackleymore School (194, slide 2845, described on p. 142), the parallel dykes 150 and 250 yards N.N.E. of M'Phun's Cairn (141), the dykes $\frac{1}{2}$ mile N.N.E. and $\frac{1}{2}$ mile east of this cairn, a thin dyke crossing the burn $\frac{1}{12}$ mile east of the "u" of "Gleann Dubh" (152 north-west), the dyke $\frac{1}{12}$ mile E.N.E. of the "M" of "Meall Reamhar" (152 north-west), and some few other places. But macroscopic observations have not generally enabled us to detect it, and we think there are comparatively few of the dykes in this district in which it is prominent: it does not seem as abundant as it is in the Sleat district of Skye.

In slide 2844, the north-west dyke which intersects the east and west one on the south side of Larach Hill (164 north-west), there are beautiful pseudomorphs of serpentine after large well-formed crystals of olivine of pyramidal habit : other little patches of serpentine with calcite probably represent altered augite. In slide 2849, from the dyke with prominent porphyritic felspar exposed on the raised beach 300 yards west of Toward

Cottages (194), pseudomorphs after olivine also occur: for description of slide see p. 136. In slide 2842, from the dyke by the big bend in Inverchapel Burn (164 south-west), there are abundant iron-yellow patches, which may be pseudomorphs after olivine.

In the dykes on the shore 250 yards south-east, and ⅓ mile S.S.E. of Glen Morag (184 south-west), the weathered surfaces of the interior portions have a peculiar spotted appearance. The spots are often about ¼ inch in diameter and project slightly from the intervening areas. They may either closely adjoin one another, or be separated by a distance equal to, or greater than, their diameters. When the rock is not too much weathered it can be seen that the spots have no sharp boundaries, and that the minerals in them sometimes extend into the adjoining areas. Their felspar seems in a much fresher condition than it is outside, where it has usually acquired a dirty yellowish-green tint. The condition of the augite is not recognisably different as far as can be made out with a hand lens, nor is the proportion different.

Thin margins or selvages of distinctly glassy rock, "tachylite," are common over the whole district, but are generally very thin, and consequently not to be found without close search. They may be seen in the following places,—the burn ⅓ mile south-east of Dun Mor (172 south-east); one of the north-west dykes in the section ⅔ mile south-west of Bodach Bochd (183 south-east); the north-west dykes in the section ⅓ mile south-east of Bodach Bochd; a thin dyke in the burn about ¾ mile S.S.E. of the Horse Seat (183 south-east); on the east side of Loch Striven, ¾ mile N.N.W. of Inverchaolain Church, and 300 yards south of Ardbeg (183 north-west); on the west shore of Loch Striven, ⅓ mile N.N.E. of Troustan; on the hillside 1 mile S.S.E. of Sgian Dubh (183 south-west), at the head of a big landslip; one of the dykes in a set of three closely parallel ones on the east shore of Loch Striven, 200 yards south of Brackleymore School (183 south-west); the dyke in the burn ½ mile E.N.E. of Garrowchoran Hill (184 south-west); the burn ½ mile south-west of Cluniter (195 north-west); the coast immediately north of Inellan Pier, and again ¼ mile north and ⅔ mile south of the pier. Even this is far from an exhaustive list.

These selvages rarely exceed ½ inch in thickness, and range perhaps from ¼ inch down to the thinnest possible film. They seem almost universally to show on their outer surfaces somewhat round-edged rhomboidal or polygonal shapes, often from ½ to 1/16 inch in length, which seem to be of a perlitic character. On breaking across one of the thicker selvages these shapes may be seen to be really only the surfaces, apparently more or less flattened against the rock intruded into, of solid forms which usually take up the whole of the distinctly tachylitic portion. When the selvages are very thin, the forms, being necessarily restricted to the thickness of the glassy area, are much thinner across the selvage than in other directions: but in other places there is no constant distinction between their breadths in different directions, and we may find more than one perlitic form as we cross the selvage from side to side.

In the most northerly of the north-west dykes in the section ⅔ mile south-west of Bodach Bochd, the shapes are seen to be separated from one another by extremely thin lines, which project slightly on the weathered surface, and are marked by a central line of depression. We think that similar lines are not uncommon elsewhere.

A slide (2852) from the glassy margin of the dyke ½ mile E.N.E. of Garrowchoran Hill is thus described by Dr Hatch: "Opaque tachylite. With minute colourless microlites (felspar)."

Besides selvages of tachylite we occasionally meet with broader masses of the same substance within and generally near the centres of dykes. The best examples occur on the east side of Glendaruel, and are not very far from one another : we do not know whether this may not depend on the relation of the sources of the dyke material to this locality or on some other condition.

A N.N.W. dyke ½ mile east of the Clachan of Glendaruel contains, besides the tachylite, two other types of rock. The inner part has a distinctly finer matrix than the outer, and shows small porphyritic felspars distinctly : it has no joints peculiar to itself, but is crossed by those which affect the rest of the dyke, and seems in places to pass into the coarser rock gradually and in others abruptly. The tachylite is only found within the porphyritic or inner portion, and itself shows more or less conspicuous porphyritic felspars : it can be seen within this inner portion, every here and there, along the length of the dyke for a distance of over ½ mile. The tachylite may form several thin bands in the breadth of the dyke, and these may vary considerably in hade and direction, and be connected with one another. But where it is most developed it occurs in a central band, which sometimes attains a thickness of 5 or 6 feet. Where it is thick it is usually easy to distinguish at some distance, as it is not affected by the same joints as the rest of the dyke, and projects with a black glossy surface as if more able to resist weathering. Satisfactory hand specimens are hard to obtain for the rock is either excessively hard, or else breaks under the hammer into numberless small cuboidal pieces along joint lines. What we may call the tachylite area is not all glossy to the eye, but is divided up into more or less round-edged rhomboidal or oblong forms of glossy rock, separated from one another by bands, sometimes 3 inches thick, which are not so glossy and weather with a somewhat rusty colour, or else a paler grey than the rest of the rock. These bands are prominently jointed across their lengths.

The above bands no doubt correspond to the "sheaths" of the pitchstone of Eskdale, figured and described by Sir A. Geikie and Mr B. N. Peach ("The Pitchstone of Eskdale," *Proc. Roy. Phy. Soc. Edin.*, vol. v., 1880), and the interior parts to the "cores." The figures and description in this paper correspond in most respects with what we observe in the Clachan dyke, but Fig. 1, showing prominent prisms across the breadth of the pitchstone from side to side, and the description on p. 242, where the sheaths are mentioned as occurring along the sides of these prominent cross prisms, do not apply to this case. On the contrary, a peculiarity of the tachylite centre is, that there is no regular set of joints which have a clear relation to the sides. The rhomboidal forms are seen whether we look at vertical or horizontal surfaces of the rocks, and we do not think that they have generally a greater length in one direction than another, though they may have locally for a little distance. Instead of comparing them with the cross prisms of ordinary dykes we would liken them to the perlitic shapes which, as already stated, are of almost universal occurrence in the tachylite selvages, though not unnaturally in the broader tachylite bands the shapes may attain a much larger size, a length of 9 inches being not uncommon. The sheaths have often, perhaps usually, a median suture as described in the Eskdale rock, but sometimes there is no suture and at other times three or more sutures in one sheath. The sheaths generally project slightly on the weathered face, and show small porphyritic felspars just as the cores do. We have never seen the glossy rock in contact with the outer coarse-grained dyke rock : it is always the sheath rock that is next to it.

Specimens of a glossy core and of a sheath from this locality were examined by Dr Hatch, and are described by him as follows. The core (slide 3268):—" Andesitic glass. Consists of isolated porphyritic crystals of fresh felspar imbedded in a chocolate-coloured glass, containing microlites, often forked, of felspar, long slender needles or microlites of a pale yellowish augite (large extinction angle), characterised by a fibrous appearance and cross-jointing and studded with minute granules of magnetite. The glass is clear in the neighbourhood of the augite and magnetite, elsewhere it is cloudy from the presence of minute belonites, the aggregation of which has, thus, evidently formed the augitic microlites and iron-ore. The porphyritic felspar is in part striated. The crystals are much corroded and contain inclusions and inlets of the glassy base." The felspar microlites are often as long as any of the porphyritic crystals, and the augite microlites longer. Each of the larger augite microlites has sometimes shorter augite microlites arranged rudely at right angles to its sides.

The sheath (slide 3269) :—" Andesitic glass, of same general character as 3268, but more devitrified, being crowded with skeleton crystals of felspar and trichites of iron-ore. The rock is somewhat decomposed as there is a quantity of brown granular matter (limonite?) present in the slide."

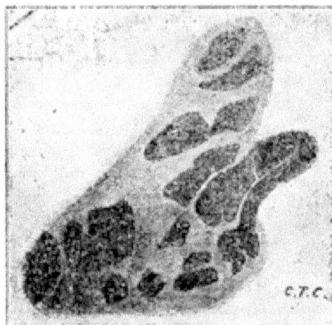

FIG 34.— ×⅟₂. Half a mile slightly W. of S. of Cruach Mhor (162 and 172). Ground plan of a portion of the centre of the dyke. The darker parts are the more glassy, black, and sounder. The paler parts are the more grey or yellow-weathering.

In another N.N.W. dyke, about ¼ mile south-west of Cruach Mhor (162 south-east), a glossy portion can be seen near the centre for a distance of nearly a mile, and its breadth in places is as much as 6 or 7 yards. The dyke occasionally makes sharp twists, but the glossy portion still keeps near the centre. The tachylite differs somewhat from that in the Clachan dyke in that it displays less commonly porphyritic felspars ; in some places it is as porphyritic, but more usually the felspars are evenly dispersed and closer together, except in certain small well-defined and usually spherical areas with diameters of ¼ inch or less, which to the eye show none at all. The outer sides of this dyke consist usually of a uniform medium-grained basalt, but occasionally it varies slightly, even in parts near to one another, and acquires a much finer matrix speckled with small porphyritic felspars. The glossy portions are divided into sheaths and cores, and the remarks made on the irregularity of these forms in the Clachan dyke apply equally to this dyke. It has also been noticed additionally that the sheaths in this appear to the eye sometimes as glossy as the cores, e.g., on the south-west side of the dyke at a point about ¼ mile south-west of the Ordnance Station on Cruach Mhor ; but they more commonly present rusty surfaces to the spectator owing to their numerous cross joints allowing readier access to water, etc. The sutures in some places are filled with thin milky-blue opalescent quartz strings. In some places, portions of the dyke that are not glossy, are themselves cut into corelike shapes by thin bands, which weather lighter grey, and have opalescent quartz strings down their centres. About 30 yards south-east of the places where the sheaths are at times seen to be glossy, more or less incomplete spherical forms, with no apparent radial structure, are observed within the glossy area. They may be compounded together, or lie one within another, and have no sheath-like bands at their margins. They occur in the cores and sometimes attain a diameter of 3 inches.

A slide (3270) obtained from near the centre of the Cruach Mhor dyke is described by Dr Hatch as follows : "Andesite (Tholeite type). Perfectly fresh rock, consisting of short lath-shaped crystals of colourless striped felspar and granular aggregates of pale yellowish-brown augite, imbedded in a base of clear chocolate-brown glass. Porphyritic felspar in isolated crystals." There seem indications of irregular perlitic cracks in the thinnest portion of the slide when examined by ¼ inch objective.

A slide (3271) obtained from the centre of the same dyke in a part of its course further north-west, about ⅔ mile south-east of Lower Duillater,

FIG. 35.— × ¼. One and a half mile S. of Craigendaive (172). Ground plan to show the pattern in one of the glassy bands in the dyke. The strings left white in the sketch weather in relief and with a yellowish tint : the longer straighter ones are approximately parallel to the length of the glassy band in which they occur.

is thus described : "Andesite (Tholeite type). Same as 3271, but with larger felspars and augites, and glass more devitrified, being full of cloudy matter and numerous barbed and torked trichites of iron-ore."

A fine-grained north and south dyke, 5 or 6 yards wide, close to the east side of the north to south burn, 1½ mile south of Craigendaive (172 east), shows, here and there, near the side, and crossing diagonally in an irregular way so as occasionally to join one another, black bands of a glossy or resinous appearance. In breadth they are somtimes as much as 14 inches. Their weathered surfaces show a perlitic pattern of dark portions inclosed by other yellowish

bands, which are often about ⅛ inch wide, weather in relief, and are divided by a median suture. The other parts with no resinous appearance show on close examination somewhat similar yellow bands, but at wider intervals ; and also occasional short strings of opalescent quartz, which sometimes attain a breadth of ½ inch, running rudely parallel with the dyke side.

The most north-easterly of the north-west running dykes in the section ⅔ mile south-west of Bodach Bochd is only 4 or 5 feet wide, and it has, besides the tachylite margins already described, an inter-mixture of a glassy with a duller fine-grained rock for most of its breadth. A specimen of this type was examined microscopically (slide 2850) by Dr Hatch and is thus described :—" Tachylite (Basalt Glass). The lighter coloured portions (in thin sections) are of a pale yellowish-brown colour, and contain isolated granules and microlites of felspar. Penetrating these are streaks of a blackish-brown glass, in some cases quite opaque. The line of division between the 2 glasses is marked by a layer of doubly refractive substance. In other parts of the lighter-coloured glass are local accumulations of blebs of a dark brown pigment, uniting here and there to opaque patches."

Some intrusive strings on the south side of the thick basalt $\frac{2}{3}$ mile E.N.E. of Inverchaolain Church (183 south-east) show a somewhat similar admixture (slide 2851), and distinct perlitic structure.

In a quarry (now disused, 1893) in the porphyritic dyke east of Toward Cottages there is a band, about 4 or 5 feet from the east side of the dyke, which is said by the quarrymen to take three times as long to bore as the rest of the dyke. The band is 2 or 3 feet wide and runs with the length of the dyke. To the eye it has a more glossy look than the rest. A specimen was examined microscopically by Dr Hatch, and is thus described (slide 2849):—" Porphyritic basalt. Abundant large grains of plagioclase felspar visible to the naked eye. Under the Microscope : —large porphyritic crystals of plagioclase and isolated serpentine-pseudo-morphs after olivine, embedded in a ground-mass made up of felspar-microlites, grains of augite and abundant glassy base, the latter powdered over with magnetite-dust which here and there accumulates to extremely dark patches."

It has been pointed out (J. J. H. Teall, *Brit. Petrog.*, p. 402) that we may plausibly explain the occurrence of glassy bands near the centres of dykes like those described, by an application of "Soret's" principle. " Suppose a mass of molten matter be injected into a fissure and to remain stagnant for a considerable length of time, the mass will be cooled at the margins, and the compounds with which the solution is most nearly saturated will accumulate in the marginal portions, leaving the centre richer in those which play the role of solvent medium."

Other dykes divided in a way which reminds us of the sheaths and cores of the Glendaruel dykes are not uncommon, though there is no glossy or resinous appearance to be noticed in them. A clear exposure of such a dyke occurs on the west shore of Loch Long, $\frac{1}{2}$ mile slightly east of south of Stronchullin (174 north-west). The dyke runs slightly west of north, and is about 7 or 8 feet wide ; in the more central decom-posing part frequent hard thin bands occur of a breadth varying from 3 inches to $\frac{1}{2}$ inch. They are all more or less cross-jointed ; sometimes the joints go right across their breadth, at other times there is a median suture from which joints proceed on either side. The joints that go all the way across are, however, in this exposure the commonest. These bands are disposed in a most irregular pattern, without any reference that could be discovered to the side of the dyke. Sometimes they go right up to the sides of the dykes. Sometimes they form circular or oval shapes which are quite isolated as seen in plane section. Such shapes may be seen inclosed in one another, or a band at one side of a shape may divide into two portions, or twist into sharp complicated curves. The dyke in the burn $\frac{1}{3}$ mile north-west of the " D " of " Dunoon " (184 north-west) shows a somewhat similar pattern.

Fig. 36.—Half a mile slightly E. of S. of Stronchullin (174). Some shapes of the cross-jointed strings within the dyke on the shore. On a reduced but not uniform scale.

Excepting as regards the glassy central portions and the porphyritic felspars, very few variations in the character of individual dykes have been noticed, which do not evidently depend on the distance of the surfaces of cooling, the coarseness of grain varying as the distance, as already described. But occasionally we do meet with thin alternating wavy stripes near their sides, which are not of quite the same character ; e.g., in a dyke running nearly north to south in the burn 1 mile S.S.W. of Gleumassan (163 south-west). In the most north-easterly dyke of the closely parallel set ⅔ mile north of Glenstriven (183 north-west), there is throughout its whole breadth, about 3 feet, a banding parallel to the side, and this is particularly distinct for a few inches next the side. Some of the bands are more amygdaloidal than others, and these, as seen at a little distance from the weathered surfaces, appear darker than the others. The bands near the side are often about ½ inch thick. In one place the dyke changes direction sharply, and the bands partly change also, but some of them become gradually cut through by the side of the dyke. The bands remind us of those mixtures of slightly different types which so often indicate fluxion structure in surface lavas.

Other cases on a larger scale may be supposed due to intrusions of one rock into another before the older rock had got thoroughly cooled and consolidated. In these cases we cease to find the difference in the joints of the two rocks, and the ready parting along the side of one of them, which are such helps to the first recognition of intrusions. We need not be surprised, then, at having observed only a few such cases. They are probably only a small part of those which actually exist. The difference in texture does of itself lead to a difference in smoothness of the weathered rock surfaces, but unless it be unusually great it is not likely to be observed at any considerable distance.

An instance of this kind occurs on the hilltop ⅔ mile E.N.E. of Ballie-more (173 north-west), where, within a north-west dyke of medium and uniform grain, marked with conspicuous, flaggy cross-joints, there are much finer-grained strings, sometimes as much as 4 inches thick. They run irregularly, hading at various angles and are sometimes almost horizontally, but in the main they may be said to be rudely parallel to the dyke side. They have no joints peculiar to themselves, but are crossed by those which affect the rest of the dyke.

A still better instance occurs in the dyke with tachylite centre on the east side of Glendaruel near the Clachan. Rather more than ½ mile slightly north of east of the Clachan, and about 100 yards from the small burn that runs through the Clachan, there is a good exposure of the dyke. For a space of from 2 to 4 yards from the sides there is a closely-flaggy cross-jointing basalt of uniform and somewhat fine grain : then comes on, quite gradually, an interior portion, in which these joints are replaced by others which cut the rock into cross prisms. On hammering the prismatic rock it is seen that the outsides are of the same rock as the flaggy sides, but there are bands within this which show a much finer matrix speckled with porphyritic felspars. These bands do not stand out from the rest in a dyke-like way, do not separate readily on hammer-ing along their junctions, and they are crossed by the same joints as the rest of the rock. The junctions are by no means so straight as those usually seen in basalt intrusions with joint differences, and some bands, as seen in plane section, are of extremely local occurrence. Yet, at this particular locality, the outer rock must have been partly cooled before the porphyritic rock came up, for, immediately at the junctions, the matrix of the latter is finer than it is further away. A little distance away in a north-west direction the same two varieties of rock occur again, and again share the same joints, but there does not seem to be now any sharp line between them. This is also the case in the two quarries in this dyke by the roadside 300 or 400 yards south of Auchateggan ; in both, the interior rock, for a breadth of 8 or 9 yards, is composed of a very fine-grained matrix speckled with a varying abundance of porphyritic felspar, and the outer 2 or 3 yards on either side is of a more uniform, coarser rock. Though the two types occupy fairly definite areas, and are in the extreme forms quite distinct, we could not, in either quarry, see any sharp line between them, but for a foot or two on either side, there seems a gradual change, the interior rock acquiring a coarser matrix, and the porphyritic felspars continuing for a little distance in the coarser rock. In the lower quarry the entire dyke is crossed by stout rudely horizontal prisms ; in the upper quarry this is also the case excepting where the central band of tachylite occurs.

Apart from the indications of fluxion structure already alluded to, *e.g.*, the rare elongation of amygdules, stripy structure near the side, and the occurrence of porphyritic felspars in bands parallel with the side, there are one or two other modes in which it sometimes shows itself. The most common of these is in the arrangement of the felspars; when we approach the edge of a dyke we can often see that the felspars, consisting of extremely thin plates, are arranged with their greater lengths parallel to one another, so that, in sections cut parallel to the edge, a number of brightly shining flakes of felspar are observed, while, in sections cut at right angles to the side, only thin needle-like shapes can be seen. This is shown clearly on the shore nearly ⅝ mile north of Blairmore Farm, and in the north-west dyke which intersects the broad east to west one in the burn on the south side of Larach Hill (164 north-west). The microscopical structure of the last dyke, slide 2844, is described on p. 156.

Again, when the dykes contain inclusions of foreign rocks, these are often arranged with their longer axes parallel to the side of the dyke. The banded dyke alluded to on p. 137, ⅔ mile north of Glenstriven, shows this excellently : the inclusions consist of schist and vein-quartz, the latter in greater proportion to the former than we should expect ; where the direction of the dyke changes, the direction of the lengths of the inclusions also changes. The dykes just on the south-west of this show similar inclusions to a less marked extent, and a similar large proportion of vein-quartz pieces. The most north-easterly dyke but one of the set of dykes in the burn 150 yards south-west of Blairmore Farm contains pieces of altered and cracked vein-quartz which are often several inches wide, and longer than this in the direction parallel to the side of the dyke.

There are three other places in which inclusions, on a large scale, have been observed, but in which no parallel arrangement was noticed. The inner coarsely vesicular portion of the dyke ¾ mile north-west of Sgarach Mor (173 north-west) contains angular pieces of hardened schist and vein-quartz (we think also of basalt of a less porphyritic type than the inclosing rock), sometimes 6 or 7 inches in length ; in places near the side of the dyke there are calcite strings running parallel with the dyke, but we do not think the inclusions can be regarded as parts of a crush breccia of later date than the dyke. The dyke in the burn a little more than ½ mile southeast of Bishop's Seat (183 north-east) includes, for a breadth of 4 yards in the middle, blocks of schist up to 3 feet in length embedded in uncrushed dyke rock. The most south-westerly dyke in the burn nearly 1 mile slightly east of north of Bishop's Seat includes many pieces of schist and vein-quartz. A dyke on the east side of Loch Striven, ⅝ mile E.S.E. of Meallan Glaic (194 north-west), contains pieces of hardened and darkened phyllite ; strings of porphyritic basalt, an inch or two thick, occur within the same dyke.

A brief examination of the Maps shows that the majority of the dykes have a north-west or N.N.W. direction ; that they are by no means straight lines, but show many rather sharp changes of direction, which sometimes continue for a considerable distance ; that the abundance of these dykes varies greatly in different parts of the area as we traverse it from north-east to south-west, across their general direction, but much less so as we cross it from north-west to south-east, the dykes occurring in groups which continue for long distances in the direction of their length. The part of Cowal north-east of St Catherine's and Carrick Castle (1-inch Map 37) is almost free from north-west dykes, but it contains east to west dykes in moderate abundance. The peninsula, west of the Kyles of Bute, is also comparatively free from north-west dykes.

The difference in the general directions of the dykes corresponds to a difference in age. The broad east to west dykes are the earliest, and there was at the time of their intrusion a tendency for the earth's crust to split in this area along east to west lines ; the other dykes were intruded at a later time when this splitting tendency had changed into a north-west direction.

The broad east to west dyke which forms the green hillock on which the ruins of Dunoon Castle stand, and which is exposed in a clear section on the shore immediately east of the castle, is older than any of the other dykes with a north-west direction with which it has been seen in contact. This is observable in sections, and, in other places where there are no sections, it can be inferred from the relations of the features made by the respective dykes.

Hand specimens of central portions of this dyke appear decidedly coarser than most of the later dykes, and the rock in hillside exposures weathers in unusually massive exfoliating blocks which disintegrate into a rich soil. Besides the coarseness of grain, field sections occasionally show other characters which are helps in identifying the dyke where it is not continuously traceable. Firstly, it is often crossed by thin red jasper-like strings, and secondly, thin hard bands of rather less coarse, and usually reddish, rock may occur in some abundance within the dyke, and confined to it. Such bands are exposed in the shore section at Dunoon ; they lie at various angles, sometimes nearly horizontally, sometimes nearly vertically, and at all intermediate angles ; they do not usually occur within seven or eight feet of the dyke side, but one was noticed to go quite to the side ; they are cut by various thin calcite and quartz strings, which also traverse the rest of the dyke ; their average breadth is perhaps from $\frac{1}{2}$ to 3 inches, and there is no difference in grain between the outer and the inner parts.

A specimen of one of these strings was examined chemically in the laboratory of the late Professor Dittmar of Anderson's College, Glasgow, and was compared with the normal dyke rock. The sample of the normal rock was obtained by mixing together two portions of the coarse dyke interior with one portion of the finer rock near the sides. The results obtained were as follows :—percentage of silica in the red bands, after drying the powdered rock at 100° C., 68·62 ; percentage of silica in the normal rock, after drying the powdered rock at 100° C.,—48·05.

In reference to similar strings in intrusive basic sills, Sir A. Geikie says ("Presidential Address," Geol. Soc. Lond., 1893, p. 179):—"There can hardly be any doubt that they represent the still fluid, and somewhat acid, parts of the mass which, owing to internal movements, were injected into portions that had already consolidated."

Similar reddish strings are seen in the section of the dyke $\frac{2}{3}$ mile south-west of Bodach Bochd (183 south-east), and attain a breadth of 9 inches. In the section on the east side of Loch Riddon there are several fine-grained strings near the interior of the dyke, which are probably of the same character essentially, though the colour is grey in place of red.

Microscopic slides were prepared from a junction specimen of the red bands (slide 2841*b*) with the darker rock (slide 2841*a*) in the locality $\frac{2}{3}$ mile south-west of Bodach Bochd, and are thus described by Dr Hatch, 2841*b* :—" Lath-shaped crystals of plagioclase felspar, more or less turbid by the separation of reddish-brown powdery material; this substance also occurs accumulated sometimes in, but more often around, the crystal, forming, in several instances, an almost opaque border. Penetrating the felspar are abundant microlites of green hornblende. This felspar must have a different composition to that in the greener rock, or else, why fresh in the one ,and weathered in the other? Patches of serpentine and calcite." 2841*a* :—" Diorite or perhaps epidiorite. Striped felspar (plagioclase) in fresh lath-shaped crystals. Bluish-green pleochroic hornblende (uralite) in fibrous masses, often with a fringe of fine needles of the same mineral, probably paramorphic after augite, of which, however, no trace is to be discerned. Scattered irregular grains of iron ore (ilmenite) and a small quantity of serpentinous matter." In the last slide there is a parallel arrangement of the constituents for a little breadth near the junction of the rocks, and this is perhaps due to slight mechanical movement that has taken place along the junction. The fact that there is no unaltered augite in the slide may be due to the same cause ; a slide of the dyke at Dunoon showed augite in some quantity.

In similar red strings in an east to west dyke which occurs near Lochgoilhead (142) Mr Teall considers that a large part of the felspar is probably orthoclase (slide 4200): the felspar forms a much larger percentage of the entire rock than in the normal dolerites : there is some quartz also.

Among the localities which show clear intersections of this Dunoon east to west dyke by later ones we may mention,—the south end of the crags $\frac{1}{2}$ mile south-east of Bodach Bochd, the section $\frac{2}{3}$ mile south-west of Bodach Bochd, and the burn $\frac{1}{4}$ mile west of Troustan (west side of Loch Striven). In all these localities the intersection is evident even at a little distance, for the older dyke weathers in its usual massive exfoliating and somewhat readily disintegrating blocks, while the later ones stick out through it as hard ribs, or are possessed of joints peculiar to themselves and with a definite relation to their margins. The later dykes in these sections consist either of porphyritic basalts with a very fine matrix, or of uniform-grained rock. In the former, the chief joints usually cross the dyke at right angles ; in the latter, the cross joints may be subordinate to others which run longitudinally. In all the sections mentioned, there are several of the later dykes to be seen, and in each, at least one of them shows tachylite selvages.

Besides this dyke, there are occasional short and thinner parallel dykes at no great distance from its side which we think are of the same age. The continuation of this dyke west of Loch Riddon can be followed at intervals nearly to Loch Fine (see p. 151).

Other dykes of the same kind also occur in the north part of the area : one passes 1 mile or so south of Carrick Castle, and runs W.S.W. through the head of Ardentinny glen and the south slopes of Beinn Mhòr : a second generally represented by 2 or more branches, runs between Lochgoilhead and Strachur : a third passes by the outlet of Loch Restil (134), and comes to the shore of Loch Fine 1 mile E.N.E. of Ard na Slaite : two more run by Achadunan (126), one on the south and the other on the north side of it. Details observed along the courses of these dykes are given elsewhere, pp. 155-160.

The later dykes with a general north-west or N.N.W. direction are certainly not all of exactly the same age : a good number of them are intersected by others, which by their joints or fineness of grain at the junction show that the earlier ones were thoroughly consolidated before the intrusion of the later. We cannot say that, among these later dykes, rocks of a certain character are older or younger than others of another character. This apparent want of distinctness of age, combined with their common direction, may indicate that the differences of age among them are not great.

One-quarter of a mile west of Troustan a porphyritic basalt running locally nearly east to west is cut by a more uniform-grained rock which runs north to south. On the other hand, 300 yards north-east of Tom Conchra (162 south-east), a porphyritic basalt intrudes into a uniform-grained dyke. In the burn 1 mile S.S.W. of Glenmassan (163 south-west), a thin fine-grained sharply-jointed dyke, running slightly south of west, cuts across a coarser dolerite running north-west. In the north to south burn east of An Leacann, about $1\frac{1}{4}$ mile south of Craigendaive (172 east), a dolerite of much the same character cuts through a fine compact aphanite-like rock. In a burn nearly $\frac{2}{3}$ mile north-east of Dun Mor (172 south-east), there are two dykes running locally east to west (the south one presenting a glassy margin to the other), and two north to south dykes : the west north to south dyke cuts the north-east to west dyke, and is distinctly cross-jointed while doing so, but it is not clear that it is later than the south east to west dyke. We do not regard the east to west dykes in this burn as being of the same age as those early east to west dykes we have just been speaking of : the east to west direction is only local, such as we often find in certain parts of the courses of many of the north-west dykes.

Each of the two neighbouring dykes $\frac{1}{2}$ mile W.S.W. of Cruach nan Capull (183) consists of two parallel bands : in the east dyke the west portion sends an irregular thin intrusion into the other portion, which is coarsely porphyritic.

Even without intersections it is often possible to prove that dykes are of different ages by finding their chilled margins in contact. Closely parallel dykes are common, particularly when they are running along some previously existing line of weakness, such as a fault line, the dykes having come up, first one and then another, along this line of weakness. The dykes first intruded have chilled against schist, or the crush breccia of the fault : the later ones partly against these rocks and partly against the earlier basalt. Good instances of such companion dykes occur in the following places,—the north to south burn $\frac{3}{4}$ mile south-east of Bodach Bochd, the depression in the hills $\frac{1}{4}$ mile west of Craigendaive, the thick band in the burn $\frac{2}{3}$ mile W.S.W. of Troustan (183), the thicker band in the burn $\frac{3}{4}$ mile S.S.W. of the top of Leacann nan Gall (183), the craggy band $\frac{1}{2}$ mile slightly east of north of Inverchaolain Church (184), and on the east shore of Loch Striven 200 yards south of Brackleymore School (194 north-west). The four first-named sets are certainly running with faults.

In the section south of Brackleymore School there are five parallel north-west dykes without any separations of other rock. The most westerly dyke is much decomposed, longitudinally banded, and about 7 feet thick. Next comes an even-surfaced harder basalt with sharply defined margins on either side, the west margin being glassy and the east margin approaching this character. The third band, about 9 feet thick, is like the one first described. The fourth, about 12 feet thick, is prominently porphyritic, and has sharp junctions on both sides. The fifth is like the first and third: it is 6 feet thick in one place, but is cut out by the porphyritic band as it goes south-east. The most south-westerly band (slide 2846), the harder band (slide 2847) next to this, and the porphyritic band (slide 2845) have been sliced, and are described by Dr Hatch as follows. 2846 :— "Basalt (would be called a dolerite by the Germans, i.e., a coarse-grained basalt), composed of plagioclase felspar in small crystals, giving lath-shaped sections, and grains of augite, allotriomorphic with regard to the felspar, but without a well-characterised ophitic structure. Separating these constituents by thin films or occurring in small wedge shaped masses is a dark-coloured glass. Olivine does not appear to be present." 2847 :— "Glassy Basalt. Consisting of plagioclase felspar and small grains of augite, irregularly distributed in an abundant glassy base. The glassy base is of a dark-brown colour and perfectly isotropic. It contains abundant spicules and trichites of magnetite which have united to produce the most fantastic skeleton forms, or have attached themselves by one end to the four sides of a microlite of felspar. Isolated microscopic fragments of limestone (effervescing with acid) appear to have been caught up by this rock while in the molten state. By the corrosive action of the magna they have been rounded to almost perfectly spherical forms. The disposition of the felspar microlites around the periphery of one of the bodies in question excludes the idea of their being secondary amygdules." 2845 :— "Porphyritic Olivine Dolerite. With the naked eye crystals of striped felspar and grains of green olivine can be made out. Under the microscope are to be seen : 1. Large porphyritic crystals of felspar (labradorite) of ideal freshness, showing polysynthetic twinning on the albite and pericline types : 2. Isolated grains of unaltered olivine. These porphyritic constituents are embedded in a microcrystalline ground-mass, made up of microlites of felspar, grains of olivine and augite, and granules of magnetite confusedly aggregated. The porphyritic felspars contain numerous inclusions apparently of a dark-coloured glassy material, arranged either centrally or in rows, marking zones of growth. The crystals are also somewhat corroded at the periphery, by the action of the magma when molten."

The influence of previously existing lines of weakness, faults, prominent joints, etc., in modifying the direction of both the early east to west dykes and the later north-west ones is extremely marked, and owing to this we find individuals of each of these classes assuming, for some distances, directions almost the same as those which normally belong to the other class. There may hence appear to be some confusion between the two classes, and it is not always possible to say to which class a particular dyke belongs. But where dykes can be traced for a distance, through several changes of direction, there should be no doubt as to which set they belong to : for in each set the thickness of the dykes is found to be appreciably less when they are proceeding along an abnormal path.

The course of the Dunoon Castle dyke between Loch Striven and Loch Riddon shows this fact well. No less than three times in this distance of about 4 miles it is distinctly seen to make sharp changes in direction, nearly at right angles to its ordinary course, and in all cases there is good evidence that there are north to south faults where the change occurs. We cannot say that the dyke is always continuous in the changed direction ; indeed, we think that in two of the cases it is not, but in all cases it is clear that the dyke exists, and in a state not markedly crushed, in a portion of the changed course, so that it is impossible to suppose that the change is due to a throw by a later fault. In all cases, too, the dyke is distinctly thinner in its changed course than in its normal east to west path : instead of 30 yards or more, breadths of 8 or 12 are now to be seen. In one case the dyke continues in its changed course for as much as ¼ mile. In each case the west portion of the normal path lies to the south of the east portion.

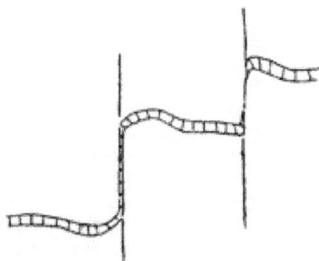

FIG. 37.—Diagram to illustrate the behaviour of dykes when coming to pre-existing fault lines which run almost at right angles to their usual direction. In this district it applies to east and west dykes meeting with north and south faults, and to north-west dykes meeting with north-east faults. Cross hatched bands are dykes. Single lines are faults.

The faults do not appear to be large ones, as far as we can judge from the few bands in the schists that are of value as horizons, but they make crush breccias of some thickness hading generally to the east. In the most westerly instance, where the change continues for ¼ mile, the fault probably shifts the outcrops of the schists about 60 yards. This is at all events the displacement seen along the same line of disturbance about 1 mile further north : the beds on the east side of the fault lie to the north of those on the west, which is as we should expect when beds dipping steeply south-east, as these, are cut by a fault hading east. We have never noticed the Dunoon Castle dyke to have any marked hade, and in the best sections it is practically vertical : and on the supposition of a later fault of vertical movement there should, of course, be no lateral shift of a vertical dyke. A later fault of lateral movement would, on the other hand, effect a lateral shift in a vertical dyke corresponding to its own movement. But the finding of the dyke in its normal condition along a portion of the changed path does, in itself, refute any explanation depending on the possible action of a later fault.

Cases like these naturally remind us of others where, in areas with less clear sections, dykes appear to be suddenly shifted for considerable distances, and it would seem likely that, as a general rule, if there is a fault at all along the apparent line of shift, that this is in part of earlier date than the dyke.

In places, e.g., in the north to south burn near Ardbeg (183 north-west), in the north to south burn south-east of Bodach Bochd, and in the burn east of Toward Hill (194 north-east) there are unusually broad bands of basalt running north to south, with directions different from the general direction of the dykes in the district. These broad bands are really compounded of dykes of different character and age, which oppose chilled margins to one another, and they are running along lines of fault, or of sets of faults. Sometimes we can see distinctly that some of the intrusions are later than some of the crushes, for unbroken tachylite margins occur close against crush breccias, but it is also common to find crushes breaking up the dykes to some extent, or slicken-lines within them, so that we must suppose there have been repeated movements along the same line of fracture, the first movements, before any intrusion, and these acted as lines of weakness for the dykes to come up along; and other movements again after the intrusions.

In the higher part of the burn near Ardbeg some of the basalt bands seem much more crushed than others, and are probably of a different age.

When N.N.W. or north-west dykes change into a north to south direction there can only be a difference in direction of from 22½° to 45°, and in these cases we have not noticed any marked alteration in breadth when diverted from their normal paths. There is, in all probability, some alteration, but it did not attract our attention when first surveying the area, and we have not had an opportunity of further examining this point. When the north-west dykes meet previously existing fault lines at an angle more nearly approaching a right angle, they act in just the same way as we have already seen the Dunoon Castle dyke to do: they are distinctly thinner than usual as long as they keep the abnormal path, or even cease to come up to the surface at all along it. A good instance of this on a small scale occurs on the east shore of Loch Riddon, about ¾ mile south of Springfield House: a locally running north to south dyke seems to stop at an east to west crush, and the first impression is that it has perhaps been thrown by the crush, but a little search reveals the dyke in a much thinner, but still perfectly sound condition, running east to west along the crush: so that it is clear that the crush is not later than the dyke, but earlier.

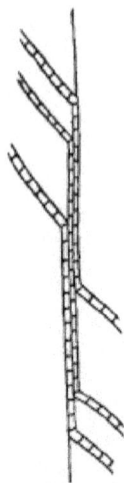

FIG. 38.—Diagram to illustrate the behaviour of the north-west dykes when meeting pre-existing north and south faults; and the consequent common occurrence of compound dykes along such faults. Cross hatched bands are dykes. Single lines are faults.

In the burn a little more than ⅓ mile N.N.E. of Knockdow (194 north-west) a north-west basalt about 4 feet thick comes from the south-east against an east to west fault or joint line, and then runs west along it: when going west it is only a few inches thick.

A burn section about ¾ mile south-west of Caol-ghleann (162 north-east) shows still more clearly the thinning of several N.N.W. dykes as they twist out of their usual course along a thin crush or carbonate of iron vein, which runs N.N.E. to S.S.W. In the most northerly of these the thickness when running N.N.W. varies from 1½ to 2 feet: the abnormal direction assumed is N.N.E., and it extends in this direction 3 yards: it either ceases to come to the surface at all for this distance, or is thinner. Another dyke, about 20 yards further up stream than this, has a thickness, when running normally on the south-east side of the stream, of about 4 feet, but when running along the crush, which it does for a distance of 20 yards in a N.N.E. direction, it is only 1½ or 2 feet thick: it is clear that the dyke is not crushed along its altered course. On the north-east side of the burn this dyke goes off in two parts: the first part that goes is, on resuming its normal direction, about 2 feet thick: a 6 inch part still keeps on in a N.N.E. direction, but after a little way suddenly turns at right angles, and assumes its normal direction again, attaining there a breadth of about 15 inches.

The broad north to south bands running with faults are not usually traceable for more than about a mile, for the dykes will not be diverted from their regular paths for more than a certain distance, and so the different dykes of which the band was compounded leave it one by one, on either side, to resume their normal directions. The occasional great thickness of these compounded bands depends on the size of the north-west dykes which the fault line has diverted into itself, and also on the closeness of the crowd of them : when the fault is traversing an area with less numerous north-west dykes, the band becomes thinner, or even ceases altogether.

The abnormal north to south directions of the east to west dykes have not been noticed to extend for more than $\frac{1}{4}$ mile. This is less than the north to south paths of the north-west dykes are often observed to be, and we may suppose that the greater the angle between the normal and abnormal paths the shorter are the abnormal paths likely to be.

In the Maps the dykes which are accompanied with crushes are marked with fault lines along their sides. But we are not able by this means to indicate whether the crushes are earlier or later than them, and so add here the particulars which have been observed about their relations in age. In the burn nearly 1 mile west of Glenmassan (163 south-west) a nearly horizontal crush crosses a north-west dyke, and perhaps throws it slightly : the crush is extremely thin and is not shown in the Map, but it is interesting as an example of horizontal crushes which are not uncommon in the district, and as an indication of the probable age of, at all events, some of them. The dyke that crosses the road slightly west of north of Ardtarag (173 south-west) has a 1 foot breccia on its side which contains basalt pieces, and it must therefore be earlier than part of the crush. Three-quarters of a mile E.N.E. of Ardtarag various north-east running crushes, hading south-east, go through a north-west basalt : one of them shifts its north-west side toward the north-east for 3 feet. A dyke, $\frac{5}{8}$ mile W.S.W. of Glenlean (173 south-west), has a hardened breccia rib within it, and there are parallel crushes on the outside also. There are thin uncrushed basalt strings within the north-east running crush $\frac{1}{4}$ mile east of Balliemore (173 south-west): frequently they run in the same direction as the crush. One-half mile slightly west of north of Ardbeg (174 north-west) there is a 3-foot basalt with reddened vein-stuff on either side which seems hardened by the dyke. Some of the breccias running with the basalts $\frac{5}{8}$ mile E.N.E. of Ardbeg (174 north-west) are altered by them. The two thin north to south dykes in the burn, nearly $\frac{1}{4}$ mile north-east of Cnoc à Mhadaidh (174 north-west) are in places quite full of slickensides, running parallel with them and the breccias on the sides : calcite strings also occasionally occur within the dykes. The most westerly of the two dykes $\frac{1}{3}$ mile east of Stronchullin (174 north-west) is quite crushed to pieces : the width is about 7 feet. The breccias close at the side of the basalt at the head of the burn nearly $\frac{1}{4}$ mile south of Blairmore Hill (174 north-west) contain pieces of decomposing basalt, and are therefore in part, at all events, later than the dyke. One-third mile N.N.W. of Blairmore Farm, there are two basalts, each from 2 to 4 feet thick, the east one of which is crossed by an east to west crush. On the west side of the north to south basalts 1 mile east of Feorlean (182 north-east) there is a hardened crush breccia on the west side of the most easterly basalt band : the basalt itself is not crushed. The dykes of basalt in the broad band in the burn $\frac{5}{8}$ mile S.S.W. of the top of Leacann nan Gall (183 north-east) are not crushed, but various breccias run between them and are presumably earlier than them. The compound basalt band $\frac{5}{8}$ mile W.S.W. of Troustan is crushed parallel to its length. The basalts in the burn $\frac{1}{2}$ mile north of

Glenstriven (183 north-west) are broken by crushes alongside of or within them. A crush in the burn $\frac{5}{8}$ mile north-east of Bishop's Seat (183 north-east) contains distinct basalt fragments. The north to south dyke in the burn $\frac{3}{4}$ mile south of Corlarach Hill (183 south-east) has altered vein-stuff at its side. In a burn $\frac{3}{4}$ mile south of Corlarach Hill, a W.N.W. basalt changes into a direction slightly north of east along crush lines : there are breccias on the sides, and also another, $1\frac{1}{2}$ foot thick, within the basalt : the sides of the basalt are chilled against the latter breccia, and, apparently, against the former also. In one of the north-west dykes in the burn nearly 1 mile south-west of Glen Morag (184 south-west) there is a breccia, several yards thick, of altered phyllite pieces, which sometimes attain a length of several inches : there is also basalt within the breccia, but this is often in large and irregular shapes, and may be parts of intrusive veins. The N.N.E. basalt in the burn $\frac{1}{2}$ mile east of Loch Loskin (184 north-west) is slightly thrown by a fault hading south. On the shore 240 yards N.N.W. of Dunoon Pier, a N.N.E. dyke, about 20 feet thick, has a rim, 2 to 3 feet thick, of hardened crush breccia sticking close to its north side, and there is another breccia at a distance of 2 feet. One hundred and fifty and two hundred and twenty yards north-west of Cluniter (195 north-east) several crushes cross a thick north-west dyke almost at right angles : the crushes hade south-east. The schist-boundary disturbances are not known to affect any basalt : near the east gates of Toward Policies two dykes occur which are clearly not crushed: 200 and 270 yards south of Inellau Pier others occur which are equally unaffected.

We have not noticed many instances of basalts running in a sill-like way, i.e., running roughly along bedding or foliation planes. Those that do occur have, we think, taken advantage, not so much of these planes directly, as of the thin crush lines which so often occur running nearly between these planes.

Near the west end of the east-west dyke $\frac{3}{4}$ mile south-west of Meall Buidhe (Arg. 173 south-east), a 2-foot sill-like band branches off in a south-west direction. One-half mile north-west of Clachaig (173 south-east) there is a basalt connection between two N.N.W. dykes : the connection is almost horizontal, and as the dip of the foliation is steeply south-east, the band cannot be along the foliation : hæmatite-stained strings run parallel to the lower edge of the sheet, and some thin basalt strings also, which have suffered from crushing. The most westerly dyke in the burn $\frac{3}{8}$ mile slightly west of north of Blairmore pier is crossed by a thin crush on the south side of the burn : the crush runs nearly with the foliation ; on the south side of the burn there is a basalt band 3 to 6 feet thick, which runs also east and west, and is traced from the east edge of the dyke to about 10 yards west of the west end. One-third mile south-west of Braingortan (183 south-west), a 1-foot basalt runs with an east to west crush which hades south : this band is only seen near a north-west dyke, so that we may conclude it is connected with it. The most westerly dyke in the burn a little more than one mile south-west of Glen Morag (184 south-west), has a thin band running from its east side for a little way along the foliation : it is crossed by a thin crush line. One-third mile N.N.E. of Corlarach (194 north-east) a north-west basalt, about 12 foot thick, and porphyritic with felspar, gives off a thin sheet on the south side.

The hades often vary in different dykes close to one another or even in the same dyke. We have not noticed one direction of hade to be more common than another. In the broad dyke $\frac{3}{4}$ mile north-west of A' Chruach (172 south-east) the hade is sometimes not more than 45° west.

Occasionally, impressions of the schist layers which have been intruded into are shown by the sides of the dykes, the sides showing prominent parallel lines dipping the same as the schists, *e.g.*, in the most southerly of the two burns ⅝ mile east of the Badd (184 north-west).

The indications of the age of the dykes which are furnished by this area, apart from the particulars of their relations already given, may be summarised in the following statements.

Throughout the schist area intersections of lamprophyres by north-west basalts are numerous (see p. 111), and it is also certain that the east to west basalts are later than the lamprophyres.

All the trachyte dykes that are known are later than the north-west basalts with which they have been seen in contact.

In the adjoining portion of Bute both north-west and east to west dykes cut through the rocks of Upper Old Red Sandstone Age : these dykes are of the same kind as those in Cowal, and in some cases clearly continuations of them.

The district was much broken by faults before the time of their intrusion, much more so than it was before the time of the lamprophyre intrusions. This remark applies to all the basalts, but in the most marked degree to the north-west ones. Notwithstanding this, faults and crush movements of later date than the later basalts are not uncommon.

The presence of more acid cotemporaneous crystalline strings has not been noticed except in the early east to west dykes. In them it is common ; they occur, *e.g.*, not only in many exposures of the Dunoon Castle dyke, but also in the two parallel dykes to the north of this in sheet 37. These strings seem closely akin to those found in many intrusive sheets of trap in the midland valley of Scotland, and it is suggested that the rocks in which they occur may be of the same age. The segregation veins in the Tertiary sheets do not seem to be of quite the same character : these (Sir A. Geikie " History of Volcanic Action," *Trans. Roy. Soc. Edin.*, vol. xxv., part 2, pp. 113–115) appear to differ from the rest of the rocks in which they occur, mainly, in the larger size and more definitely crystalline form of the constituents in them, which are the same as those in the inclosing rock. According to Sir A. Geikie ("Presidential Address," *Geol. Soc. Lond.*, 1892, pp. 140–143), the sheets in the midland valley of Scotland are probably of late Carboniferous Age.[*] If we suppose the early east and west dykes to be also of late Carboniferous Age, the interval of time between them and the north-west dykes must be immense ; for there can be no doubt that many of the latter belong to the same set as those which, still with a north-west direction, are seen in the island of Mull to intersect the bedded basalts of Tertiary Age. It is supposed by Sir A. Geikie ("History of Volcanic Action," *Trans. Roy. Soc. Edin.*, vol. xxv., part 2, p. 182) that some of the north-west basalts are of the same age as the bedded basalts, and the source of them ; but these also are tertiary.

The amount of denudation that has taken place even since the north-west dykes must be enormous.[†] If any of them flowed out at the surface, these outflows have now been entirely destroyed. If none did outflow, the existing glens must have been formed since their intrusion, for the dykes, which now mount up to the highest hilltops, would otherwise have overflowed into one part or another of these glens. In some places we can see the upward course of dykes stopping very near the present ground surface, *e.g.*, in the burn ⅝ mile S.S.W. of the top of Leacann nan Gall, where a dyke, some 20 yards lower down stream than the broad basalt

---

[*] It is probable that a sheet in the south part of the Isle of Bute, near Garroch Head, is also of this age.—W. G.

[†] *See* Sir A. Geikie, " Scenery of Scotland, 2nd edition (1887), p. 149.

band, is overlain by schist on the south side of the burn. No doubt many other instances of this kind occur, but it is unlikely that the dykes which cross over the high hills would all have stopped at low levels in the glens, just before reaching the then existing surface.

A basalt on the hill 1 mile W.N.W. of Glen Morag shows a very fine-grained surface sloping south-east, *i.e.*, nearly at right angles to the general direction of the dykes, and this surface may represent approximately an original surface of contact with the schists. A dyke in the burn ⅝ mile W.N.W. of Glen Morag is in the same line as the above exposure, but it has a flat top at its north-west end, and so is probably not connected with it at the surface.   C. T. C.

In some areas it appears as if the dykes considerably diminish in number as the elevation of the ground surface increases, and if the hilltops were clear of superficial deposits perhaps they would be found to contain fewer dykes than are revealed at the coast line.   There is a coast section between Newton Bay and Lachlan Bay (151), about 1½ mile in length, which contains no less than twenty-three dykes, some of them occurring along fault crushes. Most of these dykes pierce the outside edge of a great epidiorite mass.   About ½ mile from the coast the epidiorite forms a big ridge, the highest part of which is 570 feet. This ridge is for the most part quite bare, still retaining the smooth outlines that were left on it by the ice sheet. Along this ridge only three dykes have been observed. While this area illustrates an extreme case of the diminution of dykes as elevation increases, the adjoining tract from Lachlan Bay to Otter Ferry shows similar phenomena, but in a less marked degree. It is perhaps possible that the very compact nature of the epidiorite would itself lead to a diminution in the number of dykes which succeeded in piercing into its interior.   J. B. H.

The schists in the neighbourhood of the dykes are, for a distance of several feet, generally distinctly hardened, break less readily than usual along the foliation planes, and have not such a marked micaceous lustre. Many small specks of sulphide of iron are developed in them, and their lenticles of vein-quartz acquire many additional short lines of joint or crack, which form a distinct feature macroscopically.   The greater hardness of the altered schists frequently causes them to project beyond the general level of the other schists, and form more or less prominent ridges at the sides of the dykes : this is seen on the south side of the Castle Dyke in the Dunoon shore section.

The green or grey-green phyllites on the coast a little south of Dunoon are frequently changed to a dark grey or purple tint for a distance of several feet from the sides of the dykes. A good example occurs ⅛ mile south-east of Glen Morag.

In some places the phyllites and more micaceous schists have colour differences developed in thin bands, running generally with the foliation, which are not evident at a distance from the intrusion : some bands are rendered much darker than the normal tint, others an opaque grey or greenish-grey. This is seen in the altered schist lying between the north-west basalts a little more than 1 mile south-west of the top of Beinn Mhòr (163), and ⅔ mile east of Socach (141 south-east) close to the N.N.E. basalt, where it is running along an earlier fault. Under the microscope the rock of the first mentioned locality (slide 4737) no longer shows the original characters of the constituents, but appears made up of alternations of snow white or pale green opaque bands with other darker green almost opaque bands. The dark green bands appear black in the hand specimen: they sometimes traverse the pale bands, and in this case are bordered by a still darker rim, which contains a considerable amount of a black opaque substance with a metallic glitter. There is generally a good deal of the black opaque substance within the dark green areas, and but little of it in the paler.

In the albite mica schists the micaceous matrix is darkened and speckled with iron pyrites, while the albites acquire a reddish tint. This is seen at the north side of the east to west dyke 1¼ mile slightly west of south of Carrick Castle (specimen 4738), by the east to west dyke ⅓ mile W.S.W. of Drimsynie (142 south-west, slide 4734), by the east to west dyke on Beinn Tharsuinn (142 south-west, slide 4733), etc. Slide 4733 is thus described by Mr Teall : "The felspars (albite) make up the greater portion of the mass. The inclusions of magnetite have been more or less stained by oxidation, so that they appear red by reflected light and give a reddish tinge to the felspar as a whole. The albites are usually separated by a small quantity of brownish material, which evidently represents the original matrix in which the albites were formed and consisted largely of white and black micas with which chlorite was probably associated." Examination of the slide with a hand lens shows clearly that it is only the central parts of the albites that are reddish, the inclusions of magnetite having, as usual, been confined to these parts of the crystals. In slide 4738 there is much black opaque matter, partly with metallic glitter, lying in patches of different size between the mica flakes. These end abruptly against the albites. In slide 4734 the magnetite inclusions near the centres of the albites are also seen to be altered into ferrite. The hand specimen shows a thin pegmatite vein with felspar, and the felspar in this is the same colour as the albites of the rock mass.

The "green beds" often have their green colour changed to dark grey or black for a distance of a foot or more off the sides of the dykes. This is seen clearly in the following places,—near the east to west dyke ¼ mile west of Socach (141), ½ mile E.S.E. of the bridge at the head of Loch Fine (126), near the foot of Auchateggan burn (172 north-west), in a burn ¾ mile N.N.W. of Cnocan Sgeir (172 north-east), and in the burn ¾ mile slightly east of north of Bishop's Seat (183 north-east). A specimen from such a blackened "green bed" was obtained from the last locality, at a distance of 1 foot from the basalt, and was examined microscopically (slide 4736). Mr Teall describes it as follows : "Very dark, massive compact rock showing slight traces of banding. Under the microscope alternating coarse and fine bands— the latter containing a considerable amount of opaque material. Definitely recognisable constituents quartz and felspar. Brown mica appears to have been present, but the flakes have been modified so that the mineral can scarcely now be recognised. Micro-flaser structure. Original rock probably a very fine-grained felspathic grit or sandstone." Some parts of the opaque matter have a slight metallic glitter by reflected light. The parts of the slide that are snow white by reflected light are almost opaque to transmitted light. These differently coloured parts are no doubt the same in character as those in slide 4737.

In the section ¼ mile west of Socach the blackened green beds contain also a quartz vein with a black carbonate : the vein cannot be traced far from the dyke. On the shore 216 yards east of Glen Morag (184 south-west) the carbonate in a quartz vein 1 foot from a basalt dyke becomes a similar black colour. A string of black carbonate also occurs close to the side of a north-west basalt in Glen Finart, near the "e" of "Glen" (164 north-west). The veins referred to effervesce briskly with dilute hydrochloric acid. We do not remember any carbonate veins of this colour at a distance from intrusions.

A north-west basalt in the burn a little over $\frac{1}{3}$ mile W.S.W. of Glen Morag seems to slightly bleach the schistose limestone on its east side.

In a little burn $\frac{1}{8}$ mile N.N.E. of Glenmassan (163 south-west) a close jointing is set up in a lamprophyre in a direction parallel to the sides of a north-west basalt which intersects it. The jointing extends for a distance of several yards from the basalt.

The porphyritic basalt east of Toward Cottages alters the colour of the red sandstones on its west side for a distance of about 20 yards, changing it into a dirty yellow: the breadth of the dyke varies from 15 to 20 feet. A thinner N.N.W. dyke on the shore about $\frac{1}{8}$ mile west of this effects a similar change for a distance of about 6 feet.   C. T. C.

CHAPTER XIII.

THE IGNEOUS ROCKS (*continued*).—DESCRIPTION OF BASALTIC DYKES IN PARTICULAR DISTRICTS.

District of Tighnabruaich.

This district includes the part of Cowal which lies west of the lower part of Loch Riddon and the west Kyle of Bute, and south of a line that runs east to west between Eilean Dearg and Drum Point.

East and West Dykes.—The east and west dykes occurring in the most southerly part of Cowal at Ardlamont are not seen to be crossed by any running north and south. They are, however, almost certainly the continuation of the two large dykes which have been traced across the Island of Bute : one from Bogany Point to the south side of Ettrick Bay, and the other and largest, from Ascog Point to Watch Hill. The Bogany Point dyke is clearly cut by later tertiary dykes, running north and south, and there is no doubt that both these dykes are of a much earlier date than the Tertiary Period.

The most northerly of the Ardlamont dykes is 20 yards in width and appears on the shore at An Gnob, where it projects into the sea. It is generally a coarse dolerite but is finer at the edges where the schist in contact with it is considerably hardened. The dyke may be traced westward to the Fort at Camp Cottage beyond which it is hidden beneath beach-gravel and till. It is visible again west of Ardlamont House, somewhat to the north of its line of bearing, but soon disappears again under beach-gravel to come out again in the old sea cliff on the west side of Ardlamont Bay, some 200 yards south of where it might be expected. From this point it can be traced westward for $\frac{3}{4}$ of a mile to Rudha na Peilige where it is nearly 50 feet broad. It seems broader than this $\frac{1}{2}$ a mile to the eastward but it probably has a hade southward nearly coinciding with the slope of the bank, and so has a wider outcrop.

The large dyke at Ardlamont Point is also coarsely doleritic and is on the average about 50 yards wide towards its east end but widens westward in Port a Ghobhlan to 70 yards or more, where it ends off so abruptly as to suggest a fault running north and south. On the west side of the little bay we find it again but shifted 100 yards south of its course, and it is here also 70 yards wide. It takes a somewhat curved course westward, gradually thinning however, till on the other side of the peninsula where it meets the sea it is only 10 yards wide. About 50 yards to the south of this, there is another dyke running in the same direction 18 yards in width, and this is probably a branch from the main dyke. The schist between the two is much altered.

On the west side of Loch Riddon is a large dyke belonging to this class, being the continuation of the Dunoon dyke, which is on an average about 100 feet in width here. It can be traced almost uninterruptedly from the lochside westwards for nearly a mile, but is lost before reaching the Allt Dubh burn. Other dykes in the same general direction and probably of the same age can be traced for short distances on either side the main dyke. This is encountered again 300 yards west of Allt Dubh and $\frac{1}{4}$ mile south of its former course. It can be followed with interruptions, but somewhat diminished in width, to the head of the south branch of the Acharosson Burn, where it includes fragments of altered schist, and a branch may be observed going in a south-west direction for more than 100 yards. It now turns first north-west and then W.N.W. following generally the course of the burn for about a mile and being accompanied by veins of quartz and ferriferous carbonate. There are gaps, however, where it is not seen, and near the sharp northerly bend of the burn it is much thinner than usual, and the alteration it effects in the schist is very slight. We lose it for 600 yards and find it 300 yards south of where it might be expected, first going north-west and then westwards till it strikes the burn and disappears. Again we find it some 200 yards to the south, and trace it W.N.W. till it crosses the burn and is lost. This time we find it again 350 yards to the south and it runs westward for 600 yards to the burn. In this length it is 75 to 100 feet wide, a coarse dolerite much decomposed. At the burn opposite Acharosson it seems nearly 50 yards wide and contains veins of calcite and quartz. A little further west a dyke is seen in the burn, but it is much thinner, and again 400 yards south of the sharp bend in the burn there is a basalt dyke about 20 feet wide, ranging E.N.E. The dyke has been quarried near Drum, on the west side of the road, where it is 12 yards wide and amygdaloidal ; and beyond this no trace of it could be found. Nearly $\frac{1}{2}$ a mile to the northward of Drum a thin dyke ranging a little north of west crosses the limestone and epidiorite bands. This dyke seems to be either a branch of the Dunoon dyke or a parallel dyke of the same age.

Probable Tertiary Dykes.—The basalt dykes of probable Tertiary age which run approximately in a north-west direction occur very sparingly in the area west of the Kyles of Bute and Loch Riddon, except in the neighbourhood of Tighnabruaich and Callow. One of the more noticeable of these is shown on the Map on the west side of Tighnabruaich Burn or Easan Donn. In the southern part of its course it follows the burn and crosses it several times. It is about 20 feet wide and hades eastward 60°. In one part of its course it is a double dyke, enclosing a strip of schist, and at the top of the wood close to Dun Beag it divides into two parts which cannot further be traced. There is a large blue-grey dyke north of Callow which, after running north-westward, bends round in a N.N.E. direction and behaves as a sill dipping in a south-east direction nearly with the schist. There is a thinner and blue dyke to the west of this which starts in a N.N.W. direction nearly vertical, but in following it along its course we find that it eventually bends round nearly parallel to the preceding and behaves like it. W. G.

District of Kilfinan and Stralachlan.

This district is bounded on the east side by Newton Bay, Caol Ghleann, Glendaruel and the upper part of Loch Riddon. On the south side the boundary is a line running east and west through Eilean Dearg and Drum Point.

Basalt dykes of the north-west type are more or less abundant over the whole area; but if the district be divided into three divisions with boundaries running north-west and south-east, the dykes in the central division will be found to considerably outnumber those in the northern and southern divisions, in fact about one half of the dykes are found in the central portion. These zones of relative dyke abundance are persistent on the north-west shore of Loch Fine where this central zone has been traced as far as Loch Awe. There is also a diminution in the number of dykes on the more elevated parts of this area in consequence apparently of some of the dykes not having reached so far as the present ground surface. This is described in detail on p. 148.

The basalt dykes frequently come up along lines of fracture. In some cases the faulting has been continued after the irruption of the basalt. A good example of this is seen in the Lephinmore burn (162). A north and south fault runs along the burn, and there is a dolerite dyke parallel with the fault which has in part been brecciated along its length. The dolerite has evidently come up along the fault line and later movements along the same fault have crushed and brecciated the dolerite. The dolerite is coarsely crystalline throughout and contains clear glassy crystals of porphyritic felspar which by their rounded edges are evidently earlier than the injection of the dyke. Along the line of brecciated dolerite a fine-grained crypto-crystalline basalt about 4 feet wide has been intruded between the brecciated and the undisturbed portion of the dolerite. The later basalt does not contain the porphyritic felspars of the dolerite and its junction with the brecciated dolerite is clearly seen.

Another good instance of a basalt being affected by a later fault occurs at the western boundary of the serpentine of Glendaruel. At this locality the boundary between the serpentine and the mica schist is a fault which cuts a basalt dyke, and throws the basalt for the distance of several feet.

The courses of the burns are often determined by the presence of basalt dykes, and in these cases the gorges excavated are deeper and more precipitous than when the burns cut their channels out of the schists. Amongst the basalts themselves the coarsely crystalline varieties have yielded to erosive agencies more readily than the finer-grained varieties and have given rise consequently to the deepest gorges. At the Enacher (162) a tributary of the Lephinmore Burn a fine gorge has been eroded along the course of a massive dolerite. At the head of the gorge merely three small dykes occur, two of which can only be traced for a very short distance. The larger dyke gradually widens till it attains a width of 20 to 30 yards. The burn before the dyke is met with is quite an insignificant stream, but it deepens rapidly after meeting the dyke and runs in a deep precipitous gorge. After running for half a mile along the bottom of the gorge the dolerite is joined by another dyke to which it has been gradually converging. They run side by side for some distance separated only by a few feet of schist till they finally coalesce into a single dyke of about 40 yards in width. The dyke gradually decreases again to a breadth of 20 yards and a mile from where it is first seen at the head of the gorge disappears altogether, or is represented by a series of smaller parallel dykes separated by schist.

Another instance of the sudden deepening of a stream bed into a gorge, on the coincidence of a basalt dyke with the bed of the stream, may be seen in the valley of Stralachlan, within easy reach of the roadside, about ¼ mile from the junction of the Eas Dubh burn with the Stralachlan river at the village of Stralachlan.

Although the coarse dolerite dykes attain great breadths the finer-grained basalts may be still wider. The broadest of these latter are those which traverse the serpentine area of Glendaruel. Here we have a system of dykes extending over an area 1½ miles in length, which instead of following the usual rule of running along more or less parallel straight lines and then dying out, have a habit of deflecting from their course and coalescing with one another. After continuing as single dykes for some distance they again split up into branches, each of which may again subdivide, and one or other of them will either again meet and coalesce with another, or they may meet and coincide again with the main trunk. The whole somewhat resembles the assemblage of streams in a delta. While the individual dykes are irregular in their course, the system taken as a whole has a general north-westerly and south-easterly trend; corresponding to the general direction of the basalt dykes in the area. The different dykes of this system are fine-grained basalts with a strong tendency to split up into plates. The small branches are often about 3 or 4 yards in breadth, and the main trunk sometimes reaches a width of 30 to 40 yards or even more. Close at hand a dyke of similar composition is met with which belongs to the same phase of irruption. At its south-east portion near the serpentine two dykes are seen to approach one another and coalesce. At the watershed this dyke still remains as a single dyke, but on crossing the watershed it almost immediately splits up into two dykes running close alongside one another, the northerly dyke being about 50 yards, and the southerly dyke about 30 yards, in width. The ground here gets so covered with peat that it is difficult to say for certain what takes place, but it looks as if they coalesced again about ¼ mile further north-west—the entire width being then about 50 yards. For the next ¼ mile the peat is too thick to enable the dyke to be traced at all, but after that distance has been traversed, what appears to be the same dyke comes up again on the same course, but dwindled to a width of about 12 yards. From this place it increases gradually to 20, 30, and finally 40 yards, and then, just before entering Loch Fine, splits into 4 or 5 smaller parallel dykes separated from one another by a few yards of schist. In its habit of dying out, it corresponds to the coarse dolerite previously described; indeed, the two dykes die out in this way quite close to one another, so close that if they had been prolonged a little further they would have cut one another. These basalts have often produced great contact metamorphism on their passage through the serpentine, far more so than has usually been observed where they have pierced the sedimentary rocks and epidiorites of the district. Moreover, the serpentine in turn seems to have produced marked effects on the basalts. Serpentinous products seem to have penetrated into the joints of the basalts and have, as it were, partially serpentinised them.

Porphyritic dolerites and basalts are exceedingly common in the district. They contain porphyritic crystals of felspar which were evidently in existence at the time the dykes were injected into their present position. These porphyritic felspars are found in the glassy margins as well as in the more crystalline interior of the dykes. One of these porphyritic dolerites occurring on the hillside above Glendaruel House, was submitted to Mr Teall for microscopic examination. It was a dolerite with porphyritic crystals of a glassy felspar in a dark fine-grained matrix. Mr Teall describes it as follows:—"(4014) Microscope:—Porphyritic felspars allied to bytownite in a ground-mass composed of lath-shaped felspar, brown ophitic augite, small grains of olivine more or less serpentinised and magnetite."

Some of these porphyritic felspars exceed a square inch in size.

The finer-grained basalts show a great tendency, to split up into fine plates. In the bluish finely crystalline dykes, this platy structure is more often set up across the dyke than along it, though sometimes both types are met with in the same dyke. The tendency is to split parallel to one of the faces of the prismatic blocks that cross the dyke. The plates sometimes dip in opposite directions, owing to the tendency to split parallel to two faces of the prism. The structure is obviously due to an excessive development of prism faces in one or more directions. The fissile character prevails to its greatest extent at the edges of the dyke. If the dyke happens to be of larger size, such as 20 feet or 30 feet across, the inside plates are more in the nature of slabs. A good example of this structure may be seen in the Allt Buidhe, Lephinmore, Loch Fine.

In this district the spheroidal type of weathering has been met with in the coarser dolerites more than in the finer-grained basalts. The coarser dolerites often show massive prismatic jointing.

There is another type of dyke, with a glassy resinous lustre. This glassy type is the least common of the types described. In size these dykes approach the biggest dolerite dykes of the district, and can be traced continuously for as great a distance, and have, like them, a north-west direction.

The following are the localities of the principal dykes of this class.

1 mile south of Glendaruel House (172), and the course of the stream to north-west for ¾ mile. When the dyke leaves the burn it continues in a north-westerly direction slightly south of Cruach Chuilceachan and is met with along the same course for a mile from that hill. The place where it is finally lost sight of is 2¼ miles from where it emerges from the alluvium of the Glendaruel Valley to the south-east.

One-third of a mile north of Lephinchapel (162). The dyke occurs on the coast and can be traced up the hillside in a south-easterly direction for about ½ mile.

Half a mile north-west of Lower Duillater, Glendaruel (162). This dyke can only be followed for a short distance on the hillside.

1 mile south-west of Tom Soilleir (162). A dyke about 20 yards in breadth cuts the Lephinmore burn; after following a south-easterly course for about a mile it is lost.

These glassy dykes have already been treated of in the general description. Next to their glassy or resinous lustre, their most striking characteristic is their tendency to split up into more or less rounded or rhomboidal segments separated by divisions of lighter-coloured material. The divisional planes between the segments vary from ⅛ inch to ¾ inch across. A series of divisional planes transverse to the dyke appears to be accompanied by another set parallel to the dyke which gives a reticulated appearance to the mass. When one set predominates over the other set a kind of banded structure is set up, consequent on the difference in character of the main mass from the paler divisional planes. Sometimes the reticulated and banded conditions are seen in the same dyke. Near Cruach Chuilceachan (172), the dyke is found in a more or less devitrified condition, with tiny crystals of porphyritic felspar. It assumes the curious reticulated appearance already described, its surface being divided into innumerable segments separated by partings of paler material. The segments are sometimes perfectly rectilineal, the larger exhibit a tendency to subdivide into smaller segments.

Besides this structure, the dyke appears to be divided into longitudinal bands about 1 foot in thickness of dark-black and yellowish-white colours, which alternate with one another. Some of these inside stripes are quite glassy. The dyke assumes in weathering both a reticulated, and a longitudinal banded appearance. Sometimes these glassy portions of the dykes occupy one-half the width of the dyke, the outer parts being quite crystalline, and sharply divided from the glassy variety, so that first sight suggests an intrusion of one dyke within another.

A typical specimen of the glassy dyke was examined microscopically by Mr Teall, who describes it as follows :—" 4016. Black rock with pronounced resinous lustre, similar in macroscopic appearance to some of the Cheviot hypersthene-andesites. Microscope—The porphyritic constituents include plagioclase (probably labradorite) and a pyroxene. The plagioclase is much honeycombed with inclusions of the ground-mass. The pyroxene possesses the pleochroism of hypersthene, but as there are only two or three sections, and these are not cut in favourable directions, it is impossible to determine the nature of the mineral with certainty. The ground-mass consists of long microlites of augite, microlites of felspar and much pale-brown glass containing minute opaque (?) hair-like bodies and grains. Pyroxene-andesite."

The dykes do not always preserve a uniform increase of coarseness in passing from the outer edges to the interior portions. Occasionally, coarser and finer material may be seen alternating in the same dyke. A good instance of this is seen in a burn section ½ mile south-west of Balloch-andrain, Glendaruel, within 200 yards of the roadside. This dyke shows good longitudinal jointing.

Original upward terminations of dykes are but rarely met with. A good instance, however, of the roof of the dyke passing under the neighbouring rock, occurs in the Lephinchapel Burn, rather more than ½ mile S.S.E. of Lephinchapel, Loch Fine. Here, in the deep gorge, a dolerite dyke can be clearly seen passing underneath a quartz schist.

The upward passage of a dyke through the schist is often irregular, and frequently accompanied by small side-veins of basalt which pierce the adjoining rock. A good example of the uneven passage of a dyke may be seen in the Barnacarry Burn, Stralachlan. A dyke 2 or 3 feet in width passes through a mass of hard quartzites alternating with micaceous beds. The junction of the harder quartzites and softer micaceous beds has served for planes along which faulting has been produced, and the dyke on meeting one of these planes has been deflected, and runs on for some little distance as a sill before continuing its upward course.

The general trend of the dykes is north-west and south-east. When they take other directions they do not as a rule continue for any great distance, except perhaps between Ormidale and Upper Callow (Loch Riddon) where a north and south direction is most prevalent.

The older type of east and west dykes does not occur in this area.

J. B. H.

District of Strachur and Lochgoilhead.

This district includes all that part of Cowal which is contained in 1 inch Map 37, with the exception of the area which lies west of Newton Bay, Caol Ghleann and Glendaruel.

The Carrick Dyke.—The W.S.W. dyke that passes a mile south of Carrick Castle is of the same character, and no doubt also approximately of the same age as the Dunoon Castle dyke (p. 139). Just west of the peat moss 1½ mile slightly west of south of Carrick Castle, it contains some thin reddish strings of apparently the same character as those

described in the Dunoon dyke. The main dyke probably lies a little
seawards of the point between Loch Goil and Loch Long, but there is a
thin dyke, from 2 to 6 feet thick only, which runs through the point
with the same direction as the main dyke. The locality west of the peat
moss just mentioned is also of interest from showing the contact alteration
induced in the albite schists on the north side of the dyke: the albites
are reddened, and the rock loses its micaceous lustre to a large extent,
and the fissile tendency along the foliation planes.

One-sixth of a mile east of the "n" of "Cruach an Draghair," the south
side of the dyke runs north and south for 10 yards along a thin crush line:
the dyke itself does not seem crushed: we suppose it to be running out of it's
normal course, to take advantage of an earlier line of weakness, p. 142.
A similar change of path is inferred just on the west side of the Cruach:
the dyke feature is quite distinct as it crosses the top of the Cruach, and
for some distance down it's west side, until it comes to a north to west
fault feature: the dyke certainly does not continue on the west side of
this feature in the straight continuation of it's old path, and it must either
cease to exist at the surface for a little distance, or run along the feature
in a thinner form.

In the burn 200 yards slightly west of south of the top of Larach Hill
(164 north-west) the dyke is cut by a N.N.W. dyke from 1 to 3 yards
wide. The section can only be examined when the burn is low. The
N.N.W. dyke occupies the stream bed. There are thin strings of vein
stuff and hæmatite running parallel with the N.N.W. dyke, and crossing
the W.S.W. dyke, but they do not affect the N.N.W. dyke. The later
dyke is interesting from the clear way in which the felspars are arranged
with their thin plate-like forms parallel to the sides of the dyke. Slide
2844 from this later dyke is thus described by Dr Hatch. "Porphyritic
olivine dolerite. Felspars of two generations. The large porphyritic
felspar gives striped lath-shaped sections, extinguishing with a large
angle, and is therefore a lime-soda felspar near the anorthite end of the
series, probably labradorite. Beautiful pseudomorphs of serpentine
after large well-formed crystals of olivine (of pyramidal habit). Other
little patches of serpentine with calcite probably represent altered augite.
The porphyritic constituents are embedded in a plexus of minute
microlites of felspar, interspersed with granules of magnetite and perhaps
augite, separated by thin films of dark-coloured glass." A slide of the
earlier dyke (2843) from the same locality, is described as follows.
"Dolerite. Plagioclase in fresh crystals giving lath shaped sections.
Augite in irregular grains, allotriomorphic with regard to the felspar.
The crystals sometimes show twin lamellation. Magnetite in large
skeleton patches and grains. Apatite in slender needles. Secondary
minerals are calcite and patches of serpentinous material containing
blades of brown mica. The presence of brown mica suggests contact
metamorphism (see "Allport's researches on the Metamorphosed Dolerites
in Cornwall and Devon)." Wedged in between the crystalline constituents is a little brown interstitial matter (glass) with dendritic
devitrification products." The mica might have been attributed to
contact alteration induced by the N.N.W. dyke, but for the fact that
such flakes are not uncommon in the dyke in other places also, some
distance off later dykes, e.g., $\frac{1}{5}$ mile south-east of the first "r" of
"Garrachra Glen" (163 west). In slide 4200 (see p. 157) from the
Lochgoilhead east to west dyke, biotite also occurs in small quantity.

Half a mile south-east of the top of Beinn Mhòr (163 north-east) a
north-west dyke cuts the W.S.W. one: the actual junction of the two is
not visible, but the difference between them is shown distinctly by the

main joints, which in each dyke are approximately at right angles to the sides.

Only the north end of the change of path into a north to south direction on the east side of Garrachra Glen is seen. The section here is complicated by the presence of 2 or 3 fine-grained yellow weathering north to south lamprophyre bands, and a thin-crushed N.N.W. basalt: an east to west fault seems to throw the lamprophyres: it seems possible to infer that the broad basalt cuts the lamprophyres.

The most westerly place at which this dyke is distinctly seen is the burn section about ½ mile south-west of the "D" of "Meall Dubh" (163 south-west): it is in two parts, the north, and wider one of the two, crosses the burn 60 yards north of the other.

The Lochgoilhead Dyke.—The Lochgoilhead east to west dyke is seen well in and near the burn above Donich Lodge, from the "h" of "Donich" for a distance of rather more than ½ mile west. One-twelfth of a mile west of the "h" the dyke hades north, and there are thin crushes with barytes strings running parallel at the side. In the little burn nearly ¼ mile E.N.E. of the same letter a strong crush, also hading north, runs parallel to the dyke at a distance of 20 yards on the north side. The dyke itself has calcite strings within it, and a branch going off in a south-west direction has a crush in the middle: this crush hades south-east. The albite schists near the dyke are much hardened and altered: the albites themselves are reddened (see p. 149), while the rest of the rock is darkened and pyritised. Three-eighths of a mile east of the "h" in a little stream on the north side of the burn, the dyke is represented by two parts on the east side of a north to south fault line. The south branch is not quite in the straight continuation of the dyke on the west side of the fault, but some few yards south of it. There are several thin inconstant strings of basalt running with the fault, and these connect the north and south bands. These strings vary from a few inches to a few feet in thickness: the total thickness of them all does not exceed 4 or 5 feet. They are not crushed more than the dyke generally is in other places, and in one place one of them is seen to be distinctly chilled against the altered schists at its side. After going along this fault for about 30 yards, these north to south strings change their direction and widen out to form the north branch of the dyke.

Near the middle of the dyke on the west side of the fault, there is an 8-foot band of reddish colour: it is somewhat sharply defined from the blacker rock at the sides, but has no chilled margin and no joints separate from those of the rest of the dyke. The south branch on the east side of the fault shows the same reddish band for a certain distance east of the fault, but at a place nearly ¼ mile east of the fault, we could not find it, though there is a good section on the north side of the burn. This red band is of the same type as the red more acid strings in the Dunoon Castle dyke (see p. 139). The microscopic character of a slide (4200), from ¾ mile east of Donich Lodge, is described by Mr Teall as follows: "A dark coarse-grained rock containing a considerable amount of pinkish felspathic material. Rocks of this character are often found associated with the coarser carboniferous dolerites. The minerals are often, as in this case, in such an imperfect state of preservation as not to admit of very satisfactory determination. Felspar, more or less decomposed and stained with ferrite, forms a much larger percentage of the entire rock than in the normal dolerites. It is probably in large part orthoclase. In addition to the felspar, we find in this case augite and green alteration products, magnetite (fairly abundant), and pyrite; also quartz." The normal dyke rock (slide 4199), is described by Mr Teall as follows : "Dark massive, medium-grained rock, essentially composed of plagioclase, augite, and iron ores. The plagioclase occurs in forms giving lath-shaped sections. They are striated and the extinctions are high, indicating labradorite or an allied species. The augite is pale brown in thin sections and ophitic, though not in large masses. Diallagic striation may sometimes be seen. The iron ore occurs in irregular plates, often very ragged. They are sometimes seen to be moulded on the felspar. A small quantity of very fine micropegmatite occurs in places as interstitial matter. The accessory minerals include apatite, pyrite, and some extremely thin hexagonal plates of a brown colour (titaniferous iron mica). Dolerite." There seems also to be a little biotite present.

West of Lochgoilhead there are two main dykes running roughly parallel to one another at distances of from $\frac{1}{8}$ to $\frac{3}{4}$ mile, and each of these, but particularly the north one, may subdivide into branches. The altered albite schist near the south branch $\frac{1}{2}$ mile west of Drimsynie is exposed in a section on the hillside : slide 4734 from this locality is referred to on p. 149. The supposed change of path into a S.S.W. fault 1 mile slightly south of west of Drimsynie is only inferred : the dyke is seen in a burn 130 yards west of the fault, and also on the hill close on the east side of the fault, but the intervening ground is drift covered.

The north dyke is represented by three branches in the burn N.N.W. of Drimsynie, but the most northerly of these is only thin. The north and the middle branches are seen to be crushed by the fault running N.N.W. in the burn : the north branch is not seen close on the west side of the fault, but the middle branch is—it comes on a little south of its exposure on the east side of the fault. The south branch is crushed by the same fault in an exposure a little west of the burn ; the exposure west of the crush lies a little south of the exposure east of it.

The microscopic character of the altered albite schist near the dyke on Beinn Tharsuinn is described on p. 149.

The southerly dyke on the south side of Beinn Tharsuinn seems to be accompanied in places with a downthrow to the north of about 30 feet.

In the burn called Liogau, $\frac{1}{2}$ mile south-east of the "n" of "Liogan," a 1-foot basalt dyke lies about 20 or 30 yards north of, and parallel to, the dyke : this thin band is seen to end as it goes west.

There is a good section of the north dyke in the burn $\frac{1}{4}$ mile slightly south of west of Socach. The dyke is, as usual, divided into blocky cross prisms which run at right angles to its sides. Just on the south side of, and sometimes touching, it, there is a thin greenish dyke, apparently a camptonite. The "green beds" on the north side of the dyke are turned quite black (see p. 149). There are only two exposures of the dyke west of the above section on the east side of Loch Fine : one is in the burn $\frac{1}{2}$ mile E.N.E. of the "u" of "Strachur," and the other in the burn $\frac{1}{4}$ mile N.N.E. of the "S" of the same word : the hillsides near these sections are thickly covered with drift, and the low ground west of Strachur is composed of raised beach and alluvial material.

The south branch, as already mentioned, p. 111, is seen to cut a mica-trap sheet in the river Cur, $\frac{1}{4}$ mile E. of Strachurmore. In the burn $\frac{3}{8}$ mile slightly south of west of the "S" of "Strachur" it cuts another lamprophyre band running north and south. On the shore of Loch Fine it is exposed about $\frac{1}{4}$ mile slightly east of south of the "B" of "Strachur Bay."

The Loch Restil Dyke.—The Loch Restil east to west basalt is represented by two parallel bands, the south one the wider, near the little stream 160 yards west of the outlet of the loch : there is a crush between the bands. The dyke makes a conspicuous gulley in the steep crags on the north-west side of Beinn an Lochain : at the sides of this, $\frac{1}{2}$ mile west of the loch outlet, there is evidence (see p. 111) that the basalt is later than a lamprophyre dyke at its sides. Another gulley is formed by it in the crags at the south-west side of a landslip $\frac{1}{4}$ mile north-west of the " B " of " Beinn Lochain." A little west of these crags it is accompanied by a basalt dyke a few feet thick, which runs parallel to its north side at a distance of about 16 yards.

A sudden change of path for a distance of 40 or 50 yards along a north to south fault, with ferriferous carbonate strings, seems to occur $\frac{1}{4}$ mile north-west of the " B " of " Beinn an t-Seilich," but the basalt is not actually seen in a north and south course, and possibly it ceases to come up to the surface. One third of a mile W.N.W. of the same letter there is a good section of the dyke in and on the west side of the middle burn shown in the Map : two lamprophyres, one a coarse rock with much black mica, are seen on either side of it only a little distance off ; they do not cross the basalt, and are therefore inferred to be crossed by it.

In the burn $\frac{1}{4}$ mile south-east of the " k " of " Ardkinglas Ho," the dyke is split into blocky cross prisms : the hade of the dyke is south-west, and therefore the sides of the prisms incline somewhat north-east : there is a thin crush at the north-east side and a thin parallel dyke. In the burn $\frac{3}{4}$ mile south of the " k " of " Ardkinglas " the dyke is unusually thin,—not more than 6 or 7 feet,—and perhaps it does not exist at the surface at all in some places. In the burn $\frac{3}{4}$ mile south of the " A " of " Ardkinglas," it is not less than 10 or 11 feet broad, and there may be more than this as the south side of the dyke is not seen. On the shore, 1 mile E.N.E. of Ard na Slaite (133), the dyke is not less than 14 yards wide. In a slide (4756) of the rock from the burn nearly $\frac{3}{4}$ mile south of Bathaich ban Cottage, no olivine was detected.

Other East and West Dykes.—A dyke, seemingly a basalt, is seen along the line of an east to west fault, nearly $\frac{1}{3}$ mile south of the "oi" of " Upper Clasheoin " (126). It is only seen for a few yards, and perhaps does not belong to the set of early east to west dykes.

The east to west basalt south of Achadunan is seen at the east side of the raised beach $\frac{1}{6}$ mile slightly east of south of the bridge at the head of Loch Fine. One-sixth of a mile further east the dyke is much better seen : it is 16 yards wide and marked with blocky cross joints. The colour of the "green beds" near it is changed from green to black (see p. 149). One-eighth of a mile south of the "r" of "Garbh Allt Mor," there is an 8-foot branch on the south side, which is only seen in the burn sections close by. There are two branches in the course of a big north to south burn, not shown in the Map, $\frac{1}{4}$ mile south of the " R " of " KILMORICH " (parish): which of these it is that continues east is uncertain : the south one is the widest, and is seen in another burn 100 yards east of the burn just mentioned, while the north branch is not seen. The dyke is seen in all the burns east of this which are shown in the 1-inch Map, but its path between these burns is drift covered.

The basalt dyke north of Achadunan is finer-grained and thinner than the others described, and seems in places, *e.g.*, in the burn $\frac{1}{4}$ mile north of the " d " of " Achadunan," to contain macroscopic olivine. In the other east to west basalt dykes of the Dunoon Castle dyke type, no olivine has in Cowal been determined. There is nothing to show the relation in age of this dyke to the north-west dykes. It has a direction somewhat north of west, and perhaps it should be classed with the later north-west dykes. In the burn $\frac{1}{8}$ north of Achadunan it is 6 feet thick. In Eas Riachain it is in two or three different bands, each cross jointed and

about 5 feet thick. It is seen on the hill on the east side of the "g" of
"Lochan Mill Bhig," and is drawn in the Map from there through an
obscure morainic area in an E.S.E. direction for ½ mile, to join an exposure
in a little burn, not marked in the Map, ⅓ mile S.S.W. of the "M" of
"Meall Beag."

North-west and N.N.W. Dykes.—The north-west or N.N.W. basalts
are extremely abundant in the south-west part of the area, but rare north-
east of the line joining Carrick Castle and St Catherine's. In fact only
one is known that runs in this direction on the north-east of this line :
this is the dyke that crosses Glen Kinglas near the "g" of Kinglas." It
is seen along the south-west margin of a landslip on the south side of
the water, with a thickness of 5 or 6 feet and hading south-west. The
section in Kinglas Water does not show it, but it appears again in the
different burns on the north side of the alluvium : in these a crush rock
runs along with and is altered by it. In the burn ⅜ mile S.S.W. of the
"B" of "Binnein an Fhidleir," it makes a slight change of direction along
an early N.N.E. fault line, and in the changed path is thinner than usual.

Along the course of the N.N.E. fault a little east of Cruach nan Capull
(133) there is a basalt, of perhaps the same age as the north-west dykes,
that is taking advantage of an earlier line of fault as a ready direction of
intrusion, but we cannot be certain that this may not be connected with
the early east to west dykes, like the north and south strings connected
with the east to west dyke in Donich Burn near Lochgoilhead. That
the crush rock and vein stuff along this line are earlier than the basalt is
shown conclusively in several places, *e.g.*, ½ mile E.N.E., and a little
more than ⅛ mile east, of the "b" of "Socach" (241), and in the burn
by the last "o" of "Coire No" (133), by the great amount of alteration
it has undergone, while the basalt is uncrushed. The exposures of
basalt along the line are generally isolated by considerable stretches of
drift-covered country, and it is not certain that they are all connected.
South-west of the Coire No, all the clear sections of the fault show also
basalt accompanying it until we come within less than ½ mile of Cruach
Bhuidhe (152 north-east); and further south-west still, various north-west
basalts can be seen to shift out of their ordinary course, and run along
the fault for little distances. The dykes are often in several parallel
parts, *e.g.*, in the section ⅓ mile east of the "h" of "Socach" there is a
band 13 feet thick on the west side, then another 2 feet thick, and further
east still a third, a few inches thick. In the burn section ¾ mile slightly
east of south of the "r" of Cruach nan Capull the dyke is in 2 bands :
the west one is 8 feet thick.

The N.N.E. faults in the area where north-west dykes are abundant,
are, as we should expect, much more commonly accompanied by basalts than
they are in the area where these dykes are scarce. We will mention a few
sections which show the relation in age between the dyke and the fault
rock. One quarter of a mile south-west of the top of Meall an-T (163 north-
east), the vein stuff on the west of the N.N.E basalt is distinctly hardened
and altered : a band of basalt can be traced running along this fault for
5 miles, partly in 1-inch Map 37, and partly in 1-inch Map 29 : there
is a parallel fault, also accompanied with basalt, near the north end of the
line.

In the burn ½ mile N.N.W. of the "M" of "Maol Odhar" (162 north-
east) there are 2 basalt dykes which change their course along a thin
N.N.E. crush. In both cases the south-east part of the dyke advances
north-east along the crush—not south-west; the dykes in their N.N.E.
courses are quite uncrushed and sound, and thinner, see p. 144, than
when running north-west. The most south one is in its normal path 7

or 8 feet wide ; in its changed path it is at first $3\frac{1}{2}$ feet, but gradually diminishes to $1\frac{1}{2}$ feet, and finally ceases to appear at all.   We did not see either of these dykes resuming their old direction on the north-west side of the burn.

On the shore $\frac{1}{4}$ mile north-east of the " L " of " Leak " (151 north-east) a basalt and a fine-grained lamprophyre dyke run together in a N.N.E. direction.   The basalt is chilled against a vein of ferriferous carbonate.

On the shore 100 yards W.S.W. of Mid Letter (141) a 6-foot basalt changes into a north-east direction on coming to the side of a north-east lamprophyre band.   The lamprophyre is rather crushed in a direction parallel to its length, while the basalt is not so.   The basalt dwindles to 3 feet in width while it is running north-east ; it cuts the lamprophyre just before resuming the old north-west direction.   The same dyke contains chlorophæite amygdules (see p. 130).

The N.N.E. dyke in Caol Ghleann (152) is accompanied with parallel crushes.

On the south side of the road nearly 200 yards above Leanach (151) there are several basalts near together, running nearly north and south, and with parallel crushes.   The most west band, itself a compound one, is crushed, but the one next to this, a hard cross-jointed dyke which has been quarried, does not show distinct crushing ; yet it would seem that there must be a fault with it, for the graphite schist at its side is striking against beds which come below the graphite schist in the burn $\frac{1}{4}$ mile north-east of Leanach.

An altered crush rock occurs in the burn $\frac{1}{8}$ mile S.S.E. of the " S " of Stroudavain " (162).   It is prettily veined with red and green strings, and has a pronounced splintery fracture.   The north-west basalts which intersect it were not noticed to be crushed themselves.   Vein stuffs altered by north-west basalts have also been noticed on the shore $\frac{1}{4}$ mile south-east of the " C " of " M'Phun's Cairn " (141 north-west), and by the north-west basalt in Glen Kinglas (see p. 160).

Sections which show north-west basalts themselves considerably affected by crushing parallel to their sides, occur in the following places,—$\frac{1}{8}$ mile W.S.W. of the " B " of " Ballimore " (141 south-west), $\frac{1}{2}$ mile north-west of the " C " of " Cruach Bhuidhe " (152), $\frac{1}{2}$ mile north-east of the " a " of " Meall Dubh " (163 north-west), and $\frac{1}{4}$ mile N.N.E. of the " C " of " Creag Tharsuinn " (163 north).

Intrusions of basalts through lamprophyres are seen on the shore $\frac{1}{4}$ mile N.N.W. of the " L " of " Leak " (151), and $\frac{1}{3}$ mile west of the " S " of " Sgor Coinnich " (153).   The relations of the two sets of rocks can be inferred with confidence in many other cases, from the way the hill features—gulleys, &c.,—formed by the basalts, cross the features made by the lamprophyre sheets.

The " cracked " state of the quartz veins of the schists, which is due to contact alteration by the basalts (p. 148), is particularly well seen in a burn $\frac{1}{8}$ mile W.S.W. of the " B " of " Beinn Ruadh " (164 south-east), in a strip of schist 2 feet wide which occurs in the middle of one of a close set of parallel or bifurcating dykes.

The basalt band in the burn $\frac{1}{4}$ mile N.N.E. of the " C " of " Conchra " (162 south-east) is composed of three different dykes, all running in the same direction, and all hading slightly west.   The east part consists of an uniform medium-grained rock with a 4 inch intrusion of porphyritic basalt with fine-grained matrix.   The medium grained rock is not chilled against the porphyritic rock, and therefore the latter must have been intruded into the former.   The west part is of porphyritic basalt much like

L

the porphyritic string just mentioned, but both this and the medium-grained rock oppose chilled margins to one another; hence we suppose the medium-grained rock was first intruded and became chilled against the schists, and then the porphyritic rock subsequently came up along the chilled edge of the other, and became chilled against it in turn.

Amygdules of a white radiating zeolite occur in the dykes in the following places,—$\frac{1}{4}$ mile north-east of Tigh na Criche (141 south-west, slide 5435), 1 mile and a little more than $\frac{1}{2}$ mile south-west of the same place, by the "a" of "Clach Beinn" (163), in the burn $\frac{1}{4}$ mile below the "S" of "Strath nan Lub" (163 south-west), and elsewhere.

A quarter of a mile north-east of the "D" of "Dunans" (162 north-east) a flattish basalt surface crosses the burn in a north-west direction. This shows many irregular dark fine-grained strings, from $\frac{1}{8}$ to 1 inch in breadth, which run parallel to the direction of the surface. The strings project slightly on the rock face. There are some porphyritic crystals of felspar in the dyke, and these are in places crossed by the fine strings. The parallel strings are in some parts exceedingly numerous, and also joined diagonally by other strings of the same appearance, so that the fine-grained matter may almost equal the normal rock in bulk. Besides the dark fine-grained lines there are other yellowish-red lines, which also run in much the same direction. These last probably represent thin crush lines, and perhaps the other dark strings do also. Two feet from the north-east side of the surface, on the south-east side of the burn, is a bit of altered schist that lies horizontally on the flat surface of the basalt. Five feet further north-west of this side of the burn, and about 5 feet from the south-west side of the surface, another flat strip of schist is seen adhering to the top of the basalt exposure: and in various other places it can be seen that the basalt, apart from the fine-grained strings just mentioned, is much finer in grain at the top of the surface than it is a little below, so that it would seem that all the top must be close to an original cooling surface. Both banks of the burn are drift covered.

In the burn $\frac{1}{4}$ mile N.N.E. of the "T" of "Creag Tharsuinn" (163 north-west) a north-west basalt seems either not to exist at the surface for a space of 20 yards along its general direction, or to be very much thinner than at either side of this space.

Tachylite selvages have not been noticed so abundantly as in 1-inch Map 29. Instances occur on the shore of Loch Fine, 250 yards N.N.E. of M'Phun's Cairn (slide 5437 belongs to this dyke, but not to the tachylite portion of it), and $\frac{1}{4}$ mile south-east of the "C" of the same "Cairn." These dykes are also noteworthy from containing olivine.

C.T.C.

CHAPTER XIV.

THE IGNEOUS ROCKS (continued).—INTRUSIVE ANDESITE (?). THE SITH AN T-SLUAIN DOLERITE. TRACHYTIC DYKES.

The rocks described in this chapter have no connection with one another, but they are grouped together in one chapter, as the information we possess about each of them is but scanty, and does not take up much space.

Intrusive Andesite (?).

The rock to be described occurs in several small exposures on the coast about 1 mile north-west of the Perch near Inellan : the exposures are only seen at low tide, and are so covered by sand and seaweed that their margins are often very obscure.

The most north-westerly exposure, only about 12 yards long and 6 broad, is surrounded by the Upper Old Red Sandstone on all sides, and it seems as if the north-east exposure was also so surrounded, excepting in a seawards direction. The exposures on the south of these may be connected along a line running west or W.N.W.

It is most probable that these small outcrops are of an intrusive character, such as are not uncommon near the base of the andesite flows of the south end of Bute, but the sections do not allow us to assert this with confidence, and it is possible that they belong to these flows : their place of occurrence, a little below the limestone series, is considerably below the geological horizon of the general base of the flows in Bute.

Close flaggy joints characterise the rock of all the exposures. The directions of these joints are various : in the north-west exposure they seem arranged in a syncline ; in the north-east exposure they dip south ; and in the south-west exposure, at the north end, they are vertical and parallel to the length of the exposure, but seem to flatten in the middle.

Macroscopically the rock is of a purple or reddish-purple colour, and too fine-grained to allow us to distinguish the different constituents. A specimen from the north-west exposure has been examined microscopically (slide 2837), and is thus described by Dr Hatch : " A plexus of interlacing needles of a colourless, straight extinguishing mineral of low double refraction together with spicules and granules of iron-ore (magnetite). A small quantity of interstitial matter richly sprinkled over with opaque granules. The needles have an average length of ·52 mm., and a width of ·06 mm. They show no cleavage. A portion of the section after being etched with HCl gave no definite staining-reaction with fuchsine (test for nepheline). The mineral is perhaps oligoclase, although no twin-striation could be detected. This felspar occurs frequently in basaltic and andesitic lavas in lath-shaped microlites, giving approximately straight extinction. Resulting from the decomposition of the needles are little patches of calcite, causing effervescence when a drop of HCl is placed on the section. [N.B. One would hardly expect oligoclase to weather into calcite.] "

The Sith an t-Sluain Dolerite.

The Sith an t-Sluain (152 north-west) dolerite forms a prominent and steep rocky hill, somewhat oval in shape, with a length of about 300 yards, running slightly west of north, and a breadth of about half this. The north, and part of the east and west sides, are almost inaccessible.

A view of the south side from a little distance shows a prominent band at the edge of the rock which is marked off with some sharpness from the central portion ; we see at once that it is in a sounder condition and possessed of sharper joints, and that the main joints are parallel to its length, hading steeply slightly south of west. Our first impression at a distance was, that this might be a dyke quite unconnected with the rest of the rock on the hill. But this certainly is not so, for both at the north and the south ends of the hill, and some other places, a similar edge rock is seen, excepting that the joints in it are not always so prominently longitudinal, but sometimes are cross joints and the same in direction as, but closer than, those in the coarser interior rock. The prominent joints on the north side hade east at 20° to 50° ; the outside margin of the rock is vertical or hades steeply north.

The junctions between the edge and the interior rock are not usually clear, but it is certain that the former is not chilled against the latter. The edge rock is comparatively fine in grain, but not so fine that it cannot be seen macroscopically that the structure is ophitic—white needles of felspar clearly penetrating an abundant black mineral, augite. Small yellowish grains of olivine, sometimes the size of a large shot, also occur. The interior rock is in most parts very decomposed, disintegrating readily into rough gravel, which rolls far down the slopes below the outcrop of the dolerite, and supports a growth of short sweet grass. The same minerals occur in it as in the edge rock, but the individuals are larger. It is abundantly mixed with still coarser but sounder bands, which usually project slightly on the weathered face. These bands are often from 4 to 6 inches thick ; they may run either horizontally or vertically or obliquely to the horizon : or we may, in plane sections, find rudely oval or circular masses of similar rock, sometimes as large as a football. They are in many parts prominently ophitic, white felspar spears penetrating the black augite individuals and cutting them into strips, which are orientated similarly, in some cases for a length of 3 or 4 inches. In other parts the felspar and augite seem to be intergrown in a pegmatitic manner. We have noticed no olivine in these strings. Not uncommonly cavities, apparently amygdaloidal, filled with circular or fan-shaped radiations of some white or bluish-white zeolite, occur in these coarse veins ; they sometimes attain ½ inch in breadth, and abut against the coarsest crystals. Similar zeolite cavities also occur in places in the edge rock. The coarse veins are not chilled at their margins ; they pass directly into the normal rock of the interior.

On close examination it is seen that the edge rock at the south end of the hill is not separable by any sharp line from the interior. There is a continuous exposure from the edge rock, and probably from not far off the outside of it, into the interior rock, and there seems quite a gradual passage from the one into the other. At no place anywhere round the edge could we find a rock much finer than the ordinary type of the edge rock, though at the north side altered schist is seen only a few feet from a rock feature, vertical or hading steeply north, which must be close to the original cooling surface. In the edge rock generally there is considerable variability in the relative proportions of augite and olivine, and in places olivine may scarcely be seen, while in others it is unusually prominent. The parts specially rich in augite are generally rather coarser in grain : they occur either in strings, or in small patches which are not much longer than broad as seen in section : areas about the shape and size of marbles are common, and these sometimes pass into short strings, about 2 inches long and ½ inch broad. Such irregular mixtures of coarse augitic parts with other finer-grained parts rich in olivine are particularly well seen near the south end of the hill. The augitic parts in the edge rock have not their constituents in such large size as the coarse veins in the interior rock. They sometimes contain zeolites like those in the interior veins.

Close to the south side of the hill, the edge rock in one place shows a banded appearance, due to varying abundance of the white felspathic material, and the varying size of the olivine specks. Along the margin of two bands comes in in one place a row of zeolites, averaging ¼ inch in diameter. The conspicuous banding is only seen for a length of 2 or 3 yards, and a breadth of 1 yard. It seems to strike roughly parallel to the edge of the intrusion, but may hade slightly into the hill instead of outwards : we suspect the edge of the intrusion, judging from the prominent joints in it, inclines outwards. A loose piece gathered

on the east side of the hill showed similar banding. Some parts are
rich in olivine, and, taking the whole of the edge rock together, it is
possible it is slightly richer in this mineral than the interior rock is. But
it is hard to come to any confident conclusion about this, owing to the great
variability of both rocks. Besides the well-marked coarse strings and parts
in the interior rock already described, there are numberless other patches,
from ¼ inch to 1 inch long, of rock specially rich in augite, which are
mixed in irregularly through the rest of the mass. This is seen particularly well along the west side of the hill; in one square inch of surface
there may be little else than felspar and augite; in another square inch
little else than felspar and olivine. The olivine is generally decomposed
into a rusty tint, while the augite remains bright black. The portions
specially rich in augite have a habit of projecting somewhat on the
weathered face.

In one of the well-defined coarse veins, about 3 inches wide, the outer parts
seemed nearly all of felspar, and the inner part was specially rich in augite.

Slides of the interior rock (5438), of the edge rock (5439), and of the
coarse veins in the interior (5440), are thus described by Mr Teall,
5438 :—"Coarse-grained greyish rock. Basic plagioclase with a tendency
to idiomorphism, and olivine (fresh) are the two principal constituents.
The olivine is often allotriomorphic with respect to the felspar. In
addition to the above there is a little ophitic augite wedged in between
the felspars and iron-ores (scarce). The rock is intermediate between
olivine gabbro and troctolite." As already stated, however, there are many
parts of the interior rock, which are much richer in augite than the specimen sliced. The felspar is certainly of earlier consolidation than either
the olivine or augite. The olivine is of earlier consolidation than the
augite. A large part of the iron-ores is later than the felspar, and later
also than some of the olivine; but in other parts crystals of magnetite
are inclosed within felspar. There are many bits of iron-ore inclosed in
augite. 5439 :—"Medium-grained dark coloured rock. Same minerals
as in 5438, but possibly containing more of the ferromagnesian constituents, especially augite. Olivine-dolerite." In other parts of the edge
rock, however, the olivine is, as already described, more prominent than
the augite. 5440 :—"Very coarse-grained rock composed of white felspar
more or less idiomorphic, and black ophitic augite. Augite shows a kind
of radial grouping in places. Under the microscope, plagioclase and
ophitic augite with a few patches of iron-ore. This appears to be a kind
of basic pegmatite." The proportion of felspar does not seem greater
than in the other two slides, nor is the proportion of iron-ore less.

The variability of the rock is very interesting, but it does not seem
that there is any recognisable law regulating the distribution of the
different varieties. The whole of the rocks described belong to one period
of intrusion, and we cannot say that the edge rock is more basic than
the interior. The coarse pegmatite-like veins may be considered to
represent the residue of the magma left after the consolidation of the other
parts; they contain no olivine, and in the other varieties of rock, olivine
was seen to be of earlier consolidation than the augite.

It has already been stated that the main joints on the west side hade
steeply slightly south of west, i.e., outwards from the hill: on the north
side also, the apparent original surface is either vertical or hades steeply
north, i.e., outwards from the hill; on the south-east side the prominent
joints hade south-east, outwards from the hill. Now we suppose these
main joints to be parallel to the edge of the intrusion. The evidence
therefore strongly favours the idea that the intrusion is part of a boss,
gradually diminishing in size upwards. It may be that the present top

of the hill is not far from the original surface of the intrusion, with little more than the edge rock denuded off, and that the portions of the interior now exposed are none of them far from the edge rock.

No clear junction of schist with the dolerite is seen, but schist is observed to be much hardened and altered at various points within a few yards of the edge rock, e.g., near the "n" of "an," at the top of the "S" of "Sluain," and just on the north side of the "2" of "1428." The altered schist no longer splits with readiness along the foliation planes, and these planes, when the rock can still be split open, do not show their customary micaceous lustre. A good many bands of it are dark grey, and must in some way have been darkened by contact alteration (see p. 148). The quartz veins of the schist are now, too, much more readily broken by the hammer, breaking up into a kind of sugary mass along a set of numberless little joints (see p. 148). The colour of the quartz is in places rather purplish or pink, colours unknown in the veins of the unaltered schist.

Near the "n" of "an" the schist is for a few yards from the edge of the intrusion, tilted up in a direction almost at right angles to the usual strike in the neighbourhood, so as to run parallel to the side of the intrusion and dip at a steep angle into it. Near the "2" of "1428" the schist is, for a distance from the dolerite of a few yards only, in a vertical position with a strike parallel to the edge of it. These evidences of the mechanical distortion effected by the intrusion confirm the supposition that it belongs to a boss, not a sheet.

The supposed Tertiary, or north-west running set of basalt dykes are not uncommon in the neighbourhood of the intrusion, but none are seen cutting it or in close contact with it. The generally uncrushed state of the rock rather suggests that it also may be of Tertiary age, and the character of the pegmatite-like veins in it, so different from that of the red acid strings of the early east to west dykes, and dolerites of the midland valley of Scotland (see p. 147), agrees with this supposition.

It may be mentioned that the name Sith an t-Sluain (The Fairy Hill of Sluain) is given to the hill in consequence of the old local belief that it was a favourite haunt of the fairies. The hill appears to possess somewhat unusual acoustic properties. We were informed by the late Mr Rhynd of Strachur, that he was once on the top of it, all alone on a clear calm Sunday evening, one of those evenings when everything seems unusually distinct and vivid, when he heard some voices apparently close to him. So close they seemed that he concluded they proceeded from some people who were climbing up the steep hidden slope just below, and who would soon join him at the top. But time passed on, and they never came. Still the voices continued. He listened again more intently. Now he could make out clearly what was said. It was a mother talking to and chiding her children, and mingled with the words came the splash of oars. So downwards he looked to the loch side, and there at Newton Bay, about 1¾ mile away in a straight line, he saw the people and the boat from which the sounds ascended. It was Mr Rhynd's opinion that some such incident as this may, in time long past, have started the idea of the hill being a fairy haunt : sounds and whisperings were heard about it which could not be readily explained, and were therefore attributed to the Spirits of the Hill.

Trachytic* Dykes.

In this area only five intrusions have been recognised as trachyte. These are all dykes ; one running N.N.E. in the burn ⅝ mile east of

* It is doubtful whether this is the best term to apply to these rocks.

Cruach Mhor (162 south-east), one also running N.N.E. in the burn 200 yards west of Craigendaive, a third running N.N.W. in the burn ¼ mile south-east of Dun Mor (172 south-east), and two more running north-west in the burn ¼ mile east and ⅛ mile south-east of the "D" of "Dunans" (164 south-east). But as they are, on hasty inspection, not easy to distinguish from many of the basaltic dykes, and occur in much the same way in the field, it is possible that some dykes coloured as basalt may properly belong to them.

In the interiors of these dykes the colour of fresh fractures is pale greyish brown, less dark than most of the basalts, and the texture a uniform medium grain. But as their margins are approached the texture grows distinctly finer, and for about the last ⅜ inch the colour also changes, and becomes dark grey or black. A black glassy selvage is seen in the Cruach Mhor, Dun Mor, and one of the Dunans dykes, but in the first and last does not always exceed $\frac{1}{30}$ inch in thickness.

As already stated, two of the dykes run N.N.E., a direction which is not very common in the basaltic dykes, but we cannot, on this account, conclude that there is any difference in general habit of direction between them and the basalts, because in each of these two localities the basalts are also running N.N.E. The general direction of the basalt dykes is N.N.W., but they often vary from this in order to take advantage of some pre-existing line of weakness, such as a crush line, and the directions of the Trachytes are clearly subject to the same influences.

In the burn east of Cruach Mhor there is a strong crush, stained with hæmatite, hading slightly west, and running in the same direction as the trachyte. The schists near are much reddened and traversed by hæmatite-stained strings, and so also is a small exposure of lamprophyre in the bed of the burn, about 200 yards north-east of the broad basalt dyke. It has not been possible to show this small exposure in the 1-inch Map. The trachyte itself is not reddened in the slightest, nor are the basalts. It is clear, therefore, that the red crushes must be of earlier date than they, and that they both have locally taken a N.N.E. direction so as to take advantage of a line of weakness. It seems probable that some crushing may also have taken place at a later date after the intrusion of the trachyte, for a dark grey-green crush is seen at one place close to its side, but the dyke itself does not show any crushing. A little south of the most northerly exposure of trachyte in the burn two exfoliating basalts come in from the north-west direction, but at the crush they change direction, and run parallel with it and the trachyte, keeping on the west side of the latter. Just north of the reddened lamprophyre another basalt comes in from the north-west. What becomes of this is not clear; it certainly does not cross the trachyte, neither could we see any sign of it twisting into parallelism with it. It is hence suggested that the trachyte is of later date.

The dykes near Dunans are clearly seen cutting through north-west running basalts, and the facts observed in other localities in no case disagree with the idea that the trachytes are later than the basalts.

In the burn east of Dunans the dyke lying most to the south-east is distinctly more blocky than most of the basalts of the neighbourhood. The main joints run parallel to the length of the dyke, but there are also others almost at right angles to these. On the west side of it, on the east side of the burn, there is a prominently cross-jointed dyke about 6 feet wide, of medium-grained exfoliating basalt. On the west side of the burn close down to the water edge nothing is seen of this basalt, but the west side of the trachyte is opposed to schist. A little way up above the water, however, the basalt again appears between the trachyte and the schist. The line between the basalt and the schist meets that between the basalt and the trachyte at an angle. There is a thin crush along the former line. The latter line is not easy to examine, being partly covered with dripping moss, etc., but it is clear that the trachyte is chilled at it, and is not crushed. The east side of the trachyte on the east side of the burn is also uncrushed, though the rock next it is, in places, a much crushed and decomposed basalt. The east side of the trachyte on the west side of the burn is traceable for about twenty-five feet in a straight line, and is chilled at the margin; it cuts distinctly through a rather coarse-grained basalt dyke. The other trachyte dyke lying on the north-east of this locality is, on the west side of the burn, in close contact with a somewhat crushed basalt dyke on its north-east side. The margin of the trachyte is seen to be chilled against the basalt wherever it is distinct, and in some places at the very edge is glassy.

In the burn on the south side of Dunans one of the above trachyte dykes appears again, ⅛ mile S.S.W. of the " S " of " Strondavain." The best section is that on the south side of the burn. The dyke is about six yards wide. It breaks into large blocks, the main joints running roughly parallel to its length. Both margins are distinct, and clearly chilled. On the north-east side of it is a rather coarse-grained basalt; the joints in this do not give rise to such large blocks as those of the trachyte. The basalt is considerably crushed, and veined with many thin calcite strings, while the trachyte is in a quite sound uncrushed state throughout. The main joints in the basalt, which run parallel to the dyke side, are clearly seen to strike against the side of the trachyte, and to be cut across by it one after the other, while the corresponding joints of the trachyte are parallel to this side.

At Craigendaive the trachyte opposes a chilled margin to a decomposed somewhat coarse basalt, which runs parallel with it on its east side, but the basalt has also a chilled margin against the trachyte, and so it is impossible to say which is the earlier; one of them must first have been chilled against schists, and then the later of the two afterwards chilled against the earlier. Another basalt dyke occurs close to the east side of the trachyte, and again parallel with it, so that there is evidently a tendency for all the dykes in this locality to run out of their usual course. There is a flat feature on the hillside just to the north, running

N.N.E., and this is perhaps due to some line of weakness. The trachyte dyke is about five yards wide; it is not seen except in the burn.

The Dun Mor dyke has a coarse exfoliating basalt dyke on either side also, and running parallel. It shows glassy selvages against these, and lines of flow and elongated amygdules parallel to its margins for a breadth of several inches.

Fig. 39.— × 2. From near edge of dyke 200 yards W. of Craigendaive. Part of a natural section at right angles to dyke side: radiate spheru-litic structure unusually distinct.

All these trachytes are distinguished by the possession, near their margins, of marked spherulitic structures which can be discerned by the naked eye. This is one of the most readily recognised points in which they differ from the basaltic dykes of the area. We will especially describe those of the Craigendaive dyke. In this there are many planes running parallel to one another and the sides, at intervals of about ⅛ inch, for a distance of several inches from the sides. Most of these planes show, on their surfaces, close parallel rods, which may run in different directions on the different planes—sometimes horizontally, sometimes diagonally, and sometimes vertically, but always parallel to the others on the same plane. On examination these rods are seen to be made up of polygonal bodies, the angles of which are quite sharp, but their sides often slightly curved. We may suppose the bodies to have resulted from the mutual pressure of forms which, with more room, would have assumed a spherical shape. Even to the naked eye cross sections of these bodies often disclose a radiate structure. At other times we cannot distinguish any rods, though there may be polygonal forms. The breadth between the planes and the size of the polygonal forms in them grow gradually less and less as we approach the margins of the dyke—three or four inches off the forms are about the size of peas, and from this become less until finally they cease to be visible to the eye. At a distance of a foot from the side, in the quite coarse rock, indications of spherulitic fibres can often still be seen, but the polygonal forms are not conspicuous.

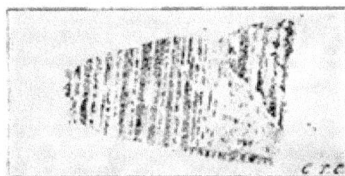

Fig. 40.—× 1. From near the side of the dyke ½ mile E. of Church Moor. The coalesced spherulites increase in size in proportion to the distances of the planes on which they lie from the dyke side. In the sketch the plane at the right hand top corner is the furthest from the side, and the spherulites on this plane are the largest.

Under the microscope the spherulitic portion of the rock (slides 3921 and 2962) is seen to consist of a pale brown glass full of minute depolarizing spicules, granules of magnetite, and brown granular matter. The spicules are sometimes aggregated into stellar groups. It is not clear that all the spherical patches are true spherulites—bodies possessed of a radiate structure. Some may be accumulations of a dark coloured ingredient of the rock, but certainly near the outer edge there are true spherulites, both singly and in groups aggregated together with length parallel to the general flow structure, and it is possible, that if the slide had been thinner, the darker spherical bodies would also have shown a radiate structure. The spherulites are often seen to be crossed by the spicules.

Several slides of the Dun Mor rock (slides 2961, 3922a, 3922b, and 3922c) were prepared, in order to show the gradual alteration in size of the spherulites as the edge is approached. The still glassy selvage (3922a) is thus described by Dr Hatch :—" Clear yellow glass, only containing a few streams of granules, some isolated long spicules, a few prisms of augite and abundant most perfectly-developed little spherulites, presenting a fibrous structure, and sometimes with a central nucleus. Between crossed Nicols they give a black cross." A part further away (slide 3922c) consists of " brown granular matter filled with long slender depolarising spicules which often radiate from a small granule of magnetite. There are also present large crystals of magnetite, also furnished with a fringe of spicules. A pale coloured augite also occurs in isolated prisms." In the slide 3922b, which comes between the two already described, the brown granular matter in the glass seems often to have accumulated around the stellar aggregates of spicules, leaving small interstitial patches of clear coloured glass. The aggregates become smaller and smaller as the original surface is approached, and in 3922a they are so minute before the distinctly spherulitic portion is reached that they appear merely as dark structureless granules under the $\frac{1}{4}$ inch object glass. These still are disposed in rows roughly parallel to one another, and show beautiful instances of fluidal structure: at times they show sharp folding, or adjacent rows separate and bend round some early formed crystal, which has been caught up in the glass. They are often seen to go straight through the spherulites without any distortion, as if the latter had not, in these cases, been developed until after the flow motion of the rock had ceased.

The slide of the Cruach Mhor rock (2960) is also a spherulitic glass. A specimen from an inch or two off the side of this dyke gave, with Walker's balance, a specific gravity of 2·34. It is possible there are some small hidden cavities within the specimen : small air bubbles continued to escape for some time while it was being weighed in water.

The central well crystalised portion of the dykes in the burn $\frac{1}{8}$ mile south-east of the "D" of "Dunans" has been microscopically examined. This (slide 3452) is composed mainly of small lath-shaped crystals of ortho-clase, together with scales of brown biotite, and a few prismatic crystals of pale, somewhat altered, pyroxene, and scattered granules of magnetite. Interstitial glassy matter appears originally to have been present. This rock was kindly analysed by Mr Player with the following results :—

Silica,	.	.	.	.	56·4
Alumina,	.	.	.	.	19
Ferric oxide,	.	.	.	.	3·5
Ferrous oxide,	.	.	.	.	4·8
Lime,	.	.	.	.	2·6
Magnesia,	.	.	.	.	1·5
Soda,	.	.	.	.	4·5
Potash,	.	.	.	.	5·0
Loss on ignition,	.	.	.	.	2·6
					99·9

The specific gravity of a specimen from the same part of the rock was determined by Walker's balance as 2·48.

The shape of the amygdules forms another conspicuous point in which these dykes differ from the basalts. In the latter, as stated, p. 129, the amygdules are almost universally rudely spherical, whereas in the trachytes they are either very irregular in shape, or prominently elongated parallel to the dyke sides. At Craigendaive and Cruach Mhor they consist chiefly of calcite with occasional specks of sulphide of iron near the centres, and some are an inch in length. Close to the surfaces of the prominent planes there is a tendency for extremely thin rust-covered hollows to show themselves on weathering : the sides of these hollows seem formed by the sides of the polygonal spherulites, and so have a crenulate aspect. At Dun Mor the amygdules sometimes attain a length of about $\frac{1}{4}$ inch with a breadth not exceeding $\frac{1}{10}$ inch, and they are composed of an undetermined greyish-brown substance with gummy lustre, which is quite soft to the knife. Possibly this is the chlorophæite of Maculloch ("Western Isles," vol. i., p. 504 ; see also Heddle, *Trans. Roy. Soc. Edin.*, vol. xxix. p. 84). In a section by the road $\frac{1}{8}$ mile slightly north of west of the "D" of "Dunans" the elongated vesicles, weathering in hollows, are especially abundant at a distance of about 18 inches from the north-east side of the trachyte : their greatest lengths are parallel to the side, and usually nearly horizontal : their dimensions vertically are often rather more than their breadths measured at right angles to the side of the dyke. One vesicle was noticed at least $1\frac{1}{2}$ inch long, though less than $\frac{1}{4}$ inch broad, and lengths of $\frac{3}{4}$ inch are not uncommon. C.T.C.

CHAPTER XV.

GENERAL GEOLOGICAL STRUCTURE OF PARTICULAR DISTRICTS.

District of Dunoon.

The district to be now described includes all the area inclosed by Loch Striven, the Clyde, the Holy Loch, Glen Lean, and the lower part of Glen Tarsan with the exception of the small area near Inellan and Toward which is occupied by rocks of Upper Old Red Sandstone Age. It extends from " the anticline " down to the south-east boundary of the schists. The small area of Upper Old Red Sandstone is described separately on pp. 92-96.

The schistose greywacke horizon in the north-west of this area is the same as that mapped in the areas north-east and south-west. For a little space west of Ardtarag the marginal lines have not been engraved. There are plenty of rock exposures in this space, and the impression conveyed by examining them was, that the bed was much folded and the margins repeatedly "v" shaped. Even in the sections elsewhere the margins are not sharply defined—we can only say that after passing an interval of about 6 or 7 yards we reach a schist in which the proportion of pebbly rock is certainly greater than it was on the other side,—and in this area they seemed still less defined. We think the south-east margin is generally more definite than the north-west. On the south-west side of Glen Lean we were in doubt whether we should not take the north-west margin about 30 yards north-west of the line engraved.

On the east side of Loch Striven, $\frac{1}{2}$ mile north-west of the Craig, the same band of coarse pebbly schists occurs as that described more fully on the other side of the loch, pp. 186-187. It has not been traced up hill to the north-west.

Crossing over Mid Hill is a band of fine-grained hornblende schist containing many crystals of magnetite and small specks of ferriferous carbonate. On the hilltop the outcrop is 8 or 9 yards wide, but near the south-west end of the line engraved for the band it is only between 1 and 6 feet. The band is on much the same horizon as those near Ardentinny.

There are three outcrops of hornblende schist a little north-west of the lowest band of green beds. The one on the west side of Inverchaolain Burn a little south of the "G" of "Glen," is only seen obscurely : it overlies a thin-foliated quartzose schist, but the line separating them has rather the aspect of a thin crush : the ground around is obscured by drift and the slipped material from a large landslip, and it is uncertain how far the bed extends. The band near the "G" of "Glen Kin" is exposed in both the burns which unite near the "G," but cannot be traced away from them owing to the thick drift covering. The outcrop north-west of Strone Saul makes a prominent show on the hillside, its edge generally sticking up in little crags over the surrounding smooth ground. There is a good section of pebbly schist and phyllite in the little burn on the east of the exposure, but no hornblende schist.

The most north-westerly broad band of green beds on the south-east side of Inverchaolain Glen is not shown on the shore of Loch Striven, the ground it should occupy being all covered by drift or raised beach deposits. But it is seen in the little east to west burn near Tighnuilt : some bands are coarsely pebbly : a little above the base some schists occur which do not belong to the green bed type. Further north-east we see in several places thin bands of green beds a little below the base of the lowest broad band : e.g., there are two such in the burn which comes into Inverchaolain Burn from the south-east by the "n" of "Inverchaolain" : one ¾ mile north of the foot of the above little burn : three on the south-east of Bealach na Sreine and the head of Glen Kin Burn : and one about ½ mile south-east of the "G" of "Glen Kin." It is not known whether these bands are partial repetitions of the broad band due to folding, or bands on a slightly different horizon. Much the greater portion of the broad zone consists of typical green beds, but here and there, e.g., near the top of the band ¾ mile south-east of Strone Saul, schists without any green tint are intermixed. A little above the top in the burn ½ mile west of the "B" of "Bochd" two thin bands of green bed occur. In the same burn the top of the broad band is drawn for a little space at a considerable angle to its usual direction : the dip of the foliation here is to the north-east, having probably been twisted into parallelism with some line of movement : a crush breccia is seen running with the foliation strike, but it is not necessary to suppose a large throw.

The north to south fault that runs with the basalt through the "n" of "Knowe," causes the foliation on its east side to twist into parallelism with it for a considerable distance, for a breadth of 6 or 7 yards. Such twists of strike are not common in this district, even near the faults which effect a great displacement. The shift at the surface in this case is about 110 yards, the west side to south. A little west of this fault is another parallel fault, also accompanied with one or more basalt dykes. This shifts the outcrop in the same direction but to a greater extent, about ⅛ mile. These faults with their accompanying dykes both make prominent gulleys through the crags, and give rise to the wildest bit of scenery there is within a few miles of Dunoon. One-third of a mile further west are two more faults which may unite before the top of the green bed is reached : the sum of their displacements is rather more than ⅓ mile, the west side again to south. We suppose all the faults mentioned in this paragraph to unite in a south direction, and to be together in the lower part of the burn ½ mile south-east of the "d" of "The Red Well." South of this, the fault seems to pass along the west side of Corlarach Hill, and runs up the burn on the west side of Toward Hill, in company with various basalt dykes, shifting the Bull Rock greywacke nearly the full width of its outcrop. In a north direction from the Giant's Knowe, the faults cannot be traced well owing to the wide-spread covering of drift. In the burn ¼ mile W.S.W. of the "l" of "Glenkin" there is a crush face, hading east and with horizontal slickens, which we suppose belongs to one of the more easterly faults.

One-quarter of a mile west of the first "B" of "Bodach Bochd" there is a north-west fault, causing a shift in the opposite direction to the faults last mentioned. The displacement at the surface is about 200 yards. This too is accompanied by a basalt dyke. The evidence for the throw is best seen a little more than ¼ mile W.N.W. of the "B" mentioned, where there are on the north-east side of the basalt good sections of green beds, and on the south-west side quartzose schists. We have not been able to trace this fault in either direction.

The supposed shift in the green bed along an east to west fault about half way between Tighnuilt and Invercholain is not actually seen, but there are two strong, nearly east to west, crushes a little east of these localities, one in the burn ¼ mile E.N.E. of the " t " of " Tighnuilt," and another in a burn, not in the 1-inch Map, a little on the south side of this burn. These both hade south, and shift the margin of the south-east green bed, the south side to the west. The combined displacement is about 200 yards, but this is in a direction diagonal to the strike. Nearly ½ mile north-east of Blar Buidhe there is a crush in the line of these, and hading in the same direction, in the course of a little burn not marked in the 1-inch Map. The throw cannot be ascertained, but we suppose it to be down south, in accordance with the hade and the appearances described on the west side of the hill.

The N.N.W. fault that throws the green beds on the south of Fiubracken Hill (174) is sometimes made up of several branches, and the amount of displacement due to each is not always evident. The base line is shifted by the most easterly fault for about 260 yards : this fault makes a crush 10 yards wide in the burn ¼ mile west of the " A " of " Ardnadam " : near where this fault, going south, is made to join the next fault on the west, some phyllite and quartzose schist are seen in the burn a little above the base line taken. The movement has to some extent been horizontal, for horizontal slicken lines are seen in one of the fault breccias in the burn ¼ mile W.N.W. of Ordnance Station 697. Another exposure of the same breccia contains small strings and specks of copper pyrites which have been partly converted into the green carbonate. On the south-east side of this compound fault the green beds make a broad spread. The boundaries of it are not definite. No other rocks but green

beds are, however, seen on the west side of the Ardnadam 100-foot beach, and there are several small rock islands in this beach which consist of green beds, generally of a pebbly character. The area east of the beach is drift covered for ⅛ mile or more, and the nearest rock exposures on this side consist of phyllite with alternations of more quartzose schist, which show repeated folding along north-east to south-west axes. We suppose that the green bed base cannot be traced north-east along the foliation strike, owing to the great number of folds by

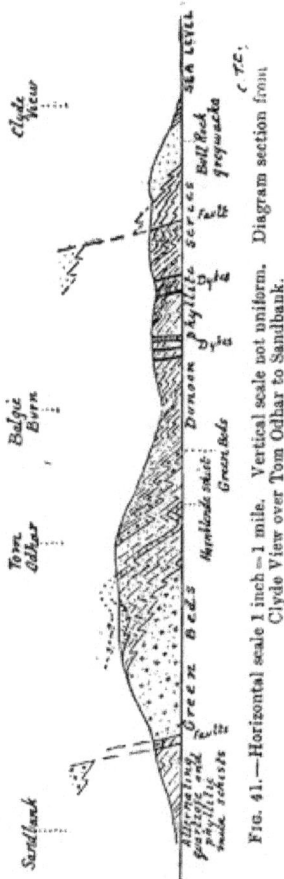

Fig. 41.—Horizontal scale 1 inch = 1 mile. Vertical scale not uniform. Clyde View over Tom Odhar to Sandbank.

which it is affected. The even run of the hornblende schist by the foot
of Loch Loskin seems to preclude the idea of there being any large fault
shifting the green beds. It is clear, too, that the top line of the green beds,
though not visible in section, must cross the foliation strike. There is
no feature on the hillside which could belong to a fault bounding the
green beds, and we conclude the appearances are due to a repetition of
small folds. The individual " v's " of the boundary drawn in the Map
are not seen on the ground, but they are inserted to indicate the style of
structure we suppose to exist.

The green beds are seen in the branches of the burn that run into
Loch Loskin. An east to west crush occurs in the west branch, and a
thin crush hading east is also seen in places in the east branch, running
near to, and parallel with, the basalt : but it is not clear how the boundaries
of the green beds are affected by these. Close to the east side of the
burn the green beds seem lost, and there is hardly room even for a
thin bed to run north-east as far as the edge of the peat at the head of
Loch Loskin. Nor could we see any further indication of green beds on
this horizon further north-east. We cannot say for certain, however, that
such do not occur, for the ground is largely obscured by drift and raised
beach deposits, and the coast near Hafton is destitute of rock exposures.

It is suggested that possibly this lower band of green beds is really an
infold, along a synclinal, of the overlying band. If this is so, both margins
of it represent the same line, the presumed base line, and the beds on either
side of it are the same. But why then do not the hornblende schist
bands, which occur near the south-east margin for so long a way, also
get repeated near the north-west margin ? There is a band of this schist
near the latter margin on either side of the burn that runs into Inverchaolain Burn between the " h " and the " a " of " Inverchaolain," but it has
not been traced for a distance of ½ mile. It appears too, from facts to
be shortly stated in connection with the green beds that come next to the
Dunoon phyllites, that the green beds may vary rapidly in thickness or
may die out altogether.

It is not possible to suppose the north-west green bed the same as the
band next the Dunoon phyllites, brought on again in the north-west limb
of an isoclinal anticline with both limbs dipping south-east, for in this
case the beds west of the north-west green bed would be the Dunoon
phyllites, and this they cannot be. There is a much greater proportion
of quartzose pebbly schist west of the north-west green bed (see, e.g., the
section in the burn on the south-west side of Bealach na Sreine) than in
the phyllite series, and also an absence of limestone.

The hornblende schist near the south-east margin of the lowest broad
green bed is represented by two bands on the west and north of Bishop's
Seat, but only one band, the lowest, is known west of the fault that
runs north to south through the " G " of " Giant." It should be noticed
that this schist does not appear to keep strictly to the same geological
horizon, being on the south-west side of Bodach Bochd well within the
green bed, while to the north-east it is at the top or some little way above
the top. Only one band is seen in the burn ½ mile north-east of the
Bishop's Seat, and this is only thin. It could not be traced east of the
burn. It is probable, however, that the folded hornblende schist near
Loch Loskin is on much the same horizon. This band is readily traceable
on the hillside by a succession of rocky knolls separated by peaty hollows
and flats, which have a general N.N.W. direction. The shape of outcrop
at once suggests folding, particularly when it is known that in the area
east of the apex of apparent fold the band has kept approximately
parallel to the north-west margin of the Dunoon phyllites, and has there-

fore a sill-like character. The north extremity of the band, north of the burn that flows into Loch Loskin is, however, of such irregular shape that we can hardly suppose the sill-like character is constant. No hornblende schist is seen where it would be expected to cross the above burn, but there is an obscure area where it may cross. The band is bare on the raised beach at Hunter's Quay, and makes a prominent ridge for about 120 yards inland.

The exposure of hornblende schist ½ mile north-east of the Badd probably represents part of the apex of fold seen in the band 1 mile N.N.E. of the Badd, this portion being cut off and shifted by the same fault as that which shifts the green beds on the north. The vein stuff of the fault is seen in the burn at the north end of the exposure, and again at the south end : the hade in both places is east.

In the shore section 250 yards south of Hunter's Quay an outcrop of hornblende schist about 6 yards wide occurs. This is a little north of the line taken as the boundary of the Dunoon phyllites. East of high-water mark it meets a small north to south fault, which advances the east side about 2 yards to the north. Just south of the outcrop there seems a small folded inlier of it. In the old cliff of the raised beach we could see nothing of the bed. In the wood ¼ mile south of the "n" of "Loskin" another exposure of similar rock occurs among pebbly schists : the thickness cannot exceed 10 feet : the horizon seems much the same as that of the band south of Hunter's Quay. Near the top of the Horse Seat (183) there is also another exposure on the same horizon.

The green bed on the north side of Brackleymore may perhaps be regarded as a folded outlier of the band on the south-east. It is not seen on the sea-shore, but there is a section along the old cliff of the 20-foot beach. The band on the south-east certainly advances its margin nearer to the north-west band the nearer the shore is approached,

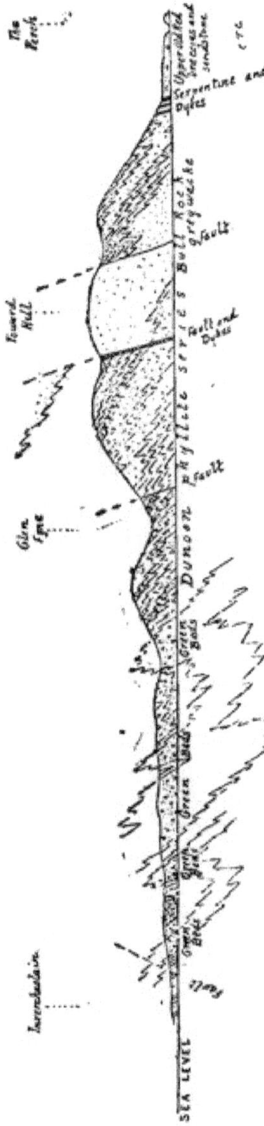

Fig. 42.—Horizontal scale 1 inch = 1 mile. Vertical scale not uniform. Diagram section from the Perch over Toward Hill and the phyllite exposures on the shore S. of Brackleymore School to Inverchaolain.

presumably by a succession of small folds, and the interval still remaining between the two bands on the shore is partly occupied by thin bands of green beds, which may well represent limbs of further folds. On the shore, 40 or 50 yards north of the north-west margin of the north-west green bed, another thin outcrop of green beds occurs. This is not shown on the map.

The area coloured as green beds between Brackleymore and the south-east slope of Kilmarnock Hill is about 1 mile in breadth across the strike of foliation. This great breadth is no doubt mainly due to folding along axes agreeing roughly with the foliation strike. Bands of green, grey, and black phyllite, like those of the Dunoon series, occur along the shore between outcrops of green beds, and run for some distance up the hill north-east. These we suppose to represent infolds of the Dunoon series lying on the green beds, and we cannot take the original thickness of the latter to be even so great as that of the north band of green beds between the school and Brackleymore. Though the phyllite is gradually lost as we ascend the hill it is more likely they are resting on the green beds than under them. The latter supposition would imply that the longer limbs of the pre-anticline folds in this locality were the north-west limbs, whereas in the coast sections on the south-east side of " the anticline " the south-east limbs are the longest almost universally (see p. 13). The " v " shape of the detached phyllite areas is not actually seen on the ground : we can only say that as we go up hill to the north-east their breadths become gradually less, or split up by bands of other rock. These other rocks are not always green beds : sometimes they are massive pebbly quartzose schists, like some seen between the phyllite and the green beds on the south side of Kilmarnock Hill. It is therefore probable that they represent a zone between the phyllites and the green beds, and should strictly have been separated from both. But the ground did not show sufficient rock exposures to enable us to separate them with satisfaction. To the north-east this intermediate zone seems to die out, or almost so : this is seen in Ardyne Burn ¾ mile south-west of the " C " of " Corlarach," and by the " D " of " DUNOON " (parish).

A traverse across the foliation strike, over the top of Blar Buidhe shows, as far as observed, no schist other than green beds. A little east of this hill a band of hornblende schist can be traced running almost at right angles to the foliation. This hornblende schist seems to occur approximately along the junction of the green beds with the underlying rocks, at the apex of the same fold as that which affects the hornblende schist of the Badd, etc. To the N.N.E. of this schist the green beds seem gradually to thin away, and we suppose them to fold out along the same axes as those which affect the phyllite areas west of Kilmarnock Hill, and the green beds by Ardnadam.

There are good exposures of the green beds in the burns south of the Badd, but the hillsides adjacent are obscure and do not allow us to fix rock boundaries with confidence. We suppose there is a good deal of folding, and also a fault which, east of the Horse Seat, cuts out part of the folded green beds and brings beds which are below the green beds against the Dunoon phyllites on the east. The little burn east of the Horse Seat is mostly in drift : the only rock seen, near the head of it, is black phyllite. If it had not been for this, we could have supposed a connection between the green beds on either side of Horse Seat. The supposed fault is nowhere seen in section, but there is a N.N.W. running feature, facing east, on the east side of the Horse Seat, which may be caused by it. The fault on the east side of this fault is seen distinctly in the burn ½ mile south-east of the Badd, and again in the burn between the " D " and the

" U " of " DUNOON " (parish) : in the first place there is a crush with quartz, ferriferous carbonate and specks of copper pyrites, and a subordinate breccia, hading east, some 30 yards on the north-east, and various parallel strings of ferriferous carbonate, by which the adjacent green beds have been stained an ochreous colour : in the last place the fault is in two parts, both of which hade east.

The green beds are seen well in the burn ⅔ mile slightly north of east of the Badd, but are mixed with a quartzose pebbly schist in one part. The next section to the north-east is that at the head of the burn near the " O " of " Tom Odhar," where the breadth of the zone is less than it is south-west, and there is also a considerable admixture of phyllite. Still further north-east the zone continues to decrease in breadth. In the burn going through the word " Dunloskin " it occurs in many different bands, belonging apparently to different limbs of folds, and the thickness in the east exposures is not more than a few yards. We saw no green beds on this horizon between this burn and the Clyde. There is an almost continuous rock section between Hunter's Quay and Kirn, and no genuine green bed is seen. In the want of one we are left in a little doubt as to the best line to take for the north-west margin of the phyllites, for there is no sharp change from this series to the rocks on the north-west. No doubt this want of definition would have been felt in the area to the south-west had it not been for the opportune occurrence of the green beds on approximately the best line between the more phyllitic and the more quartzose zones. It was considered allowable therefore to adopt them as the boundary without further scruple.

The two hornblende schist outcrops within the green beds area on Kilmarnock Hill (194) form prominent hillocks. The hornblende is rather darker in macroscopic aspect than is usually the case on the south side of " the anticline." About ⅓ mile south of the hilltop, close by the boundary of the phyllites, a third outcrop of such schist occurs : the thickness and boundaries of this could not be determined.

The section of the Dunoon phyllite series on the coast south of Dunoon and the structures to be observed in it are fully described on pp. 9-18. There are on either side of this phyllite series, through most of the area, well characterised bands of schist, a green bed on the north-west and the Bull Rock greywacke schist on the south-east. The opposed margins of these beds can be mapped with comparative ease, and include between them the schists which we have called, as a group, the Dunoon phyllite series. On the coast south of Hunter's Quay there is, as just said, no green bed to define the north-west margin of the series, and we feel in doubt as to where the edge of it should be taken : the line could not well, however, be made to diverge more than 20 to 30 yards from that adopted. Fortunately this is the only clear section wherein this green bed fails us (see p. 31).

There are several north-east running crushes in Ardyne Burn from 150 to 250 yards north of the " A " of " Ardyne " : we suppose these to be the same as those seen about 1 mile N.N.E. of here, one at the east side of Ardyne Burn, and the other crossing the burn that flows past Corlarach. There are two prominent features facing north-west on the north-west side of Corlarach Hill. These probably represent the fault lines seen in the burn north-east of here, between the " D " and " U " of " DUNOON " (parish), and already mentioned on this page : the north-west one probably runs on to one of those in Ardyne Burn ; the south-east one we have drawn south to join the fault which shifts the south-east margin of the phyllite series on the west of Toward Hill.

The fault that shifts the south-east margin of the phyllite series on the

M

south of Dunoon runs along a hollow in the hills west of Ardhallow. The schists in the burn are mainly phyllites, often stained with hæmatite and crushed, and intersected with many basalt dykes which seem to converge as they approach the main fault. To the north the fault makes a prominent feature facing west, as long as it has the Bull Rock greywacke on its east side, and the exposures of greywacke near it are sometimes stained with hæmatite. Further north still, the direction becomes doubtful : close to the foot of Glen Morag Burn there are a few thin crushes, and we have carried the line on doubtfully to join these.

The character of the Bull Rock greywacke is described elsewhere, p. 15. Its outcrops are distinct on either side of the Ardhallow fault, forming more craggy ground than the phyllites. This difference prevails generally, but it is not so marked as one would have expected, and between Beinn Ruadh and the crags near Tor Aluinn is hardly noticed. On the east side of the raised beaches near the foot of Ardyne Burn the band is almost covered with boulder clay for a length of $\frac{1}{2}$ mile along the strike, but on the west of the burn it appears again in four small island-like exposures within the gravel area. In three of these it has been quarried. The shore at Ardyne Point, where the greywacke should occur, is covered with gravel.

This bed makes one of the best building stones of the district, and there is a large quarry in it, a little north of the Bull Rock, which provides most of the Dunoon building material. In the quarry a good many quartz veins occur, and, on looking over some of the loose blocks derived from these, we noticed many small specks of copper pyrites, iron glance, chalybite, and small pyramidal crystals of quartz.

The fault between Toward Hill and Beinn Ruadh seems to effect a surface displacement in the base of the Bull Rock greywacke of about $\frac{1}{4}$ mile, the south side to west. The fault cannot be traced with confidence in either direction. One-third of a mile slightly west of north of the "B" of "Beinn Ruadh" there is a north to south bank feature facing east, which we suppose due to it. The feature can be followed north for about $\frac{1}{4}$ mile.

The N.N.W. fault drawn a little over $\frac{1}{4}$ mile south-west of Garrowchoran Hill is not established with certainty. The only evidence for it occurs near the north-west margin of the bed. Here a north-west basalt dyke has some breadth of massive pebbly schist on the north-east side, striking north-west of some phyllite exposures on the south-west side. But these exposures are not extensive and possibly they are due to phyllites folded within the greywacke area.

Between the Bull Rock greywacke and the serpentine no schists have been mapped out of the general mass for more than a short distance. The craggy east face of the Tom near Inellan is formed by a massive pebbly schist, with pebbles chiefly of quartz and white felspar. This schist is bounded on the west by black and purple phyllites, which were formerly quarried for roofing slates. The line between these rocks is drawn on the map for a short distance : on the north side of the Tom phyllites appear to strike against the pebbly rock, but whether this is owing to folding or some fault we are unable to say.

The high band of phyllite marked in the burn north of the "a" of "Chapelton" was formerly extensively worked for roofing slates in a quarry on the east of the burn. The colour is purple or green. The low band is purple or black.

The serpentine and the schists close to it, and the schist-boundary fault are described already (pp. 72-74, 95). So also (p. 30) are the peculiar gritty phyllite beds which are seen in places a little north-west of this fault.

Three changes of direction of the broad east to west basalt dyke have been suggested in this area, where this dyke meets with north or north-west faults : viz., on the east of Inverchaolain Burn, in the burn south-west of Bodach Bochd, and on the west side of Kilbride Hill. The last locality is much obscured by boulder clay, and all we can say is, that, if the dyke exists at the surface, its course on the south-west side of the fault must be considerably south of its course on the east side of the fault. In the Bodach Bochd locality, the dyke is seen within 80 yards of the burn on the west side, and 200 yards on the east side, and the exposure on this east side is so much north of that on the other that the dyke must either change its direction approximately along the burn and fault course, or else cease to come to the surface in the intervening area. The basalts actually seen in the burn are of different characters to the east to west broad dyke, but it is possible that under the drift at the sides of the burn the latter may also occur. On the east of Inverchaolain burn the south end of the changed path is seen distinctly.

District of Blairmore.

This district is inclosed by Loch Long, the Holy Loch, the Loch Eck valley, and the north margin of 1-inch Map 29.

As far as known, all the schists, with the exception of some hornblende schists, belong to the group of alternating schistose grits, phyllites, and albite schists, which occurs between "the anticline" and the "green beds" on the south-east side of "the anticline." The schists mapped out are,—the south-east portion of a massive greywacke horizon on the east of Loch Eck, a band of coarse pebbly schist on Blairmore hill, the series of thin hornblende schists which strike across Stronchullin hill, and a broader hornblende schist near Blairmore.

The schistose greywacke near Loch Eck is part of a zone which has been traced a considerable distance to the south-west. The schists on its south-east side are more thinly flaggy, more micaceous and less pebbly than it. The margins are not sharp lines, but a certain breadth of rock has to be traversed before we can see that most of the schists on one side differ from those on the other in the characters mentioned.

The greywacke is crossed, at a distance of about $\frac{1}{4}$ mile from Loch Eck, by a north to south fault. This advances the margin of the band on its east side about $\frac{1}{4}$ mile to the north : to the south the fault forms a conspicuous slack just before it reaches the superficial deposits of the valley ; to the north, outside the area being described, it also forms a slack at intervals for about $1\frac{1}{2}$ mile, and two basalts occur along with it in one place.

The Blairmore Hill pebbly schist has a quartzose matrix : this contains pebbles both of quartz and felspar (generally white), and there are also lumps, 2 or 3 inches long at times, of grey, compact, creamy-weathering felstone or hornstone-like rock, which form hollows on the weathered rock faces. These lumps are crossed by the foliation planes of the matrix. There is an outcrop, almost entirely of pebbly rock, for a breadth of at least 100 yards. The north-west margin is definite, and has been mapped : the south-east side has not been examined with much care.

Two north-east faults traverse the hill, and both advance the south-east side towards the north-west. The crushes belonging to the west fault occur in different bands, occupying in all a breadth of about 12 yards, and hade south-east : they effect together a shift of about 150 yards. The breccia of the east fault is about 6 feet wide.

The Stronchullin Hill hornblende schists are best seen in the crags on the E.N.E. side of the hill. None are observed on the sea-coast, the area where they would be expected being all covered with superficial deposits. There are at least five bands, all of much the same type: the hornblende is represented by shreds of pale green fibrous actinolite, and the felspar by saussurite: abundant specks of ferriferous carbonate also occur. They can hardly be distinguished from the schists around except on hammering,

and it is probable that there are other bands, besides those mapped, which have escaped notice. Perhaps the thickest outcrop is that in the little burn ¾ mile E.N.E. of the hill top. This is about 25 yards wide. A fault vein runs along the burn course across the outcrop, and shifts the east side 10 yards to the north: the vein is 2 feet wide and hades east.

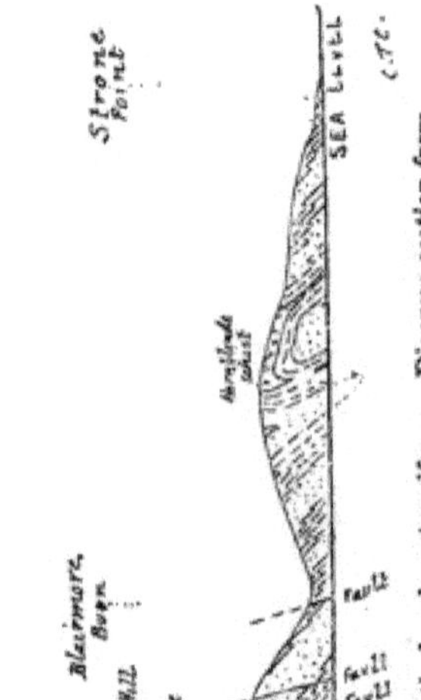

The hilltop is too obscure to allow the bands to be traced continuously, but on the south-west slope several are seen again. The most southerly band is affected by two or three faults near the little burn ½ mile south-west of the hilltop: it is very full of carbonate specks, some as large as a pea, and these give a slaggy aspect to the weathered rock. Between the basalt dykes ¾ mile south-west of Stronchullin Hill, there are at least three outcrops of the rock, but in the Map these have been generalised as two. The most southerly one is 10 feet thick.

The east margin of the Blairmore hornblende schist makes some crags facing east, which form a conspicuous contrast with the rest of the hill-side. The west margin is probably nearly coterminous with the inland edge of a flattish area, which is overlooked by a feature with abundant exposures of thin flaggy mica schist and pebbly schist. Perhaps the feature is due to some small crush line. The greater length of the outcrop trends N.N.W., at right angles to the strike of foliation. The bed is probably of sill-like character, with its greater length lying along the arch of an anticline into which it has been folded. It is not seen well in the limbs of the supposed fold: perhaps this is due to the pulling out and thinning which these received during the folding. There is, however, an exposure, much veined and decomposed, on the shore, nearly ¼ mile north of Blairmore Pier, and this continues into the cliff at the back of the raised beach. The bed is not seen on the shore south of the pier, but there is room for it. The rock is generally rather more massive in character than those of

Fig. 43.—Horizontal scale 1 inch=1 mile. Vertical scale not uniform. Diagram section from Strone Point to Blairbeg and Stronchullin Hill.

Stronchullin Hill, and the grain coarser : the hornblende shreads some-times attain a breadth of $\frac{1}{20}$ inch. Supposing the physical structure is as we have suggested, the hornblende schist should, if it continued, reappear again further north-west, in confirmity with the ordinary rule of "pre-anticline" folding (see p. 13). As a fact, no such schist is known to the north-west until we reach Stronchullin Hill.

The chief faults observed in the area run in a north or N.N.E. direction, and are doubtless continuations of those on the south-west side of the Holy Loch. Four of them have been described already, and it may have been noticed that they all advance the beds on their east sides towards the north : they downthrow therefore to the east. Apart from the evidence afforded by the schists, we often obtain great help in esti-mating the throws of faults in this district by their effect on the lamprophyre sheets : there are some good instances of this close by the north margin of the Map. The faults can also frequently be traced by the features they make on the hill. The crush rock is generally softer than the uncrushed, and yields more readily to denudation, so that it has a habit of forming gulleys in streams, and slacks or terrace shapes on hillsides.

The crush that trends north from near Finnartbeg is seen well in the burn between Finnartbeg and Cnoc à Mhadaidh. A little before it reaches the basalt dyke it is in all about 10 yards wide : the dykes coming into it twist out of their ordinary course and keep along it for some distance. These basalts a little on the north-east of Cnoc à Mhadaidh are themselves crushed, and so the fault movement must be of two ages, one before and one after the basalt intrusion.

The north and south features, facing west, which run north from the Holy Loch a little on the east side of Finnartmore are no doubt due to crush breccias. These unite in the burn due east of the "n" of "Kilmun Hill," and breccias, hading east, are seen in the burn below this point in several places. Further north still, the continuation of the crush is cut on the north side of Blairmore Burn in two places, by levels driven in search of lead ore. These trials, and a third one in a different crush on the north side of Cnoc à Mhadaidh, are mentioned further on (p. 287). On the east side of the features, near Finnartmore, there are others at intervals all the way to Loch Long, but these are not so prominent and only one has been engraved in the Map.

The best coast section is that on Loch Long from about $\frac{1}{6}$ mile north of Gairletter Point to about $\frac{1}{4}$ mile south of Ardentinny. This is described on pp. 18-22. Other good sections occur,—at Strone Point, in much folded alternations of micaceous and more quartzose thinly foliated schists ; in Blairmore Burn ; and in Stronchullin Burn, for nearly 1 mile above Stronchullin. Both these burns run almost at right angles to the folia-tion, and show repeated folds of the schists, with a steady foliation which runs approximately parallel to the axes of fold. In parts nearer "the anticline," many of the hillsides—Creag Liath and the adjoining side of Loch Eck particularly—show good sections.

District of Glen Massan.

This district is enclosed by the Loch Eck valley, Glen Lean, the lower part of Glen Tarsan, Glen Laoigh, and the north margin of 1-inch Map 29.

With the exception of a hornblende schist on the south side of Coire Athaich, all the schists appear to belong to the group of schistose grits, phyllites, and albite schists which occur at or near "the anticline." The bands mapped are the hornblende schist, an horizon of greywacke which runs across Sgarach Mor, and one or two other greywacke bands on the north-west side of Glen Tarsan.

The schistose greywackes on the north-west side of Glen Tarsan are on the north-west side of "the anticline," and the Sgarach Mor bed on the south-east side, but, for the reasons stated elsewhere (p. 86), it is not believed that they represent one another. The interval between the two subdivisions of the bed on the north-west side is but little, and perhaps it does not run so continuously as shown on the Map. It consists of very thinly foliated mica schists, and is best seen on the north and north-east sides of Balliemore, within ¾ mile of this place; an interval is seen again close by the north margin of the Map. The north-west margin of the north-west greywacke is the most definite of the lines bounding these two beds. Just on the east side of Glen Laoigh this margin appears thrown along the line of a basalt band which is compounded of more than one intrusion. The amount of displacement cannot be estimated accurately, as the ground is covered with drift for a little distance on either side. The Sgarach Mor bed is not well defined near the head of Coire Athaich, and lines have not been drawn for its margins here. The obscurity may be partly due to folds through which the margins cannot be traced, owing to lack of clear definition, and partly to the numerous faults which break through the outcrops of the lamprophyre sheet on the south-east side of Glen Tarsan, etc. Within the bed, and generally rather nearer the south-east margin than the north-west, there is some breadth of thinner more micaceous schist. This is seen on the north-east side of the River Massan; it occurs in two bands, each about 50 yards broad, and separated by the same width of more massive pebbly rock. On the south-west side of Glenmassan, near Stonefield, a north running basalt crosses the band, and, judging from the outcrops at the sides of the alluvium, it must be accompanied with a considerable amount of displacement, the west side towards the north. This fault seems to divide into two parts on the north-east side of the glen, the basalt continuing with the west part. The east part is quite distinct just outside the north margin of the area being described, throwing a lamprophyre sheet about 120 feet down to the north. On the south side of the greywacke, the fault shifts the Coire Athaich hornblende schist for about 100 yards along the ground surface, and, after this, makes a conspicuous slack on the hill. We could not see it crossing Eas Mor, but we suppose it may continue across both this and the Glen Lean burn (probably near Clachaig, where the course of the burn is through drift and alluvium) and join the fault crossing the west side of the Bishop's Seat.

On the west side of A' Chreachan there are other north and south basalts traversing the bed and accompanied with fault crushes. The crushes are at least several feet wide in some places; the east side appears to be somewhat advanced to the north, as before. These crushes are not seen in Coire Athaich Burn, but there is a space in which no rock is seen, and we have carried the line on through this to a north to south gap in the crags of Creag Capull. This gap contains a series of basalt dykes; the presence of such a series all running out of their ordinary course of itself affords a strong presumption that they are taking advantage of some previously existing line of weakness; and besides this, a crush breccia, about 1 foot wide, is seen just on the east side of the gap. There is room for more crushes on the west side also. On the north side of Glenmassan the same crushes are seen again in the burn that goes through the "l" of "Glen." At a distance of ¼ mile from the burn-foot a thin basalt dyke is seen running with the fault, and this is shortly afterwards joined by another dyke coming in from the south-east. Forty yards above this a lamprophyre dyke comes in from the same direction; this is not seen on the other side of the burn, and is no doubt faulted. The dyke mapped on the north-west side is a portion of the basalt. Another portion of it continues up the stream a little further, but in a very crushed state.

About ½ mile slightly north of east of Glen Lean the south-east margin of the greywacke is perhaps folded. It is not actually seen to be so, but the pebbly schist in a small stream not marked in the Map seems to strike against the thinner foliated more micaceous rocks in the burn a little to the north.

The section in the burn ⅛ mile north of Glenmassan shows one or two north running basalts accompanied with a crush. West of these are several thin lamprophyre sheets, generalised as one in the Map, and on the east a single broader sheet at a slightly lower level. These are clearly shifted by the fault movement.

The hornblende schist in Coire Athaich shows an apparent thickness of from 12 to 20 feet. It is so full of ferriferous carbonate that it has the aspect of a limestone. The Stonefield north to south fault displaces it as already described. A little on the east side of this another north to south fault displaces it again in a similar way, but only for a distance of about 10 yards. It is probable that this fault also is a part of the Stonefield fault. No trace of the bed was seen in the burn on the north-east side of Clachaig hill.

Apart from the faults already mentioned, it may be well to call attention to the two or three N.N.E. faults in the area between Glen Tarsan and Glen Laoigh. The throws of these do not appear to be large, but they may form prominent features, and represent the south terminations of a set of faults which have been traced a great distance to the north-east. The great Tyndrum fault is one of the set. In Sheet 37 there are indications that part of the movement along these is of earlier date than some of the lamprophyres.

The little bit of coast line in this area shows hardly any rock exposure. A great part of the burn courses of the larger glens are also covered with superficial deposits. The slopes of A' Chruach, A' Chreachan, and Sgarach Mor show excellent rock exposures.

District of Colintraive.

This district is enclosed by Glendaruel, Loch Riddon, the east Kyle of Bute, Loch Striven, Glen Laoigh, and the north margin of 1-inch Map 29.

The schists in this area extend from the "green beds" on the north-west side of "the anticline" to the "green beds" on the south-east side, and include some hornblende schists.

Two bands of the last-mentioned rock occur at the north end of the area. The base of the lowest of them runs about ⅝ mile north-west of the base of the green beds, and is apparently on a considerably lower horizon than any other hornblende schist which occurs in 1-inch Map 29 on the north-west of the anticline. A W.N.W. fault, accompanied with a basalt dyke, traverses these on the south-east side of Cruach Conchra, and advances their outcrops on the south-west side towards the north-west. The advance in the case of the lower band is about 120 yards. Smaller shifts also occur along the basalts dykes ½ mile north of Cruach Conchra. On the south-west side of Cruach Mhor a wide peat moss covers the hilltop, and neither band was seen on the far side of this.

In the Map it will be seen that a little north-west of the base of the green beds various other lines, running roughly parallel to it, have been engraved but not coloured. These were drawn in an attempt to separate out, one by one, those beds which had the characteristic green-bed character from those which had not. The attempt had to be left incomplete owing to the amount of time required. As a fact, it is probable that hardly half the beds included in the green beds zone on the north-west side of the anticline are typical green beds, the rest being grey quartzose pebbly beds and more micaceous thinly foliated schists.

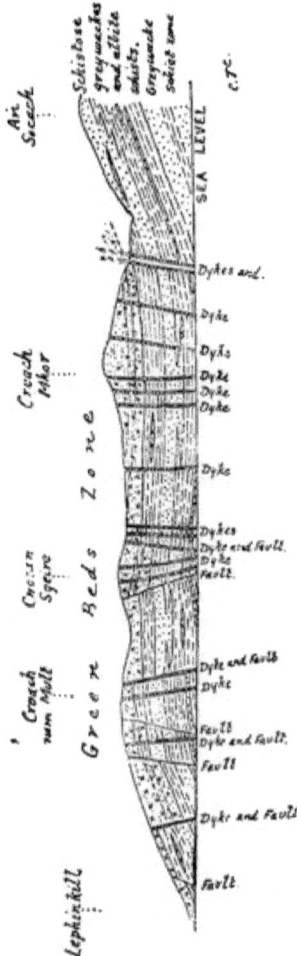

FIG. 44.—Horizontal scale 1 inch=1 mile. Vertical scale not uniform. Diagram section, nearly along the strike, between Lephinkill and An Sceach, to show the faulting.

Below the basement green bed mapped near Lephinkill there is, in the feature on the north-east side of the fault by the "h" of "Lephinkill," another outcrop of a similar character, but it is only 2 feet thick and is not shown in the Map. Such bands may be infolds of the higher lying bed, but it has not been possible to make this out with certainty in any case. The mapped base is distinct almost continuously from Lephinkill to a little north-east of Cnocan Sgeir, and by its help many faults have been picked out —eight in all. Some of these throw in one direction and some in the other, and taken together they almost compensate for one another. The general direction is north-west. The fault nearest Lephinkill is indicated by a prominent N.N.W. feature, facing west, which passes through the "h" of "Lephinkill." The amount of shift at the surface is about 80 yards, the west side to the south. The fault next this is the largest of the series mentioned. It passes ⅝ mile south-west of Cruach nam Mult. The basement green bed is very distinct on the north-east side, at a place about ½ mile slightly east of south of the Ordnance Station 1009. On the south-west side it is not so distinct, but it is probable that the amount of shift is about ½ mile, the south-west side to the north-west. This fault is nowhere seen in section. One-half mile S.S.W. of the Cruach there is an east to west feature which we suppose to mark its course for a little way. A basalt dyke runs at the base of the feature, and the schist on its south side is in one place striking nearly parallel with it, with a dip to the north. On the east side of the burn that passes by the last "n" of "Auchnagarran" there is another feature, facing south-west, which may be due to the fault. The amount

of shift in the top of the greywacke horizon cannot be made out satisfactorily. The next largest of this set of faults is the fault that passes ½
mile east of Cnocan Sgeir. It is accompanied by one or two porphyritic
basalt dykes. The basement green bed is seen near both sides of the
fault. The amount of surface displacement is a little over ⅛ mile the
south-west side to the south-east. It is clear that there must be a shift in
the same direction in the top of the greywacke horizon on the east side of
Tamhnich Burn, for between the "n" and the last "h" of "Tamhnich"
there are good sections of the thinly fissile phyllitic mica schists that
come over the greywacke, while the schist exposures striking at them, on
the east side of the burn, are nearly all massive, often pebbly, schists.
Probably the fault runs along an obscure feature about ¼ mile north-east
of the "n" of "Tamhnich."

At the sides of Tamhnich Burn the basement green bed is hidden by
boulder clay, but it is shown in the stream itself. About ¾ mile south-
west of An Socach it is crossed by a broad zone of basalt and crush rock
(seen best in the two burns north-west of Balliemore), which seem to
cause some shift in it, the north-east side towards the north-west,—the
same direction as is indicated by the top of the greywacke zone to the
south-east. Some 300 or 400 yards on the north-east side of the slack one
of the best exposures of green beds in the district occurs, sticking up in a
prominent hillock.

Good sections of typical green beds, alternating with other schist, occur
in most of the burns that fall into the Ruel from the east side above
Lephinkill; perhaps the one in the burn passing by the "O" of
"KILMODAN" may be specially mentioned.

As elsewhere stated (p. 149), the green beds have their colour changed
into dark grey or black near basalt dykes for a distance of a foot or more.
This is seen near the foot of Auchateggan Burn, and in a burn ¾ mile
N.N.W. of Cnocan Sgeir.

On the south-east side of these green beds there is a breadth of about
¾ mile which consists largely of thinly foliated micaceous albite schists ;
but within this is a band of massive quartzose pebbly greywacke which
has been mapped where possible. This is seen in a crag ¼ mile north-
east of the "k" of "Dalleik." In the crag it seems about 20 feet thick,
and a little way below is another similar outcrop which may be the upper
band repeated by folding. The direction of stretching in the crag is about
20° north of the direction of dip of the foliation. Some of the felspar
pebbles are pulled out in this direction for a length of 1¾ inch, and crossed
at right angles to the length by four or five cracks (sometimes ¼ inch broad)
which are filled with granulitic quartz. A band which is presumably the
same is seen again in the crags ¾ mile N.N.W. of Cruach nan Cuilean.

A little below the above pebbly band we come to a broad zone of schistose
greywacke, the top of which is moderately definite, and has been mapped.
This zone extends south-east across the strike, at least as far as the
"ch" of "Auchenbreck" and the "d" of "Craigendaive," but the ground
is too obscure to permit of drawing a base line with any confidence. The
fault in Tamhnich Burn is seen distinctly in the stream bed ; the hade is
west; the horizonal shift at the ground surface seems about 50 yards.
The two larger faults on either side of the burn are not seen in section,
but are continuations of breaks which are distinct in the green beds base.

The greywacke horizons on the north side of Fearna Bagh, and on the south-
west side of Dun Mor, are no doubt the same, but they have not been joined
to one another, as there is an area near Cnoc Breamanach where the marginal
lines are unusually indefinite. Perhaps this is due to folding, but the folds
could not be traced out. The Fearna Bagh greywacke is on the south-east

side of the anticline of foliation, while the greywacke above Aucheubreck
and Craigendaive is on the north-west side. They might hence be supposed
to be the same, but for reasons elsewhere mentioned (p. 86), we do not
think they are. Even on the supposition that the foliation agreed with
bedding, which of course is frequently not the case, there would be a
difficulty in supposing these horizons the same, as a greater thickness of
thin micaceous schist is seen (e.g., on Dun Mor) below the south-east greywacke, dipping south-east, than there is dipping north-west below the
north-west greywacke. Near the centre of the Fearna Bagh bed there is
some thickness of thinner more micaceous schist; this is seen well in the
Fearna Bagh section. On the north side of the 100-foot beach, inland
from this bay, the south-east margin of the bed is unusually indefinite.
The appearances may well be due to folding.

The fault crossing the greywacke $\frac{1}{2}$ mile west of Cnoc Breamanach
seems to shift the north-west margin of the bed about 60 yards, the east
side being advanced to the north. The crush is seen by the side of the
most westerly basalt dyke, about $\frac{1}{4}$ mile south of the last "ll" of " Allt
a' Chapuill." The crush rock is hardened and altered by the basalt;
the basalt itself shows no sign of crushing. We have continued this
fault south along a feature facing west, and this leads to a place where
the broad east to west basalt dyke becomes much thinner, or even dies out
for a certain space, and changes its direction so as to run along the
supposed fault line. We suppose the dyke to have come up along this
previously existing line in much less volume than usual, owing to the
line being so different in direction to its normal course (see p. 144).
Some other basalts, probably of later date than this east to west one, have
also come up along the same line a little south of this locality.

The fault that crosses the greywacke at the head of Allt Glac na maill
makes a strong hæmatite-stained breccia, occasionally accompanied by
basalt, nearly continuously in the west branch of Milton burn between
the "n" of "burn" and the "h" of "Breamanach."

One-quarter of a mile south-west of the top of Coraddie the broad
basalt changes its direction sharply. A fault is supposed to accompany
it in the changed path. No crush is actually seen, but the changed path is
in the same line as a powerful crush in the burn north of Lochan na Leirg
which can be traced up to this place by a prominent feature. The dyke
is seen in the changed path, but in a thinner form; it is not faulted nor
unusually crushed.

In the locality $\frac{1}{4}$ mile north-east of Beinn Bhreac the east to west
dyke is only seen in the north part of the changed path. It is more
crushed than the later north-west dykes that intersect it, but is by no
means a train of crushed pieces dragged out along a fault of date subsequent to the intrusion. The intrusion has originally had a nearly
north and south direction. In the burn to the north the crushing
affects some of the later dykes also, and throws some lamprophyre sheets
on the west of the burn.

On the west side of Loch Striven, about $\frac{1}{2}$ mile north-west of Ardbeg, a
massive coarsely pebbly bed is seen, and is shown in the Map up
the hill to the north-west for a little distance. The appearances in
nature are much more complicated than would be judged from the
Map, but the small scale compels us to generalise. As a fact, there are no
fewer than three outcrops of this schist in the coast section. The most
northerly is clearly along the crest of an anticlinal, and, though it is 5 or
6 yards wide at the sea level, it thins out to 1 yard in a distance of 20
feet above this. On the south side of this north portion it is seen clearly
how the first foliation which elongates the pebbles, though folded, is still

constantly crossing the original bedding. Just on the south-east of the burn which runs from Lochan na Leirg various pebbly schists come on the south-east side of the mapped band, and these may be repetitions of the same bed. In the burn itself there are various basalt dykes, and crushes running with them, and these shift the pebbly bands somewhat, the west side towards the north. On the north-west side of the burn two pebbly bands are seen ; the south-east one is the broadest and as coarsely pebbly in some portions as the north-west one. But suddenly at a point 170 yards west of the burn the south-east band is replaced, presumably owing to folding, by thinly foliated very micaceous schists. And so it seems probable that this band is only a repetition of the other one. West of Lochan na Leirg some more pebbly portions occur, both south-east and north-west of the main band, but how they come in is not clear. A massive pebbly band on the west side of the fault which runs through the "h" of "Breamanach" may be on the same horizon, but it is not so coarse.

No other schists have been mapped until we come to the green beds on the south-east side of the anticline, near Strone Point. There are two bands, both well seen in the raised beach margin on the north of the point, which can be traced west as far as a prominent feature facing west. Various exposures of basalt are seen along the continuation of this feature to the north, and in all probability basalt occurs with it here also ; but, besides this, it is necessary to suppose a fault of some magnitude, for on the west side of the feature we see nothing of the green beds, though they may occur, after a considerable shift to the south, in the drift-covered area south of Glaic ; but striking against the top of the upper green bed is a considerable breadth of black phyllite and thin quartzose schist. On the south-east side of the upper bed come some thin alternations of grey-green phyllite and quartzose schist, and then a considerable breadth of massive grey pebbly schist; after this, for $\frac{1}{8}$ mile or more, as far as the extreme end of Strone Point, the ground is drift covered, with the exception of a very small exposure at the raised beach margin. The exposure consists of grey phyllite and massive pebbly schist ; on the whole it did not suggest the presence of beds of the Dunoon series. On the east side of Loch Striven there is some massive pebbly quartzose schist between the most south-east green bed and the phyllites on its south-east side (see p. 176). Whether the beds at Strone Point belong to the same series we are unable to say. The outcrops of green beds on either side of the loch do not match in number, or strike accurately for one another.

Good sections are abundant both on the coast and in the inland scars. Several of these are elsewhere described (pp. 23-28). They are scarce on the north-east shore of the Kyles of Bute, but nearly continuous on the west side of Loch Striven. One-quarter of a mile N.N.W. of Troustan the limbs of early folds, themselves twisted by the later "anticline" folding, are well shown by the thin impure calcareous bands. The coast to the north-west is also interesting, because it shows clear instances of the first foliation crossing the stripe or bedding of the schists. C.T.C.

CHAPTER XVI.

GENERAL GEOLOGICAL STRUCTURE OF PARTICULAR DISTRICTS (*continued*).

District of Tighnabruaich.

This district forms the most southerly part of Cowal, and lies south of Kilfinan and the hills, Cruach Kilfinan, and Beinn Bhreac. Its northern boundary is an east and west line from Eilean Dearg in Loch Riddon to Drum Point on Loch Fine, and the sea forms its other boundaries. It comprises the southern part of the parish of Kilfinan,* but the only place of importance in it is Tighnabruaich, which, stretching with Kames for two miles along the Kyles of Bute, is a favourite summer resort. The only part of the area which is more than 1000 feet in elevation lies around Beinn Capuill, and this high ground slopes rapidly down to the shore at Loch Riddon and Tighnabruaich. The greater part of the district is extremely undulating, being much cut up by a number of small streams and hollows into many isolated hills, not a few of which are steep and craggy. The coast in Loch Riddon and the Kyles has a comparatively smooth outline, while in the south and west, bordering on Loch Fine, the outline is varied by a good many small bays, the shores of which are diversified by beautiful patches of native wood. The finest of these is Kilbride Bay at the mouth of Allt Osda. The shores of Loch Riddon and the Kyles are also beautifully wooded. There are two natural lochs of considerable size, Loch na Melldalloch and Loch Asgog, besides a good number of old lochs now silted up and overgrown with peat, which are coloured on the Map as alluvium.

The district embraces strips of the following sets of beds :—

    Garnetiferous schist horizon.

    Loch Tay limestone, with its associated epidiorites and hornblende schists.

    Green beds.

    Mica schists, with subordinate gritty and pebbly bands.—Nearly whole extent.

The country occupied by these bands is very narrow in the north-west, gradually widens out to about the central part of the district where the anticline is, and narrows again gradually to Ardlamont Point, the southern extremity.

Garnetiferous Schist Horizon.—This band occupies the coast from the north side of Auchalick Bay to beyond Drum Point, and is the most north-westerly of the strips to be described. It is a fine-grained and thinly foliated sericite mica schist, extremely crumpled, and traversed by very numerous quartz veins, many of which are parallel to the foliation planes. Gritty bands are met with occasionally, but none of sufficient importance to be separately noted, and towards the base on the north side of Auchalick Bay there are a few thin green bands of the type hereafter to be mentioned, besides a thin band of limestone ; and this, together with the limestone of Ach Chreagach, will be described with the Loch Tay limestone. Good sections of the beds can be seen along the rough and craggy coast and in the interior about Ach Creagach and Inveryne Barr (Δ 268), but the interior generally is not very

---

* The name has nothing to do with Fionn, the mythical Gaelic hero, but is derived from St Finan, who lived in the 7th Century.

rocky, and has a somewhat tame appearance. Several small faults or marked joints are seen on the coast at Drum Point and to the southward, which have various directions, and at Port Leathan are two more important parallel faults, running N.N.E., about 30 yards apart. These are best seen on the south side of the bay, where the rock is much stained near them, but they are also seen on the north side. A very large north-west fault comes into this bay, and this will be described in the sequel. The dip is everywhere to the north-west, but the contortions are so violent and frequent that it would be useless to give any estimate of the average amount. There is a marked feature running nearly due north from the north side of Auchalick Bay to the summit of Inveryne Barr, which may be a fault line. The rock has not been much worked. There is a quarry in it east of the road, about ½ mile north of Drum.

Loch Tay Limestone.—This is a light-coloured highly crystalline limestone with abundant plates of white mica on the foliation planes. It is generally accompanied by bands of epidiorite, and often divided by them into two or more separate bands. There are small old quarries in the upper part, W.N.W. of Acharosson, and the lower part is thin bedded, and seems more micaceous than the upper. The limestone can be traced pretty continuously from the path leading between Kilfinan and Acharosson down to the north shore of Auchalick Bay. As no good section shows the top or bottom of the limestone it is impossible to give its exact thickness, but it seems altogether as much as 30 or 40 feet. It cannot be seen for some distance on either side the main road, but south of the bog which crosses its course both parts can be well seen up to the fault that runs east and west to the south of Cor Mheall. This shifts the outcrop altogether between 150 and 200 yards to the west, and would seem to have a downthrow north of 200 to 300 feet. At Inveryne, south of the fault, both bands are seen again for a short distance, and there is a quarry in the upper one, showing that it rests on the epidiorite. Both bands soon get covered up with drift, but they must be shifted to the north-west, nearly 400 yards, by the Melldalloch fault, and if the dip of the beds is 30° to 40°, as it seems, the downthrow on the north-east would be as much as 600 feet. On the other side of this fault there are old quarries in both the bands, but they show little rock, and the limestone finally disappears under the old marine alluvium of the beach. About 150 yards to the north-west there is a thin band of calcareous schist, probably 2 to 3 feet thick at least, which has been traced for ¼ mile to the north-east, under the hill called Inveryne Barr (Δ 268 on the 1-inch Map). This band seems to be quite distinct from the Loch Tay limestone.

To the south-west of Ach Chreagach and north-west of Drum a band of limestone occurs that is quite different from that last mentioned. There are several quarries which show a good thickness of limestone and calcareous schist. The most northerly quarry is in rock of the latter kind ; in the next there is a good and thick limestone on the same line of strike. Perhaps there is a fault to the south of this quarry, for the third quarry is much further east, out of the line of strike. This last shows calcareous beds for 20 yards wide across the strike, but it cannot be traced northward. No trace of these beds could be met with to the north or south of these quarries, and their occurrence here is certainly puzzling, but it seems possible that they may be parts of the Loch Tay limestone brought up again by sharp folds.

Epidiorite and Hornblende Schist.—This, which nearly everywhere accompanies the Loch Tay limestone, has the general characters which have been elsewhere described. In this district, however, the rock is more massive than usually it is further north, and is more an epidiorite than a

hornblende schist. There are two, and in one place three, bands of the
rock, but near the main road not one of the bands could be found, and
they are not supposed to be quite continuous, but to behave as intrusive
sheets often do, die out for a time and set in again nearly on the same
horizon.

Near the Acharosson path there is one band 50 feet broad intercalated
with the limestone, and a much broader and very massive band above
the limestone. Only the upper band is seen between the main road and
the hog, but south of this two bands are visible in force near Cor Mheall,
and probably each is as much as 50 feet thick, for the upper one is quite
100 feet in width at the surface, while the lower one spreads over a pretty
wide area—seemingly as much as 250 feet wide. There is no section,
however, from which the thickness of either can be exactly ascertained.
Both bands are shifted equally with the limestone by the Cor Mheall fault,
and can be traced for some distance till they disappear under the drift. .
They seem to be thinning gradually, for when they appear again south of
the large Melldalloch fault they are much less thick, being only about 30
feet in width all the way to the bay, where another and higher band of
about the same width appears about 40 yards north-west. This can be traced
for about 300 yards inland. It is coarsely crystalline generally and pretty
massive, but finer at the edges and more properly a hornblende schist.
The rock generally contains small garnets, and the schist into which it
has been intruded is decidedly hardened along the line of contact.

Green Beds.—These are characterised by the abundance of black mica, of
chlorite and of epidote, not only in the finer portions but also in those which
are strong and gritty, and are occasionally quarried. The less gritty bands
often decompose into a very soft greenish schist, *e.g.*, in the burns which
run into the Acharosson burn from the south, between Auchanaskioch and
Acharosson, especially the one engraved on the 1-inch Map. The bedding
and foliation in these schists seem much more regular and not so
crumpled as in the ordinary mica schists, and the quartz veins are much
less numerous, except in the mica-schist bands, which occur in places
among the green beds. The line for the base of the green beds is a fairly
good one, for the lower part of the strip mapped is mainly composed
of beds of the genuine "green-beds" type, but the upper line coinciding
with the Loch Tay limestone is a purely conventional one in this district.
It marks nearly, though not quite accurately, the upper limit of the green
beds, but includes in the upper part of the strip also a good deal of mica
schist of the ordinary type, which could not well be separated out from it.
It must therefore be clearly understood that the upper half of the strip
coloured as "green-beds" is a mixture containing perhaps more ordinary
mica schist than green beds proper.

The coast between Auchalick and Ardmarnock bays gives good sections
in some of the upper beds, showing mainly grey mica schists much contorted, with the usual quartz veins, and with occasional bands of green
chloritic schist. There is a strong and marked band of the latter on the
south side of Auchalick bay, and another on the north side of Ardmarnock
Bay, close to which are three thin basalt dykes trending W.N.W. generally,
but somewhat meandering ; one of these is 3 feet wide, and another 2 feet
6 inches.* To the northward is a long curved fault, the general trend
being east of north, and west of this, and west of Carraig nam Ban,
is another small basalt dyke 2 feet wide. Strong chloritic schist occurs
about Cor Mheall, and a soft green bed, much decomposed and 10 feet
thick, may be seen on the north side of the basalt dyke quarry near Drum.

* It is probably one of these dykes that has been quarried about ¼ mile inland
in a plantation.

Alternations of green and grey schist may be seen in many places of the lower part of the Acharosson burn, and in this ground, to the north-east of Drum, occurs a band of hornblende schist about 20 feet broad, which is probably an intrusive rock.

There are good sections of the green beds proper in Ardmarnock Bay, about Black Harbour and to southward. Two small basalt dykes traverse the beds in Ardmarnock Bay in a W.N.W. direction. A strong green gritty band has been quarried by the roadside ¼ mile north of Ardmarnock House, and in the plantation east of the road, ½ mile further north, there is a marked grey mica schist band which runs north-east past the site of the old fort. There are good sections in the green beds about Melldalloch and in many other places, and in the Acharosson burn. Thin lamprophyre dykes occur in them at Auchanaskioch, and both north and south of Melldalloch Farm. The Kilfinan fault is probably prolonged southward across the Acharosson burn, shifting the green beds there, but the evidence for it did not seem clear at the time the survey was made, and the line for the base of the green beds was left unfinished. The fault passing by Melldalloch shifts the base of the green beds about 2000 feet, so that if the dip be taken as 30°, which seems an underestimate, the throw must be as much as 1000 feet down north.

Mica Schist.—This rock occupies by far the larger portion of the district, which probably contains representatives of nearly all the horizons between the two sets of green beds on the north-west and south-east. Two marked bands of strong gritty and pebbly schist stretch across it from south-west to north-east, one on either side the anticlinal line, and the whole from Ardmarnock House to Ardlamont Point, in whatever way dipping, is probably one set of beds. Very little of it seems to be a true phyllitic mica schist, and the prevailing type of fine-grained mica schist gives one the impression that the mass of it was originally a fine-grained shaly or flaggy sandstone, in cases alternating with sandy shale.

The strip of mica schist which comes south-east of the green beds falls to be described first. This strip contains here and there thin bands of strong gritty or pebbly schist, which cannot generally be traced very far, but as many as four of these were followed for some distance in the southern part of the strip and coloured yellow on the Map. The most important of these is the one west of Creag Mhor, which is 50 feet broad. These gritty bands contain less mica than the finer parts, and few of the quartz veins so common in the ordinary schist, and generally they are not so much contorted. Several of these thin gritty bands may be seen on the coast about Rudha Preasach, alternating with the finer type of schist. One of these bands south of the point can be traced inland for 300 yards. Another occurs further to the north-west, which is crossed by a fault throwing down a few feet on the west side. A marked feature, running in a north-east direction parallel to and east of this gritty band, is probably a fault feature. Sections are numerous inland in the little crags on the moors, particularly west and south-west of Creag Mhor and north of Loch na Melldalloch. The best stream sections are in the Acharosson burn and in Stowhall burn, which joins it, running near the foot path to Acharosson. Two veins or faults are to be seen in these beds in the Acharosson burn, both having a direction a few degrees north of west. The lowest of these is a breccia, with much red ferriferous carbouate, and occurs a little west of the point where a small burn joins the main stream from the north ; the other, which is a pink ferriferous carbonate, is about 600 yards further east. If the Kilfinan fault continues across this area, which is likely enough, it must pass by the west end of the crag called Sròn.Dubh, and to the west of the big moss. There are four detached bits of intrusive red lamprophyre, dykes and sills, along the western edge, south of Acharosson, which are engraved on the 1-inch Map, but perhaps not coloured, and there is a thicker mass of the same kind further east, which crosses the Stowhall burn. The width of this strip of mica schist is a little more than the shift of the Melldalloch fault, which crosses it about its centre, so that its total thickness may be about 1000 feet, on the supposition that the north-west dip averages about 30°. It certainly seems to average a good deal more on the east side of Bagh Buic, where the north-west joints are so prominent, but the crumplings are so numerous in places that not much reliance can be placed on the above estimate.

Gritty Schist of Glenan.—Under this head will be described a band of strong schist, containing numerous gritty and pebbly bands, which stretches from Glenan Bay to near the Acharosson burn. It is about 1000 yards in average width, but contains within it a good deal of the fine-grained ordinary mica schist. Lines have been drawn for the top and bottom of the mass, and coloured, and the more markedly gritty or pebbly bands, inside the area, have also been indicated by a yellow colour. The most traceable of these bands is the coarse and pebbly one 30 to 50 yards in width, which can be clearly followed for 2 miles along the west side of East Glenan burn, past Creag an Fhasgaidh and Carn Bhuilg to the Melldalloch fault. It forms a good feature in several places, and under Creag an Fhasgaidh is crossed by several faults ; two (most southerly) throwing down to south about one half the thickness of the band, while two others throw down to the north nearly its whole thickness. This is almost certainly the same band which is marked as trending northwards from the west side of the upper dam towards the Acharosson burn. At Capull Cloch, ½ mile E.N.E. of Loch na Melldalloch, the band is very conspicuous, about 70 yards broad, a conglomerate schist with much drawn out pebbles of quartz and felspar. Another marked pebbly band can be traced along the west-side of West Glenan Burn and under Creag Mhor. The upper limit of the whole group is not so clearly defined as the lower one, and in the northern part of the area is somewhat obscure. It would seem to run from near the east end of Loch na Melldalloch N.N.E. to the big peat moss, and on-ward in the same direction towards the Acharosson burn. Two sheets of red lamprophyre, or perhaps more properly of felsite, 30 to 40 yards in width, occur here in the schists south of the Acharosson Burn, and a broader band of about 50 yards on the north side of the burn, the latter dying out rapidly at either end. The large felsitic mass, which is found further east at the great bend in the burn, has been previously noticed, p. 150. A band of the same character occurs south-west of the Acharosson footpath, and seems about 70 yards broad, and another or the same is quarried for road metal on the eastern side of Loch na Melldalloch.

There seems to be a fault cutting off the coarse gritty band west of the upper dam, and there are two large faults, trending about N.N.W., between Ardmarnock House and Glenan, both of which throw down to the west and shift the beds from 100 to 150 yards. Sections of these gritty schists are also to be had in the islands off the coast,—Eilean Buic, Eilean na Beithe, and Eilean Buidhe. The dip generally seems as much as 30° or 40°.

Up the East Glenan burn there is a fairly well defined horizon of gritty schist, 200 to 400 yards east of the main band which has been described.

Anticlinal Area.—It will be convenient to treat under this head, of the schists which are found between the two bands of gritty schist coloured yellow on the map. "The anticline" in the foliation passes through the middle of the band from south-west to north-east, from Port a Mhadaidh on Loch Fine, over Cnoc à Chasteil, south of Binnein Ban, past Craignafeich, Tighnabruaich reservoir, north of Barr Liath, along a small stream and moss south of Beinn Capuill to the head of Loch Riddon. It must be understood, however, that it is not a simple anticline, but a compound one, with many minor folds, which are practically nearly flat for some distance on either side of the central line, the position of which it is not always easy to fix. Thus, for instance, north-west and west of Binnein Mor, near Loch Riddon, the foliation is nearly horizontal for a considerable distance "The anticline" seems, however, to be steeper, especially on its north side, and more simple as we proceed from this point south-west towards Loch Fine.

The rocks of this division are in general more folded than the preceding, and the dip more variable. The ground occupied by them is also rougher and more decidedly craggy. The rocks seem for the most part to be of the finer-grained variety of mica schist, but a few gritty bands are met with occasionally. The whole, however, have undergone a more pronounced metamorphism than the beds further removed from the anticlinal line.

Sections of the beds on the north side of "the anticline" may be had on the shore of Loch Fine, between Port a Mhadaidh and Lub Faochaige :— fine grey mica schists, full of quartz veins, much crumpled, and with many reversed folds. Inland these beds are well exhibited on the hill, Barr nan Damh, west of Derybruich, where they have similiar characters, and several marked features running north-east, perhaps caused by joints, may be noted. These much crumpled rocks can also be well seen on Cnoc à Chasteil, west of Asgog Loch, and on Binnein Ban, also on Creag an Fhithich, A' Chruach, and other crags and hills, especially Cnoc Uaine and around Beinn Capuill. To the north-west of the latter, and near the head of Allt Dubh, two basalt dykes penetrate the schist, running pretty close together and nearly parallel to the burn ; one of these is 5 to 6 feet wide.

On looking at the Map it will at once be seen that the horizon under description, between "the anticline" centre and the gritty schist band to the north, is the one along which lamprophyre intrusions are most numerous along the strike, in fact there is an almost continuous band of these running from Lub Faochaige, on Loch Fine, by Beinn Capuill, to the head of Loch Riddon. Red bands of this description occur just east of the summit of Beinn Capuill, and further to the north-east, from 6 to 10 feet thick. A thicker and coarser band on the west side of the hill is crossed by the Dunoon dyke, and seems cut off by a fault at its north end. It is a coarse red mica trap, much decomposed in places, and is shifted more and more westward by a series of faults ranging east and west. The first of these, north-west of Beinn Capuill, shifts it 100 yards, and at a little south of this the mass contains lumps of quartz. The next fault at Coire Ban, south head of Allt Mor, shifts it westward 200 yards, and the third about 100 yards. South of this the sill is much wider and is soon lost sight of in a doubtful way. Similar bands may be seen westward near Allt Mor and the eastern branch of the Creag-an-fhithich Burn, where one band is at least 15 feet thick, but there is a good deal of drift about the heads of these burns, and the sections are poor. In the Creag-an-fhithich Burn, about 350 yards east of the Acharosson footpath, there is a fault trending W.N.W., which seems to cut off a red lamprophyre sill on the north, and a quartz vein, 2 to 3 feet wide, runs northward in the burn. A little further east is a calcareous breccia, also running W.N.W., near which are

N

thin veins of ferriferous calcite. About 300 yards west of the footpath there is a thin grey amygdaloidal basalt dyke running along the south side of the burn. South-east from this, and east of the upper dam, is a coarse red mica trap, with large red felspars and black mica, which may be the continuation of the one last described in the burn east of the footpath. Eastward of this, and on the other side of the footpath, a quartz vein 6 inches thick is sticking to a vertical face of schist trending W.N.W.

Two parallel red intrusive lamprophyre sheets occur south of Loch na Sgine, running under A' Chruach, and the most westerly one is prolonged under Creag an Fhithich, where it is 5 feet or more. There are marked features in these two hills which seem often to run along north-east joints parallel or nearly so to the strike. Passing to the west side of the Creag-an-fhithich Burn we come on the sills again, near the main road, and one runs up the west side of Allt an Eas, west of which the strike joint features are very prominent. Another band is on the east side of Allt an Eas and north of Binnein Ban, where its course is a zigzag. West of Binnein Ban there are three of these red intrusive bands, the lower or most easterly one being 7 to 8 feet thick, and traceable, with interruptions, to Derybruich. South of Binnein Ban is a marked hollow along the strike, probably due to a fault or joint, and to the east of this three N.N.W. lines of fault. There is a fine feature running north from Alltan nam Breac which flows into Asgog Loch, to the road at Allt an Eas. Two red sheets are seen on the coast, one being opposite to Eilean na Beithe, and the other on the south side of Lub Faochaige. This last is a reddish brown band of a few feet thick, curving in direction and giving off thin strings at its eastern end. A vein of ferriferous calcite is to be seen in several places in the upper part of the Acharosson burn, accompanying the Dunoon east and west dyke there.

The north-westerly dip in this area is much greater than it is in the adjoining district to the north-east, and as illustrating this it may be mentioned that the distance of the base of the gritty group from the centre of "the anticline" is half as much again near Beinn Capuill as it is on Loch Fine. Good sections of these rocks on the south side of "the anticline" may be seen on the coast between Rudha Stillaig and Port a Mhadaidh. South of the latter they are much folded, with numerous quartz veins, and a gritty band, 30 to 40 yards broad, may be seen at the top of the old coast cliff. The dip is comparatively low, and south of Port Ghabhar, the beds seem practically horizontal in places; still they are much of the same fine character, fine mica schists, with occasional thin gritty bands. A fault here runs north-east, and can be traced to near Polphail. The coast is rocky, with only traces of the old shelf cut by the sea, on which are the stacks of the old coast line. Near the fort Creag a' Chasteil are four nearly parallel faults, trending 25° to 30° north of west, and the rocks continue to be of the same character round the head-land of Rudha Stillaig to Port Leathan, where occur two thin dykes or sills, probably of the lamprophyre type, running in the direction of the strike, about 50 yards apart.

The fine grey micaceous schist, much crumpled and quartzy, may be observed inland along the strike in many places. There is a marked strike feature 500 yards south-east of Polphail, extending for nearly ½ mile. Some coarse bands occur about Barr Iolaich, where the prominent foliation dip is 30° to S.E., and another foliation dips 60° in same direction. Rocks of a similar character are found round Loch Asgog and Craignafeich. South of the latter place, 80 yards, is a basalt dyke running east and west, which has been quarried, though only 2 to 3 feet wide. Another basalt dyke trending W.N.W. crosses the small stream to the west, and is 6 feet

wide where quarried by the roadside ; and south-west of this in the burn and running along it is another, 4 to 5 feet wide. In the field east of the powder works plantation, and west of the road leading to Kames, there is outcropping much strong grey mica schi·t, and rock of the same character appears between Achanlochan and the pier, where it has been quarried. There are sections along the upper road here (new road) showing :—
North Side,

> Basalt dyke, 12 to 18 inches.
> Mica schist, 8 feet.
> Basalt dyke, soft brown decomposed, 8 feet.
> Mica schist.

On the roadside south of Craignafeich is a soft purple mica schist. A bit of rock crops out in the moss west of Achanlochan, with much contorted nearly flat foliation, and another foliation which is nearly vertical. To the east of this, about 300 yards, the dip varies from 0° to 20° while the dip of the cleavage foliation is high. To the south and south-east of Barr Liath the beds are extremely plicated, but probably nearly flat on the whole. Good sections can be seen in the lower part of Allt Mor, where there are two thin basalt dykes, and in the branches of the Tighnabruaich Burn. At the foot of Allt Mor much quartz occurs on the shore in strings, veins, and lumps. Between the two branches of the Tighnabruaich burn a thin lamprophyre sill occurs in the schist, and a basalt dyke up the burn west of Rudha Ban. There is another basalt dyke in Allt Dubh, E.S.E. of Beinn Capuill, which runs along one of the bends of the burn, and is accompanied by a vein. Numerous sections may be had in the crags about Binneiu Mor, north of which are several lines of fault, the most important of which is W.N.W.

The schist behind Tighnabruaich is in places of a very fine-grained character, it might indeed be called a phyllitic mica schist. It may be seen in the new road south of the Established Church, and schist of a similar character crops out in the path west of Stillaig Farm.

The Stillaig Band of Gritty Schist.—This has a width across the strike of nearly 400 yards, and on the coast it forms the western part of Eilean Aoidhe. It is mostly a strong, coarse and gritty schist, with some pebbly bands, and some that are thin, fine, and very micaceous with numerous quartz veins. Beds of the latter kind much corrugated are found at the extreme western end of the island and seem separated from the main gritty mass by a fault along the strike. The south-east dip seems to average 30° or 40° throughout, and near the upper boundary occurs a fine grey dyke or sill 5 to 6 feet thick, which runs nearly along the strike, but dips in the opposite way to the schist, or north-west. It is probably of the lamprophyre type.

Strong bands of gritty schist appear in the woods to the north-east ; and about Stillaig and Asgog farms the breadth of the band continues much the same ; but the sections inland are not good though some may be seen in it, in its course past Achadalvory, Auchagoyl, and Kames farm till it strikes the shore about the pier at Kames. It seems to miss the Tighnabruaich shore but appears again in strong force at Rudha Ban, where it can be clearly differentiated into three gritty bands with partings of fine mica schist. The most southerly of these has on the shore on the east side of Port Driseach a width of about 100 yards. It is mostly a rather fine quartzose strong and massive-looking schist very finely foliated with abundant mica, and white in colour. The dip south-east seems 30° or more, and there are signs of a later foliation dipping 75°. To the northwards is a band of fine micaceous schist, 50 to 60 yards broad, with quartz veins. The second gritty band is 80 to 100 yards broad, and there follows a second band of mica schist, 75 to 90 yards broad, similar in character to the first. The third gritty band is 100 to 150 yards broad. The two foliations may be seen in several places, e.g., 300 yards south of Allt-Dubh and near the summit of the hill south of the footpath leading to Callow. In the first of these places both dip south-east, one at 35°, the other at 70°, and there are some reversed dips which are probably over-folds.

The south part of Eilean Dubh opposite Callow for a width of 150 yards is mostly gritty and pebbly schist, with a few finer micaceous bands, one of which seems to form the southern hollow crossing the island in a north-east direction, while another forms the hollow near the northern boundary of the mass. The gritty schist is white, quartzose, finely foliated and crumpled. A fault on the west side of the island trending north-west seems to throw down to south-west about a foot. About the middle of the island on the east side is a very prominent joint which dips north-west about 10°, and near this the foliation is much crumpled, ranging from nearly horizontal to the vertical. At the south-east end in some coarse pebbly schist are traces of secondary foliation. The interior of the island is rough and rocky and very difficult to traverse, being covered with tall heather and brushwood.

The Ardlamont Peninsula.—The remainder of the district forming a triangle of which Ardlamont Point is the apex may be thus conveniently designated. We will begin with the shore section at Asgog Bay. Moderately fine grey micaceous schist full of quartz veins and much plicated in places forms the island of Sgat Mhor, the east side of Eilean Aoidhe, and the west side of Aagog Bay. A N.N.E. crush crosses Sgat Mhor; two faults run east and west along the south side of Eilean Aoidhe, and one trending north-west is in the hollow which cuts off the eastern end from the rest of the island. There seems also a large fault going W.N.W. between the island and the mainland on the north. There are irregular cracks filled with breccia and quartz 100 yards west of the point bounding Asgog Bay on the west, and near the mouth of the burn is an east and west crush line. The rock east of the bay is much the same in character, round by Sgat Bheag and Port na Beiste, and numerous faults are seen, the major part of which run in a W.N.W. direction. On the east side of Sgat Bheag is a grey intrusive sill of the lamprophyre kind. It is 4 to 5 feet thick, and appears on the mainland opposite, striking with the schist, and being faulted by its faults. The most northerly of these faults shifts the sill about 60 yards to the west. A few gritty bands are found in places, and there are some lines of crush going N.N.E. or nearly with the strike, and others ranging nearly east and west. The same kind of rock forms both sides of Kilbride Bay, and is generally full of quartz veins and much crumpled, with occasional crushes in various directions. A little north of Rudha na Peilige, a gritty band occurs 25 yards in width, which may be the same band that has been quarried to the north-east and north-west of Corra, not far from the road. To the eastward of the point on the shore, the rock is mostly fine and very micaceous, but there are gritty bands occasionally; and 100 yards east of the mouth of a little burn are two thin basalt dykes running northwards close together. To the N.N.E. of this and beyond the large east and west dyke, is a very coarse pebbly band, a few feet thick, in which are quartz pebbles drawn out to an inch in length; and there is a good deal of coarse schist on the shore about the centre of the bay, with several lines of crush. Passing to the east side of the bay there are some thin coarse bands, one of which is 10 yards wide; and 350 yards further south another strong coarse band is 25 yards in width.

There are several thin dykes and sills on the shore about 600 yards to south of this, which may belong to the lamprophyre type, and one in the same direction, E.N.E., is seen in a quarry west of Point Farm. There is also a larger dyke on the shore 15 feet wide in the same direction. We pass the large east and west dolerite dykes described on p. 150, and come near Ardlamont Point on a greyish-brown fine-grained sill, a few feet in thickness, clearly intrusive in the fine mica schist, but running nearly along the strike. It does not seem to cross the big dyke, being apparently of earlier date than it, but it appears again in the bay on the north side 100 yards from the dyke, where it is 3 to 4 feet thick, and it can be traced interruptedly to the north-east running out to sea, and either widening or splitting into two near low-water mark. This is probably a somewhat basic lamprophyre. There are also some thin sills which lie to the south-east of it, fine and white, and only a few inches thick. Between this and Point Farm a few thin sills can also be traced for a short distance. In the fine grey mica schist to the northward along the shore, there is a coarse band several feet thick which thins out very rapidly along the strike in both directions. The usual kind of mica schist prevails along the shore past An Gnob where it is altered by the big dyke, and soon after we come to a smash running W.N.W. North of Blindman's Bay there are strong bands occasionally; some very coarse and pebbly are to be seen south-east of Carry, nearly vertical, among alternations of various kinds with an E.N.E. strike. Northward to Kames, rock of the usual fine character prevails with occasional alternations of a stronger kind, but the sections on the coast are poor, and no detailed description is necessary.

Numerous outcrops of the fine-grained mica schists are found in the interior on the moors about Cnoc Mor and west of Colachla. On the east side of Allt Osda there are several strong grey gritty bands south-east of Coille na Sithe alternating with the ordinary kind, and one of these bands has been traced to the north-east beyond the large peat bog. There is another south of this which forms a fine feature, perhaps partly due to a fault. There are several N.N.W. features due to joints or faults to the east of Kilbride Church, and one feature long and prominent running due south from Allt a Chaoruinn. Others occur about Cruach a Kames, one of which is nearly along the strike. Numerous sections may be observed in addition about Sithean Mor, Cnoc na Carraigh, Cnoc na Carraige, and Cnoc na Cailliche.   W.G.

District of Kilfinan and Stralachlan.

This district lies west of Newton Bay, Caol Ghleann, Glendaruel, and the upper part of Loch Riddon, and north of the Tighnabruaich district.

The schists lying within this area belong to the following groups :—

Anticlinal grits and mica schists.
Green beds.
Loch Tay limestone.
Mica schist (with green beds and other schists not differentiated).
Quartzites and Phyllites of the Ardrishaig series.

The anticlinal grits and schists occupy the whole of the ground to the south-east of a line a little north of Ormidale, passing through Cruach Camuillt, a little to the east of Cruach na Broighleag, and 1 mile to the west of Kilfinan.

This line forms the south-east boundary of the "green beds." The "green bed" zone extends north-west as far as the first appearance of the Loch Tay limestone, which passing just above the village of Kilfinan continues a north-easterly course along the western slopes of Glendaruel, and finally crosses the boundary burn of this area ½ mile north-west of Sròn Cruaich.

The next zone of mica schists is bounded on the south-east by the main mass of the Loch Tay limestone, and on the north-west by the Ardrishaig series.

The south-east horizon of the Ardrishaig series follows a line from the head of Stralachlan Glen to a point on Loch Fine about 2½ miles north of Otter Ferry.

As a glance at the published Map (sheet 29) will show, the demarcation of the "green beds" from the anticlinal grits shows the existence of numerous faults, giving a general irregularity to the boundary. The mapping out of the base line of the "green beds" was the means of detecting many of these faults, which otherwise would have escaped observation.

There are occasional bands of "green beds" to the south-east of the boundary drawn. They are not numerous, are generally of small extent, and have not been shown on the published Map. Two such bands occur about ½ mile south-west of Cruach Camuillt. The farthest of them from the main "green bed" zone is about ¼ mile distant from it.

The "green beds" are well seen in the burn sections of Allt Lean Achaidh (a north and south tributary of the Kilfinan Burn), and the main trunk of the Kilfinan Burn from the exposure of the limestone to the "green bed" boundary to the east. This last section traversing the entire "green bed" series is the most important section of the "green beds."

There are also good sections of the "green beds" exposed in most of the burns of the western slope of Glendaruel, among which may be instanced the burn that joins the Ruel just to the south of Glendaruel House, and the burn joining the Ruel near Muymore.

The Loch Tay limestone, as regards its main mass, follows a more or less even course in a north-east and south-west direction. The superficial deposits prevent its being traced continuously along the line of strike, but wherever burn sections occur along its horizon the limestone is invariably found. In burn sections more than one zone of limestone may be seen associated with beds of hornblende schist or of mica schist. It cannot be said with certainty how far this repetition of the limestone may be explained by folding. But as some dip sections contain only one exposure of limestone, it is probable, when more than one band is seen, that folding will best explain the appearances. For a similar reason, the width of outcrop in different sections varies considerably.

Bands of hornblende schist first make their appearance close to the outcrop of the Loch Tay limestone, and they may occur in any of the sedimentary zones that lie to the west of the Loch Tay limestone outcrop. They do not, however, always accompany the limestone. There are some good sections containing the Loch Tay limestone which have no trace of hornblende schist. As a general rule, however, the Loch Tay limestone is associated with hornblende schist.

In that part of the area that falls within sheet 29, limestone outcrops sometimes occur to the west of the main limestone outcrop. Numerous bands are found in the neighbourhood of Otter Ferry and Largiemore. As the associated beds are similar to those found in association with the main limestone mass, it is supposed that they belong to the Loch Tay limestone and owe their present position to folding.

Good sections of the limestone and hornblende schist in association occur in the Kilfinan burn close to the village of Kilfinan, but still better sections are met with in the numerous burns on the western slope of Glendaruel, notably the Kilbridemore burn, where repeated folding has given rise to large continuous exposures of limestone. The burn at Largiemore also affords a good section.

The sedimentary zone above the main limestone mass contains "green beds," but they are not nearly so numerous as the "green beds" to the south-east of the limestone, in fact they occupy a very subordinate position, occurring as small bands scattered over the area. There is, however, one locality ½ mile south-east of Barnacarry, in which they are strongly developed : this locality shows the most north-westerly "green bed" exposure in this area.

In this zone various sills which occur scattered through the area are similar in character to those found in association with the limestone.

A big mass of serpentine occurs within this zone, the characters of which are described on p. 75. The main mass is nearly 1½ miles in length, and in its widest part has a breadth exceeding ¼ mile. The northern extremity of the mass occurs ½ mile south-west of Kilbridemore, from which point it extends nearly 1½ miles to the south-west, occurring as a big sill similar in behaviour to the epidiorites. A fault coincides with the western boundary for a considerable distance, and in one place a north-west fault divides the mass which has narrowed at the fault. Four smaller masses of serpentine occur along the same horizon to the north-east, which makes the whole line of serpentine extend over 2 miles.

The next zone to the north-west is represented by a set of siliceous and micaceous beds of the Ardrishaig phyllite series, in which the siliceous constituents are the stronger. These rocks are well displayed along the coast section between Largiemore and Stralachlan. There are also good sections in the burns which cut through this series, the courses of which are usually transverse to the strike of the beds. Amongst them may be specially enumerated :—The burn running into Loch Fine at Lephinmore. The Barnacarry Burn at Barnacarry. The burn which joins the Stralachlan river immediately to the south of Stralachlan church.

Various sills of epidiorite and hornblende schist are met with, but they occur in greatest strength in that portion of the area that lies between Stralachlan Valley and Loch Fine. They approach, in this locality, in size and habit, the Glendaruel serpentine mass rather than the long even sills of the Loch Tay limestone area.. The big ridge of which Bàrr nan Damh (571 feet) is the highest point, is made up almost entirely of one of these great epidiorite masses. Although they attain great dimensions their characters here are still those of sills rather than bosses.

Over the entire area the folding is intense. In the Ardrishaig series, however, the limbs of the folds point in the same direction, and are rarely crumpled. Hence, in walking along a hill slope, the enormous folding of the beds would never be suspected, the beds appearing to succeed one another with a steady dip. It is only when they are seen in section that the repeated folds can be detected, and they are not well seen then unless good alternations of quartzose with micaceous bands occur.

In the zones to the south-east of the Ardrishaig series the beds are not only intensely folded, but the limbs of the folds are themselves crumpled, and the highly-folded nature of the beds is obvious in examining any rock exposure.

Good instances of these foldings of the Ardrishaig series can be seen in the Barnacarry Burn, the Lephinmore Burn, and the burn running into the south end of Newton Bay.

The valley of Stralachlan occurs along the strike of the Ardrishaig series, and owes its existence to the erosion of the zone of softer micaceous beds which occur along the course of the present valley, flanked on either side by highly siliceous rocks.

Of the faults which have been observed in this area, some follow roughly the line of strike of the beds. One of the most persistent of these is the extension of the fault from the east side of Glendaruel, which after being lost sight of under the alluvial terraces emerges again close to Ballochandrain on the west side of the glen, continues its south-westerly course for about a mile along the bed of the big stream flowing past Ballochandrain, which has excavated a deep gorge along the line of fault. For the distance of ¾ mile beyond the head of the burn, all trace of the fault is lost, the ground being covered with peat and heather. It can then, however, be picked up again continuing its normal direction along the bed of a stream which joins the Kilfinan Burn, nearly 1 mile due west of Cruach na Broighleag, where a basalt dyke has come up along the vein breccia. After leaving the Kilfinan Burn no further trace of it is seen in this area. Although this fault can be followed for several miles, there is no evidence of a big throw.

In the parish of Stralachlan, close to Cruach Mòr and Cruach Fasgach, long features occur along the line of strike, which probably owe their origin to strike faults. The surface of the ground is so thickly covered with peat and heather that nothing definite can be made out.

A strike fault also forms part of the western boundary of the Glendaruel serpentine, already described.

Numerous faults are found transverse to the strike, the large majority of such faults having a trend a little north of west. The trend of the basalt dykes often coincides with the lines of fault, as for example in the Barnacarry Burn, and in the Lephinmore Burn.

J.B.H.

CHAPTER XVII.

GENERAL GEOLOGICAL STRUCTURE OF PARTICULAR DISTRICTS (continued).

District of Carrick Castle.

This district is inclosed by part of Loch Eck, the valley west of Carnach Mòr, Curra Lochain, the lower part of Lettermay Burn, Loch Goil, part of Loch Long, and part of the south margin of 1-inch Map 37.

The schists in this area consist of albite or phyllitic mica schists, and greywacke or quartzose schists. These rocks alternate so repeatedly with one another in thin bands, that it is not usual to meet with horizons of either type, which promise from their thickness to be readily traceable for some distance. Sometimes we start on a band which seems definitely bounded and homogeneous in character, but after tracing it a little distance, it gets mixed with schists of a different class, either near the margins or in the middle, and we lose the power of identifying it from the schists at the sides. Probably these difficulties are largely due to mixture of different bands by acute folding with axes lying parallel to the prominent foliation.

The schist lines which have been traced for the greatest distance, are the tops of greywacke schist horizons, one on the north-west side of "the anticline," and the other on the south-east side. Neither of these is very well defined, but in a district like this, where better lines are scarce, it was advisable to make the most of them, and they have certainly helped to disclose some of the larger faults and folds.

The top of the greywacke schist horizon on the north-west side of "the anticline," enters the area from the north, a little above the foot of Curra Lochain. There is drift for a little distance on either side of the loch, and the position of the line can only be located roughly. From the foot of the lochain down to the N.N.E. lamprophyre dyke, $\frac{1}{4}$ mile south of the second "a" of "Curra Lochain," the rock in the burn is nearly all massive quartzose schist, often distinctly pebbly. On the east side of Beinn Bheula, the quartzose schist is overlain by thin alternations of phyllitic albite schist and quartzose schist. The separating line between this series and the under one is not definitely marked. We see merely that after passing over a certain thickness of beds, we get on to a series in which the proportion of massive quartzose schist is less than it was before. The difficulty of tracing the line is increased by the difficulty of access to some of the crags. Near the top of the line, $\frac{1}{2}$ mile E.N.E. of the last "n" of "Beinn Bheula," there are several faults coming together from different directions. There are two ranging north-east or N.N.E., but their effect is uncertain. Another runs E.N.E. and hades north. The crush in this is sometimes 3 feet wide, but liable to sudden variation. Near the hill top to the west, it throws a thin lamprophyre sheet down to the north, 40 or 50 feet. The amount of throw at the greywacke top is uncertain, the other faults largely obscuring the section.

Five-eighths of a mile south-east of the "l" of "Beinn Bheula," the line is crossed by a N.N.E. crush, and a crushed lamprophyre dyke on the east side of this, but it is not clear that either of these causes appreciable altera-tion in level. Three-eighths of a mile south-west of the foot of Lochan nan Cnaimh, the line is crossed by another N.N.E. crush. In the burn near this locality, the fault throws down a lamprophyre dyke consider-ably to the west. The shift in the greywacke schist is not clear. South of this place we have carried on the line doubtfully for more than 2 miles. There are good exposures of albite schists above the greywacke for all this distance, on Cnoc Tricriche, Sgor Coinnich, Beinn Bhreac, and Cruach a' Bhuic; but on the tops of some of the hills, thin bands of quartzose schist come on again over the main mass of albite schist. The line between the series which consists mainly of albite schists, and the underlying greywackes is too poorly defined to enable us to find out small faults and folds. We have frequently found that a certain band of greywacke or quartzose schist, that seemed in one place to act as a good top to the greywacke horizon, having little else but albite schist above it, would, on passing north a little distance, become overlain by schist, in which the proportion of quartzose schist was, for a certain thickness of section, as great as, or greater than, that of albite schist. To keep the line a consistent division between the more quartzose and the more phyllitic parts of the section, we have apparently to ascend every here and there into higher beds as we go north. The general course of the line adopted is almost level from the N.N.E. fault near Lochan nan Cnaimh to Cruach a' Bhuic, yet the impression derived from examining cursorily the dip of the foliation on the hill tops between these places is that, the general dip, except on Cruach a' Bhuic, is dis-tinctly north-east. This is probably due to the existence of many small acute folds with their limbs hading north-west, which have the effect of bringing up the top beds of the greywacke again and again as we go north. Unfortunately, none of these folds is so distinct as to be shown on the Map with confidence. The line must only be regarded as a generalisation. The horizontal character of the line forms a conspicuous contrast to that of the different lamprophyre sheets, which are mapped in the neighbourhood. These have distinct hades to the north-west. They may be considered to keep almost parallel to the prominent foliation without regard to those north-west limbs of synclines, through the influence of which the bedding of the schists is being repeatedly brought back again to its old level.

Fig. 45.—1 inch = 20 feet. One-quarter of a mile E. of Lochan nan Cnaimh (153). Contorted phyllitic mica schist and greywacke schist. Axes of folds and strain slips hade N.W.

The remarks made about the top of the greywacke on the east side of the hill-top apply also to that on the west side. In the burn ½ mile south of Beinn Dubhain, the top is thrown down west by a north to south fault, perhaps about 40 feet. West of this place the line gets quickly to a lower level. Just before it reaches the alluvium at the head of Loch Eck it is poorly defined, and on the west side of Loch Eck we were not able to map out any thick horizon until we advanced some distance down Garrachra Glen (163).

Below the top of the greywacke there is usually a large proportion of massive quartzose pebbly schist on both sides of Cnoc na Tricriche. Such is seen, e.g., on the west side, in Coire Ealt and the burn drawn through the "Point" of "Dornoch Point," and, on the east side, in the upper half of the burn that flows south-east from Cnoc na Tricriche.

East and south of Cruach a' Bhuic the structure requires further description. In the burn head ¼ mile south-east of this hill there are massive quartzose schists at a higher level than those below the albite schist to the south-west. We suppose this is due to a north to south fault. On the east side of the supposed fault there are two outcrops of albite schist. The north outcrop is about 300 yards wide. More than half this width has various outcrops of quartzose pebbly schist opposed to it on the west side of the supposed fault—a circumstance which helps to confirm the existence of the fault. The south outcrop is not so wide and is not distinct close to the fault. The dip in both these outcrops is south-east, and they both, but particularly the north one, get thinner as they are traced downhill towards Loch Goil. In the more southerly of the two burns ¾ mile south-west of Carrick Castle, the width of the north band seems only 50 yards. North-east of this section the ground is drift-covered until we come near the side of the loch. We saw no albite schist near the loch thick enough to map: a large proportion of the rock is massive quartzose pebbly schist. The south band is not so well defined as the north one: north-east from the fault the north margin of the south band recedes slightly from the south margin of the north band, and at the same time there is apparently a slight diminution in the width of outcrop of the south band. One-quarter of a mile south-west of the "A" of "Ardnahien" the south band enters on a drift area, but on the shore 120 yards N.N.E. of Ardnahien there is an outcrop of albite schist,

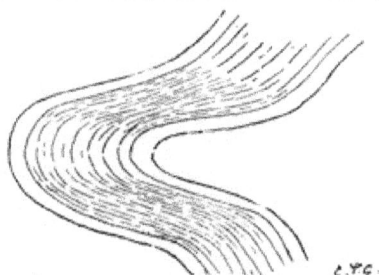

FIG. 46.— × ¼. Near Carrick Castle. Folded quartzose and more phyllitic mica schists. The more shaded parts are the more micaceous, and contain more magnetite crystals than the rest. The limits are but slightly thinned, yet, in the same bed, considerable differences have been developed between them and the crests. Axis of fold hades N.W.

about 20 yards wide, which may be a continuation of it. This is the only mapable albite schist seen in or near the shore between Carrick and ½ mile south-east of Ardnahien. The albite schists south-east of Ardnahien have been traced uphill to the W.S.W., and strike to the south-east of the Cruach a' Bhuic albite schist.

South of Cruach a' Bhuic, west of the supposed fault, the albite schist runs downhill with a south-east dip. Both margins are well-defined, and gradually approximate towards one another the lower they descend. South-east of Cnoc Fraorach the width of outcrop is 300 yards: ½ mile S.S.E. of this Cnoc it is 80 to 90 yards. At the last place there is a fault feature running W.N.W., and on the south side of the fault the albite schist is shifted to the east, and is also thinner, hardly 50 yards. Further south-west the outcrop continues to get narrower until in the burn between the "F" and "i" of "Glen Finart" it is hardly 16 yards. About 50 yards east of the outcrop, on the south side of the fault, there is another albite schist, of about the same thickness.

We seem thus gradually to lose the thick albite schist of the hills between Cnoc Tricriche and Cruach a' Bhuic, as we descend to low levels a little on the south-east side of "the anticline." Anyone who examined the section between Rudha nan Eoin and Carrick, and then that on the hills mentioned, would certainly find a difficulty in matching the two. Yet if "the anticline" had been a simple anticline of bedding the beds on the hills must have been seen again on the coast line. It seems necessary to suppose that the Beinn Bhreac, etc., albite schist has practically all died out, or else been folded out, by the time we reach the sea level. The latter alternative is much the most probable. Besides the great amount of folding always seen in good schist sections, there are other reasons in support of this supposition. Near Rudha nan Eoin we pass gradually, in a south-east direction, from a thick massive quartzose pebbly horizon on to much more thinly fissile schist, in which there is a considerable proportion of phyllitic and fine-grained albite

schists. Apart from the smaller size of the albites (see p. 42) this
series does not much resemble the schist of Beinn Bhreac, etc.   The
mica in the former is not so prominent.   It is rare to get distinct
micaceous "satiny" bands more than 4 to 6 inches thick.   There is

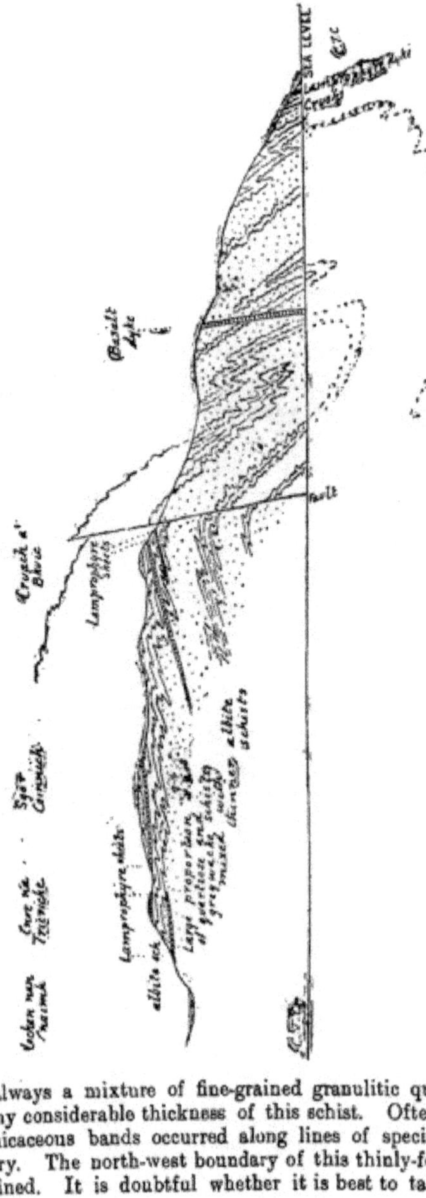

FIG. 47.—Horizontal scale 1 inch = 1 mile. Vertical scale not uniform. Diagram section between Toll a' Bhuie and Lochan nan Cnaimh. To show the supposed structure near the centre of "the anticline."

nearly always a mixture of fine-grained granulitic quartzose schist, but
never any considerable thickness of this schist.   Often it seems as if the
more micaceous bands occurred along lines of special strain and were
secondary.   The north-west boundary of this thinly-foliated series is not
well-defined.   It is doubtful whether it is best to take it a little north-

west of the Rudha, or a little south of it, the rock at the point itself being an alternating series, with a larger proportion of massive rock than there is to the south. The line adopted in the Map is the south line. This line can be traced to the south side of Creachan Mor, and it is certain that the massive rocks of this hill, and of Cruach Eighrach, correspond to those on the coast between Toll nam Muc and Rudha nan Eoin. If "the anticline" had been a simple anticline of bedding, this massive horizon should overlie the Beinn Bhreac albite schists, and we should have met with this horizon again either on the south slope or top of Beinn Bheula. But we never meet it again either on Beinn Bheula, or between Beinn Bheula and the base of the green beds to the north-west.

These are the facts which, taken in conjunction with the other important differences in the beds on either side of "the anticline" (see p. 86), have led us to suppose that very near the centre of "the anticline" there are also important folds, the axes of which have been folded by "the anticline." The top of the greywacke schist we have mapped on the north-west side of "the anticline" is not the same as that on the south-east side, but the one line may be considered the one margin, and the other the other margin of a great greywacke series.

On the top of Cruach a' Bhuic there is a somewhat flat band of massive pebbly schist, which may possibly represent part of the underlying greywacke folded over.

On the south-west side of Creachan Mor there is a N.N.W. scar feature, facing west, which develops to the south into a slack or depression. By the "u" of Cruach à Chaise, a clayey crush, 2 to 3 feet wide, is seen along this feature. The top of the greywacke horizon seems shifted 80 to 90 yards to the south on the west side. One-fifth of a mile north-west of the "C" of "Coille Mheadhonach" the same margin is again shifted by a fault, the west side to south, 80 or 90 yards. Half a mile north of this place, various parallel crush breccias are seen along the fault, hading east. The surface displacement of the line taken for the base of the greywacke is 50 to 60 yards, the west side to south. For a little way north-west of the greywacke horizon the beds are not so massive or quartzose as they are to the south-east, but we reach massive pebbly schists again before the albite schist which runs up to Cruach a' Bhuic is reached. It must not be supposed that all the schists on this greywacke horizon are greywacke schists: there are thin phyllitic mica schists as well, but the proportion of these is less than outside the horizon. Near the middle of the horizon, between the fault last described and the hilltop to the north-east, the south-east margin of a phyllitic interval has been mapped: it is overlain by an unusually quartzose schist without evident pebbles, which forms a strong contrast in colour with the phyllitic schist, so that it is easy to map the line between them. The line is much folded and often dips north-west for a little space, but the general inclination is distinctly south-east.

There are good sections of the greywacke schist on Castle Crag and the hillside to the south-east, from the first "C" of "Castle Crag" for nearly ½ mile south-east. On this hill both the north-west and the south-east margins of the horizon are unusually well-defined.

Between 200 and 300 yards east of Loch Eck the greywacke schist horizon meets with a north to south fault; the surface displacement is rather more than ¼ mile, the west side to south. There is a conspicuous slack along the course of the fault which continues to ⅞ mile south of the "T" of "Tom Soilleir." In the burn nearly ½ mile north of Coylet there are two basalt dykes accompanying the fault. The crush breccia near the east dyke is distinctly hardened and altered.

The albite schists south-east of Ardnahien have been already referred to. Three hundred and fifty yards south-east of Ardnahien there is a fairly-defined south-east margin of an alternating series of albite or phyllitic mica schists with other more quartzose schists. South-east of this comes a bed, 60 to 70 yards wide, of massive pobbly schist, and then a bed chiefly of albite or phyllitic mica schist, about 130 yards wide. This last albite schist is well-defined on both sides, but ⅓ mile south-west of the shore an outcrop of quartzose schist, 20 yards wide, appears in the middle of it. The quartzose schist may be a folded inlier of the under-lying bed. South-west of the quartzose schist the margins of the albite schist rapidly approximate ; ⅛ mile north-west of the "C" of "Cruach an Draghair" the width is hardly 30 yards. At the "C" of "Cruach" it is only 16 yards, and soon after that ceases to be traceable.

Five-eighths of a mile slightly north of east of Cruach nam Miseag there are two small patches of albite schist resting on more massive pebbly quart-zose schist. These we regard as outliers of the lower part of the albite schist of Cnoc Tricriche, Beinn Bhreac, etc.

There are two faults in this district which may represent the Ben Donich N.N.E. fault. The west fault makes a feature facing north-west all the way between the burn ¼ mile E.S.E. of the "r" of "Mullach a' Chuirn" and ¼ mile north of Lochan nan Cnaimh. The crush rock is seen close to the burn mentioned. This is the fault which throws the lampro-phyre at a place ¾ mile south-west of the foot of Lochan nan Cnaimh, as already mentioned, p. 201. The east fault is exposed in the burn ¼ mile south-east of the "i" of "Corriesyke." The hade is east. It keeps in or near this burn as far as the sharp bend to the west. A little further south-west it comes into another burn and keeps along it for ⅛ mile or more. No doubt the direction of the fault has determined the directions of both these burns as far as they coincide with it. The crush must cross the glen south of Cruach nam Miseag near the 700-foot contour. It forms a slack or depression on the hill-top south of this glen, just east of the 1000-foot contour. There is a section of the crush, still hading east, and associated with ferriferous carbonate strings, ⅜ mile west of the "C" of "Cuilmuich." We suppose this fault may be the one the indications of which we have already described near Chruac a' Bhuic (see p. 202), and in the greywacke schist ½ mile north-west of the "C" of "Coille Mheadhonach" (see p. 205). The surface displacement in the albite schist of Chruac a' Bhuic is in the opposite direction to that near Coille Mheadhonach, but it is not easy to see what can become of either fault unless they are the same. On the south-west side of Glen Finart the fault probably continues through the second "a" of "Barnacabber," and the last "o" of "Tom nan Con." In the burn just south-west of this "o" the crush is exposed : it consists of crush clay and quartz veins hading east.

North to south or N.N.W. faults run from Loch Long by Toll a' Bhuic and Knap Burn and make conspicuous features in the landscape. The fault at Toll a' Bhuic effects a surface displacement in a lamprophyre dyke of 30 to 40 yards, the west side to the south, but it cannot be traced far. It is not clear that there is any shift in the lamprophyre along the course of the Knap Burn fault.

Three-eighths of a mile south-west of the "P" of "Port an Lochan" there are indications of several N.N.W. crushes. One of these shifts a lamprophyre dyke about 3 yards, the west side to south.

The fault mentioned (p. 205) on the south-east side of Creachan Mor seems to divide into two parts ¼ mile north of "A" of "Am Binnein." The west part makes a scar feature, facing west, on the south side of Am

Binnein, and enters Loch Long on the east side of Shepherds Point. The crush on the shore is 9 feet wide and hades east : the slickensides are horizontal. The east part also forms a scar feature facing west. It runs through the "i" of "Clunie Wood" down to the "oc" of "Stronvochlan," and formed the east margin of the old Finart Bay in raised beach times. One-eighth of a mile north of the "i" of "Clunie" two or three parallel crushes are seen along the line, all hading east.

On the shore ¼ mile north-east of Toll a' Bhuic there is a gulley about 16 yards wide, running nearly with the strike of foliation. We suppose it due to a thin crush or to two thin parallel crushes: the main part of the rock within the gulley is sometimes, e.g., on the west side of Toll a' Bhuic N.N.W. fault, sound schist, but there is always a space where no rock is seen, and there may be a crush beneath this. The gulley is readily traced as far as the east side of the Shepherd's Point fault, where it is accompanied with a lamprophyre dyke.

Part of the Coire No (133) N.N.E., disturbances come into or near the north-west corner of the area being described, and may perhaps be considered the originating cause of the low valley crossing the watershed between Sugach Aodainn and Leamhanin. In Sugach Aodainn, ⅔ mile south-west of the "u" of "Sugach," a 6-foot basalt dyke is seen running with parallel crushes in a N.N.E. direction : there is a smaller basalt branch on the east side : some crushes seem to trend north, crossing over to a N.N.E. feature, along the course of which, other crushes, hading west, cross the stream due east of the "d" of Bridgend."

The foliation dip between Am Binnein (164) and Loch Long is often vertical or sometimes even dipping north-west (see p. 83). The northwest dip is particularly to be noted between ¼ mile and ½ mile south-east of this hill-top. On the north-east side of Tom Soilleir (163 north-east) there is a tendency for the foliation dip to be west or south-west, instead

S E   N W

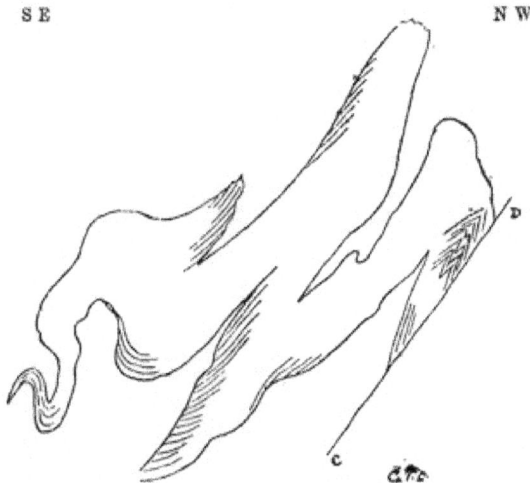

Fig. 48.—1 inch = 8 feet. Two hundred and fifty yards E.S.E. of Ardnahien (153). Plan of folded pebbly quartzose schist band in a steep rock face sloping N.E. The first foliation near the margin is indicated in several places. It often crosses the margin, and in some places is parallel to axes of early fold. Line "CD" may be a discordance.

of north-west. We could not see any clear reason for this. Some of the
exposures showing the south-west dip are a considerable distance off the
north-west depression along the road between Larach Hill and Whistlefield,
and it hardly seems feasible to connect the dip with possible disturbances
along this line.

The quartzose bands mixed with the more thinly fissile schists on the
south-east side of the Creachan Mor greywacke contain a good many
impure calcareous bands, but these are rarely more than an inch or two
thick.

There are good sections along the side of Loch Long, particularly
between Carrick and Finart Bay. The "pre-anticline" folds and
structure lines are often folded by the later folds with axes hading north-
west. Good hillside sections are also abundant throughout the district.
Plate V., near the end of the volume, represents part of a cliff section
on the roadside a little south of Coylet Inn, Loch Eck. The prominent
folds in the section are probably of "anticline" age.

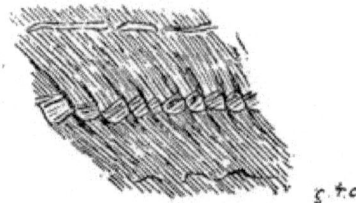

FIG. 49.— × ⅓. One-third of a mile S.E. of Mullach à Chuirn (142). Quartzose
schist band in phyllitic mica schist. Band is repeatedly broken through by
foliation-faults which throw an earlier foliation and bedding constantly in the
same direction. There are four or five times as many breaks in the phyllitic
schist as in the quartzose, and the hades of these become steeper when crossing
the latter schist.

District of Beinn Mhòr.

This district is inclosed by the Strachur and Loch Eck Valley, part of
the south margin of 1-inch Map 37, the River Ruel, Caol Ghleann, the
upper part of Stralachlan River, and the part of Loch Fine between
Newton Bay and Strachur.

All the schist zones from the Ardrishaig phyllites to the alternating
greywacke and albite schists occur within this region. Most of it lies
on the north-west side of "the anticline," but a small portion near the
south-east corner, on and near Clach Beinn, is on the south-east side. In
the north-west half of the area there are many bands of epidiorite and
hornblende schist, but south-east of a line a little below the green beds
none are known.

The sections of the Ardrishaig phyllites on the shore, and the folding
shown in this series, are described elsewhere, p. 58. The ground near
the centre of "the anticline" is more elevated and more craggy than that
to the north-east. Sections are so abundant in this rocky area that it is
not necessary to mention them in detail. The crags south of Coirantee
show very clear instances of folding of the axes of early "pre-anticline"
folds, like those described on p. 24.

We think it probable that the epidiorite and hornblende schist exposures
on the shore near Stucreach and Leak belong to one bed repeated by folding and faulting. The exposure ⅓ mile south-west of Stucreach has pebbly
green beds a little north-west of it: a fault, to be shortly described,
lies a little east of it: its west end is also cut off by another north to

south fault. Near Leak there are four or five exposures of epidiorite which are quite isolated at the surface. The two smallest near the lamprophyre on the east of the hamlet, can hardly be shown on the 1-inch Map. The exposures west of the hamlet are larger and irregularly oval in shape. In the rocky point ¼ mile north-west of the " L " of " Leak " there are two exposures, one lying north-east of the other. They are separated by a thin band of phyllitic mica schist.

The line adopted to roughly divide the Ardrishaig phyllites from the garnetiferous mica schists is the most north-westerly zone of graphite schist (see p. 48). On the shore south-west of Strachur Bay garnets are abundant in the schists to within 200 yards of Creag nam Faoileann, but they are probably north-west of all the graphite schists. The numerous thin outcrops of epidiorite and hornblende schist on this Creag (see p. 66), belong probably to the same series as those near the head of Eas Dubh (142). The schists on the shore north-east of the Creag probably occur between these epidiorites and the most north-westerly graphite schist.

Between Creag nam Faoileann and Mid Letter there are several north-west faults. The most south-westerly fault has a slight down-throw south-west, as shown by a band of pebbly green beds on the shore, but the others throw down north-east, and the general effect of the group is to advance the graphite schist on its south-west side towards the south-east.

200 yards south-west of Mid Letter there is a north-west fault in two parallel parts, which throw down to north-west. The direction of shift can be made out in the band of green beds alluded to in the last paragraph. The effect on the graphite schist is uncertain.

In the burn ⅓ mile S.S.W. of Mid Letter there are strong crushes ranging slightly west of north, and the strike of the foliation is parallel to them for a little space. These crushes probably keep a little east of the burn to the south, and cause the graphite schist exposures in this burn to be further to the south-east than we should otherwise have expected. A little below the most south-easterly graphite schist there is an obscure exposure of limestone. On the west side of the burn is another crush, seen in the burns ½ mile north-east of An Carr and rather more than ¼ mile south-west of Mid Letter. In the latter locality the hade is west. There are no exposures of graphite schist on the west side of the fault for ¼ mile or more.

In the burns ¼ mile south-east of Stucreach there are good sections of graphite schist : the most north-westerly outcrop is accompanied with a lamprophyre sheet in part of its course, but in the branch off the west burn there are prominent folds in the graphite schist, and, coincident with these folds, a gradual separation from the lamprophyre. Close to the section a small fault seems required with downthrow east, but the course of the supposed fault is not clear : in a north-west direction there are some crushes in the burn ⅓ mile south-east of the " b " of " Stucreach," but these hade west.

Between ¾ and ⅔ mile S.S.W. of the " S " of " Stucreach " there is evidence for a good many north-west faults. The effects of these are seen best in a green bed north-west of the graphite schist. The most north-easterly fault that has an important throw is the fault in the burn ⅔ mile south-west of the same " S." The hade is north-east and the down-throw, both in the green bed and an accompanying lamprophyre, is also in this direction. The displacement at the surface is about 130 yards. The strike of the foliation is parallel to the crushes for a little distance. The other faults, at least three, south-west of this, all throw down

O

south-west, and the green bed is gradually lifted to a little above its old level. There is a fault on the shore ⅜ mile south-west of the "S" of "Stucreach" which is accompanied by a thin basaltic dyke, and the foliation on the south-west side is twisted into parallelism. This may be compounded of several of the faults that are observed further south-east. The downthrow appears to be south-west, for an epidiorite, like that at Stucreach, is brought on to the shore again a little south-west of the fault. The hill-side where these faults would be expected to cross the graphite schists is almost entirely covered by drift or a patch of peat and alluvium, which runs nearly with the strike of the beds. Graphite schist is seen at a point ⅜ mile N.N.E. of the "h" of "Sith an t-Sluain." This is probably just on the high side of the fault which throws down south-west. There are no exposures of this schist further south-west until we come ⅓ mile north of the "n" of "Leanach," where the burn displays a good section of graphite schist underlying a graphitic limestone and separated from it by a strong crush hading north-west, nearly with the schist. The schist under the crush is much contorted. The different twists cannot be traced far. There is certainly more than one line of crush. Both limbs of the folds in the contorted part are generally inclined in the same direction : as we look north-west towards the exposure the axes of anticlines generally point south-west, so that there appears to have been a movement along the crush lines of the rocks on the north-west side in a south-west direction over the rocks on the south-east side. This is the opposite direction to that indicated at the crush in the graphite schist at the head of Eas Dubh (see p. 109).

Quartzose schist below the graphite schist is exposed in the burn from near the limekiln, 200 yards north-east of Leanach, up to where the road crosses the burn. Near the limekiln there are four or five basaltic dykes, running north or N.N.W., and with indications of crushing, either within or near the sides. Some of the crushes are probably accompanied with distinct brows, as graphite schist is seen close to the two most westerly dykes, considerably south-east of the strike of the graphite schist mentioned in the preceding paragraph. On the hill side ⅝ mile N.N.E. of Leanach, hornblende schist is striking at a green bed on the east. It is probable that part of the disturbances near the limekiln cross between these exposures. To the north, on the coast east of Leak, there are three or four crushes which may continue to the same place. The most westerly crush is accompanied with a camptonite dyke. Both this crush and the next three cause shifts in some irregular lamprophyre intrusions : they appear to downthrow west. The most easterly crush is accompanied by two dykes, one basaltic and the other a fine-grained lamprophyre : the former is chilled against a ferriferous carbonate vein.

On the north-east side of Meall Reamhar there is a N.N.E. running feature, facing north-west, along which a band of green beds is thrown down north-west. This feature can be traced north-east to a little beyond the "ch" of "Tigh na Criche." A little west of this feature, near the "r" of "Mid Letter," there are four north to south crushes hading west. The two middle crushes are close together, and there is an obscure exposure of green beds between them.

The Loch Tay limestone is exposed at the hamlet of Glenaluan in a quarry which has been extensively worked. This quarry and various veins and crush breccias in it have been described by Mr W. Ivison Macadam (Trans. Edin. Geol. Soc., vol. iv. 1883, p. 101), and a series of analyses is given by him of the limestone and vein rocks, etc. The details of two of these analyses are given on p. 47. None of the crush breccias and slicken-side lines seemed to us to materially displace the bed, and owing to the amount of other geological detail, they are not inserted in the map. By the "a" of "Glensluan" there is a thin exposure of limestone 10 to 12 feet above the main outcrop. A similar exposure is also seen ¼ mile further south-west. A lenticular band of phyllitic schist, a few feet thick, occurs within but near the top of the limestone. In the limestone below this there is a band of lamprophyre which gradually dies out going south-west. There are two bands of hornblende schist in the limestone : the upper band is at the base of the upper, purer, part of the limestone, and the lower at the base of the calcareous quartzite part.

In the bend of the River Cur by Tombuidhe a green bed is seen on the south side of a strong crush. The direction of the latter is doubtful : we have drawn it E.N.E. Whether this green bed is the band a little above the limestone at Glensluan is doubtful. Possibly it may be near the top of the main green beds zone and have the Loch Tay limestone above it. Between the crush and the green bed is an obscure exposure of calcareous quartzose schist which might represent part of the lower portion of the limestone.

One-half mile south-west of the last "n" of "Glensluan" the limestone suddenly comes down into the bed of the burn on the south-west side of three basaltic dykes. The two east dykes are generalised as one in the Map : they are both partly crushed. There is a band of hornblende schist within the limestone and another at its base, just as at Glensluan. The base of the limestone continues in the north-east bank of the burn for nearly ¼ mile south-west of the dykes. Then we meet with a north to south crush, hading west and downthrowing in the same direction, so that the hornblende schist within the limestone on the west side, lies against beds below the limestone on the east side.

Eighty yards further up stream there is a larger north-west fault, which shifts the limestone some distance on to the south-east side of the burn. There is a section showing the upper part of the limestone, the thin phyllitic schist above it, and the hornblende schist within it, in the little burn ¼ mile south of the "h" of "Gleann Dubh." The amount of displacement seems about 180 yards. Where the fault crosses the main burn there are two thin bands of green beds on the south-west side of it, opposed to the lower part of the limestone. On the south-west side of the fault in the little burn there are two outcrops of hornblende schist between the limestone and the top of the green beds zone. These are seen at intervals as far as the head of Stralachlan River. Before the limestone gets into the main burn again it meets with another N.N.E. fault—probably the continuation of that mentioned, p. 220, in the burn ⅛ mile east of the "s" of "Creggans." The downthrow is west but not large. On the west side of the fault there is a burn section of the limestone and the leaden coloured phyllitic schist just below it : there is only one hornblende schist in the limestone and this is a little above the base. One hundred yards further up the burn, hornblende schist occurs again in the north-east bank striking at beds below the limestone. There is a crush hading west on the south-east side of the schist, and we therefore presume that it represents the band within the limestone. Between this exposure and the burn ¼ mile E.S.E. of the last "n" of "Sith an t-Sluain" the limestone is covered by drift, but the green bed a little above the limestone is seen near the middle of this space. In the above burn there are two faults, one running nearly east to west, and the other N.N.W., and these let down the beds between them to a small extent. There is a hornblende schist below the limestone, and below this schist there is calcareous quartzite. On the south side of this burn the limestone makes a con-

spicuous scar with hornblende schist below. Two hundred yards south
of Cruach nan Capull (152) there is a prominent bank feature facing south.
To the west this twists into a north-west direction, and is accompanied by
a throw in a green bed: the south-west side is shifted north-east about
60–70 yards. The displacement in the limestone is probably greater
than this, but it cannot be accurately determined as there are no rock
exposures very near the south side of the feature.

There are many sections of the limestone and adjoining schists in the
small streams S.S.E. of Cruach nan Capull. There must be many faults
in the locality, but their exact directions and effects are not always
ascertainable, owing to the amount of drift on the hill sides. The hornblende schists in and at the base of the bed are liable to considerable variation in apparent thickness. One-quarter of a mile south-west of the second
"n" of "nau" there are two outcrops of hornblende schist, but neither seems
more than a few feet thick. South-west of here the outcrops are always
much wider, and there are sometimes 3 of them. Near the head of Stralachlan River, about ¼ mile W.N.W. of the "u" of "Sron Cruaich" the
outcrop within the limestone is about 40 yards wide, and that at the base
about 50 yards. The upper part of the limestone is the purest. In the lower
part there is a considerable proportion of calcareous quartzite, but good
sections of mottled limestone also (see p. 46). In the river ⅟₆ mile N.N.W.
of the top of the main limestone outcrop there are two more exposures of limestone. These may both be folded inliers of the Loch Tay limestone. The
larger, or south-east exposure, contains some mottled bands very like those
of this limestone, but most of it is calcareous quartzite: a hornblende
schist occurs within it: the foliation dip of the quartzite changes from
north-west to south-east near the middle of the exposure. There is a band
of green beds 50 yards north-west of the top of the main outcrop of limestone. We did not see this green bed again before the higher outcrops of
limestone were reached, though the section is a good one. Possibly this
absence of repetition may be due to the green bed being at the bottom of
a syncline. There are two or three green beds lower down the river before
the fields near Leanach are reached: a W.N.W. fault runs along the river
course and advances the most north-westerly bed considerably west on the
south side of the fault. There are also some thin hornblende schists and
epidiorites in the same reach which are at times hard to distinguish from
the green beds. South-east of the main outcrop of limestone there are four
outcrops of hornblende schist in the river, not counting that at the base.
The most south-easterly outcrop, within 70 yards of the boundary of the
parish of Stralachlan, is bounded on the west by a N.N.E. crush hading
west, but it is possible that this is a reversed hade and that the band is
really the same as that on the west side of the fault. Five-twelfths of a
mile N.N.E. of this locality there are two N.N.E. faults crossing the limestone: one downthrows west, but the other and larger one of the two,
though it hades west, has a downthrow in the opposite direction. The
larger fault is accompanied by two dykes, one basaltic and the other a
lamprophyre. We suppose this fault continues N.N.E. to the crushes
in the burn ⅓ mile S.S.W. of Mid Letter.

At the head of Stralachlan River, 50 to 60 yards south-east of the
parish boundary, we lose the rock section. Between this place and the
limestone we could see no green beds except one band, about 5 feet thick,
not far below the limestone: most of the exposure is either hornblende
schist or rather massive quartzose schist. We have therefore to draw the
top of the main green beds zone further south-east of the parish boundary.
Even after allowing for the slope of the ground being in the same direction
as the dip, this makes the distance between the limestone and the green

beds considerably greater than it usually is in the north part of Cowal. All the way from here to the burn ¼ mile E.S.E. of the last " n " of " Sith an t-Sluain " the top of the green beds zone is entirely covered by drift. This is also commonly the case through most of the area being described. We have drawn the line in the Map, so as to show where to stop the green beds colour, but it is of little stratigraphical value.

There are six outcrops of hornblende schist in Garvie Burn below Strath nan Lub. The lowest but two, ¼ mile south of the last " n " of "Strondavain," only occurs on the south side of the burn : it comes in on the crest of an anticline with both limbs hading north-west. It is probable that some of the others are also due to repetition by folding. The hill sides near the burn are obscured by drift for long distances, and it is doubtful whether the two outcrops in the bare area east of An Cruachan (162) continue to the exposures to which they have been connected in the Map.

On the south-east of An Cruachan there is a prominent N.N.E. depression partly occupied by a peat moss. North-west of the hilltop there is a basaltic dyke, with a strong crush at its side, in the same line as the depression. The crush hades west. Neither of the hornblende schists is seen on the south-west side of the depression until we advance a little within 1-inch Map 29, after passing another feature, due to a fault with downthrow N.N.E.

There are two exposures of hornblende schist in the burn ½ mile S.S.W. of the " o " of " Caol-ghleann." These probably belong to one band dipping almost with the hill slope. The horizon is apparently much the same as that of the most north-westerly exposure in Garvie Burn. The ground between here and Garvie Burn is obscured by peat on the hill top, and drift in the lower areas, and no other exposure on the same horizon has been noticed in it.

On the south-east side of Dunans Burn there are small exposures of hornblende schist south-east of the fault running along the burn, at places ¼ and ⅓ mile north-east of the " D " of " Dunans." Two other outcrops of similar schist are also seen ⅛ mile or more slightly north of east of these. All these are isolated by drift from one another. It is possible they belong to one band. The hill slope being nearly with the dip, the band may make a broad outcrop under the drift.

There is a hornblende schist in the burn 200 yards south of the last " a " of " Caol-ghleann. A little north of the burn a north-west fault is required, to account for a shift in the outcrop, towards the north-west, of about 200 yards. In a burn north-east of the supposed fault there are two outcrops of hornblende schist. It is probably a continuation of the north-west outcrop which is seen in the burn ¼ mile north of the " h " of " Caol-ghleann," intersected by many parallel basalt dykes. On the south-west side of the dykes the top of the hornblende schist is gradually covered by the overlying schist, but in the stream a little lower down it comes out again, apparently owing to a local flattening of the dip, and keeps in the burn for 70 to 80 yards. After this the base emerges out of the burn. We suppose the exposures just at the foot of the burn belong to the same band: they are cut off on the south-west by the north-west fault mentioned in the beginning of the paragraph. The fault makes a slight shift in a N.N.E. lamprophyre on the north-west side of Caol Ghleann Burn. Many other parallel faults, at least five of them, throw the same intrusion, sometimes in one direction and sometimes in another, between this fault and the alluvium at Caol-ghleann. The most south-westerly exposure of rock in the burn near the north end of this alluvium is dipping south-west, at right angles to the ordinary dip in the neighbourhood, and is crossed by red crushes parallel to the strike. This is probably close to the north-west fault, accompanied with basalt dykes, which, in the area 1 mile south-west of Caol-ghleann shifts the base of the green beds zone, the north-east side to the north-west, for $\frac{1}{4}$ mile (see p. 215). Accompanying basalt dykes are seen in a burn $\frac{1}{4}$ mile south-east of Coal-ghleann.

On the west side of the burn by the "G" of "Caol Ghleann" there are two outcrops of hornblende schist. The higher one is the wider. We suppose it to be the same band as that already mentioned in the burn 200 yards south of the last "a" of Caol Ghleann, etc. Strong crushes, accompanied in part of their course by a basalt dyke, run up the burn on the north-west side of the exposures, and continue for a long distance N.N.E. to join those with a similiar direction in Ghleann Dubh (see p. 211). In the burn $\frac{3}{4}$ mile S.S.W. of the "h" of "Tom a' Bhiorain" a basaltic and lamprophyre dyke both accompany the crush. On the east side of the same Tom there is a feature facing west, which we suppose to be along the course of the fault. We could not trace the fault satis-factorily on the south-west side of Coal-ghleann, but on the east side of Maol Odhar and An Cruachan there are features in the line of the fault, and in Garvie Burn, along the same direction, there are thin crushes hading west.

The base of the green beds zone is obscured by drift at the side of the alluvium east of Balliemore. There is a small exposure of green beds in a little stream near the bottom of the wood $\frac{1}{2}$ mile E.S.E. of Balliemore, and in a burn not quite $\frac{1}{2}$ mile south-east. This last exposure is probably near the base of the zone, for all the exposures for some distance east of it are quartzose or phyllitic mica schist. In a little burn $\frac{5}{8}$ mile south-east of Balliemore there are several thin basaltic dykes, at least four, run-ning north-west with thin schist partings between them. The schist in the partings sometimes strikes parallel to the dykes. Some of the dykes hade south-west. There are crushes near the north-east side of the dykes, and further off, green beds. These crushes and others which probably occur under the drift a few yards north-east of them, must have a downthrow north-east, for 160 yards south-east of the exposure mentioned, and on the south-west side of the direction of the crushes, we see green beds striking far east of the probable base of the green beds on the other side of the crushes. The amount of shift at the surface seems about $\frac{1}{4}$ mile. The basement green beds on the south-west side of the fault gradually rise to the south and form a prominent crag—Creag Bhaogh. A band of hornblende schist, sometimes ·16 feet thick, but not usually so much, is seen about 12 feet above the base for as long as the exposure continues quite bare. The beds under the green beds are alternations of quartzose and other more phyllitic mica schists. Near the "B" of "Bhaogh" there is an outcrop of green beds, about 1 foot thick, 30 feet below the main base.

In the burn $\frac{1}{4}$ mile south-east of the last "r" of "Glen Branter" there is a rather wide exposure of pebbly green beds, about 200 yards south-east of the main base. On the north-east side of this there is a basalt dyke hading west and accompanied with crushes. We could not see any outcrop of green bed on the north-east of the dyke. In a south-west direction the green bed is seen again in several places about $\frac{1}{3}$ mile east of the last "r" of "Glen Branter."

The hillside between the base of the green beds zone in the burn near the "r" of "Glen Branter" and $\frac{1}{4}$ mile slightly west of south of the "G" of the same words is drift covered. A north-west fault, accom-panied with a basalt dyke, is seen in the burn near the "a." It has a hade south-west, but the amount of shift in the green beds is uncertain. In the burn ¼ mile west of the first "r" of "Sròn Criche" there are two thin outcrops of green beds not quite ¼ mile up the burn from the main base : the upper one is only 1 foot thick. An outcrop in about the same position with respect to the main zone is seen further south-west, between ¾ mile slightly south of west of the "S" of "Sròn Criche" and ¾ mile south-west of the same letter. At one place the band is a good deal folded.

In the burn running from the north-west near the "L" of "Strath nan Lub" there are two basalt dykes in the general direction of the burn. Near the north-east one there is a fault hading south-west ; on the south-west side of the south-west dyke the schist is reddened and vertical and almost parallel to the dyke. On the north-east side of the upper part of the burn there are good exposures of schist ; a thin green bed is seen about ½ mile north-west of the "u" of "Lub," but the base of the main zone cannot come on until 140 to 160 yards further north-west. In the nearest good section south-west of the dykes, that in the burn running between the "a" and last "n" of "nan," green beds are seen, at intervals, down to near the top of this word. They form the whole of the bottom 100 yards of the exposure, and must be within the main zone. A lower outcrop, corresponding to that on the north-east of the dykes, occurs in the burn a little east of the top of the "L" of "Lub." The amount of shift at the surface along the line of the dykes seems about ¼ mile. The dykes and crushes are known to continue some distance both north-west and south-east. Their indications north-west, near Caol-gbleaun, are mentioned on p. 214. Going south-east they run through the Bealach an t-Saic down into Garrachra Glen, and effect a considerable displacement in the top of the greywacke schist horizon mapped on the north-west side of this glen. In the different localities traversed, the foliation near the fault is generally deflected into parallelism with it for a little space.

In the burn running south-east between the "t" and "h" of "Strath nan Lub" there are occasional exposures, chiefly of quartzose schist, up to nearly half way between the 800 and 900 contour lines. The base of the green beds cannot be below these exposures. On the opposite side of the strath, in the burn that enters the strath near the "S," green beds are seen up to about the same level. These strike considerably south-east of the base line on the north-west side of the strath. We suppose there is a fault with large downthrow west between the outcrops mentioned. The hill sides are too much obscured with drift to enable us to fix the direction of the fault accurately. The broad hornblende schist ½ mile north-west of the "a" of "nan" seems to end suddenly on the west and is perhaps shifted north by a continuation of the same fault. By the side of the wire fence on the hill top between Strath nan Lub and Caol Ghleann, there is a small exposure of hornblende schist sticking up through the peat. This is perhaps the same hornblende schist, on the east side of the fault.

In the burn ¾ mile south of the "S" of "Strath nan Lub" the base of the green beds zone is a little above the 1000 foot contour. One mile south of the same letter there are green beds considerably south-east of the strike of this line, and opposed on their north-west side to quartzose schist. There must be a fault with downthrow south-west, shifting the base line on the south-west side towards the south-east. The amount of surface displacement is 300 to 400 yards. Both north-west and south-east of this locality there are basalt dykes along the line of the supposed fault.

South-east of the green beds there are but few schist lines that have been traced. The top of the apparently thick greywacke schist on the north west side of "the anticline" comes into the area from the south on the south side of Carn Bàn, and keeps fairly well defined as far as the side of the Bealach an t-Saic fault. Below the top there is a large proportion of massive quartzose, often pebbly, schist, all the way down to the alluvium near Garrachra. On the north-east side of the fault the top is not so definite, and going further north-east becomes gradually less and less so, so that we had to cease mapping it. In the steep south-east slope of Creag Tharsuinn we cannot make out that there is any thick zone in which quartzose schists largely preponderate over the more phyllitic. We suppose the appearance to be due to a great folding and mixing together of phyllitic mica schists with the greywacke schist horizon.

The margins of various greywacke schists have been traced on the slopes of Beinn Bheag, Cruach Bhuidhe, etc., but none of them continued distinct far. Some of them lose definition in a gradual way, either by the oncoming of phyllitic schists in the greywacke area, or the oncoming of greywacke in the more phyllitic areas. This suggests that the changes are due to gradual intermixing by folding. As examples of this gradual change, we may mention the top of a greywacke schist $\frac{3}{4}$ mile north-west of Stuck, and the bottom of a similar schist $\frac{1}{2}$ mile E.S.E. and $\frac{1}{8}$ mile S.W. of the "B" of "Meall Breac." ·

In the want of traceable schists, the numerous lamprophyre sheets in the hills east of Loch Eck have been found useful in finding out faults. In this area faults are exceedingly numerous; in fact, nearly all the straight parts of the streams and more prominent nicks in crags are occupied by crush lines. The most general direction is slightly north of west, and the downthrow north. Between Cruach Bhuidhe and Bernice Glen ten faults have been mapped, all throwing down north. The fault in Bernice Glen has a downthrow to north of 80 to 100 feet.

There are also a good number of other faults with a N.N.E. direction. Two of these cross the west side of Meall an-T, and are accompanied with basalt dykes. Two more cross the top of Clach Beinn. They both downthrow east; the west one about 120 feet, and the east one 20 to 30 feet.

A conspicuous feature ranges N.N.E. in Garrachra Glen, and no doubt represents a continuation of one of the faults in Coire No (see p. 222). Near the "G" of "Garrachra" (House) there are several feet of crush rock along the feature. The hade is east. The schist east of the fault is much contorted along axes which are nearly at right angles to the direction of fault (compare the N.N.E. disturbances mentioned on pp. 109, 210). Further north-east the Bealach an t-Saic basalts, or some of them, twist into a south-west direction along the fault for a little distance. By the first "a" of "Garrachra Glen" a lamprophyre sheet affords evidence of a downthrow east, perhaps of about 40 feet. All along the east side of Creag Mholach the fault forms a conspicuous feature facing east. It crosses Glen Shellish between the "h" and "e" of "Shellish," and is accompanied in the burn by a basalt dyke. A basalt dyke is also seen with this fault $\frac{1}{2}$ mile south-east of Glenshellish farm.

On the south-east face of Creag Tharsuinn there are prominent joint planes dipping E.S.E. much with the hill slope. This is nearly opposite to the foliation dip. On the south side of Clach Beinn there are also prominent joints striking north-east, and hading south-east at about 30°. This is the same direction as the foliation dip.

District of St Catherine's.

This district is inclosed by Hell's Glen, the Lochgoilhead valley, the lower part of Lettermay Burn, Curra Lochain, the valley west of Carnach Mar the valley between Bridgeud and Strachur, and part of Loch Fine.

All the groups of sedimentary schist found on the north-west side of "the anticline" come into this area, and there are also many outcrops of epidiorite, hornblende schist, etc.

In the Ardrishaig phyllites the schists mapped are epidiorites, hornblende schist, various outcrops of quartzite schist, and a few thin limestones. The quartzite schists have not generally been traced, but are drawn in the burn sections and on the shore of Loch Fine. They are extremely abundant, particularly in the north-west part of the series.

North-east of Ardchyline there is an unusually wide outcrop of quartzite schist. The width of bare quartzite on the hillside is about $\frac{1}{8}$ mile, but taking into consideration the sections in the burns on either side, we consider there is about as much more hidden under drift. Most of this outcrop seems to end in three fingers pointing south-west. The most south-easterly finger ends 100 yards north of the "A of "Ardchyline." The middle finger ends, rather bluntly, 200 yards north of the "r" of the same word. The north-west finger is not actually seen to end, but in the burn $\frac{1}{2}$ mile N.N.W. of the "A" of "Ardchyline" it is only 40 yards wide. This fingering out is no doubt due to folding. In the most northerly of the streams at Ardchyline there are good sections of green-grey phyllite, the foliation of which strikes right at the quartzite. Further north-west there is still a considerable width of quartzite in the burn $\frac{1}{2}$ mile N.N.W. of the "A." This is bounded on the west by N.N.E. crushes accompanied with a lamprophyre dyke. The effect of the crushes is uncertain. There is no rock seen on the west for some distance. Creag a' Phuill is composed of massive quartzite schist, and this can be traced north-east for several hundred yards. In all probability it is the same bed as the one near Ardchyline. The quartzite $\frac{1}{8}$ mile east of the Creag is underlain by an epidiorite band, on the other side of which more quartzite schist appears. If we look on the two outcrops of epidiorite at St Catherine's as repetitions of one bed by folding, which, as stated shortly, is not unlikely, it may also be the case that the quartzites on either side of the epidiorite near the Creag are the same, the epidiorite being in a fold between them. The epidiorite is hidden by drift in the north-east part of its course between the Creag and Ardchyline. It is possible it runs down a dip slope and joins some epidiorite exposures at the back of the raised beach $\frac{1}{4}$ mile north-west of Ardchyline.

Five-eighths of a mile north-west of Ardchyline a little rocky knoll of epidiorite appears near the north end of the quartzite schist. The south margin is a gently curving line. Close to the side of the line quartzite schist is seen. The greater length of the margin runs at right angles to the prominent foliation of the district. We suppose the epidiorite to be part of an old intrusive sill, and the south margin to be a folded margin. The epidiorite knob is only about 30 yards wide from south to north. The north face of it is a rough crag which represents a fault line. In the burn to the north, in some places hardly 30 yards distant, there are good exposures, chiefly of quartzite schist, striking at it. The fault is seen in the burn $\frac{3}{8}$-mile north of the "A" of "Ardchyline": the hade is south; it is accompanied by a lamprophyre dyke 4 to 10 feet thick. In the burn west of the knob a similar dyke is also seen along the supposed fault. The epidiorite is shifted by the fault, the north side to west: the amount of displacement of the east margin of the epidiorite outcrop is about 170 yards, of the west margin rather more than this, the outcrop on the north side of the fault being rather wider than on the south. In the knob south of the fault, the east margin bends round and becomes the west margin. It therefore is probable that on the north side of the fault also, both margins may represent the same line repeated by folding, and we may consider the south-east band of the St Catherine's epidiorite to be the same bed as the north-west band. Whether the south-east band represents an inlier along an anticline or an outlier at the bottom of a syncline is not certain. South-east of the south-east outcrop there is a considerable proportion of quartzite schist which may correspond to that near Ardchyline. The two bands come very near one another at St Catherine's, and there is less quartzite schist between them than to the south-east : this may be explained by supposing we are dealing at St Catherine's with a less thickness of beds, the mass of the quartzite not being reached.

Near Arduagowan Cottage, and between it and Airidh a' Ghobhainn, there are bands of epidiorite which have a habit of gradually thinning and ultimately dying out to the north-east. There are wide exposures of quartzite in the middle of the space between the two outcrops south of Arduagowan Cottage, but phyllite comes on on either side of the quartzite, between it and the epidiorite.

Three-quarters of a mile north-west of Ardchyline a north-west fault shifts the outcrops of both epidiorites, the south-west side to the north-west. The north-west band is well exposed at the back of the raised beach for ½ mile north-east of the fault, but on the south-west side it is only seen on the shore, for about 80 yards. The fault in its course south-east most soon meet with the east to west fault already mentioned ⅝ mile north of Ardchyline. A little south-east of their supposed junction there are two lines of fault, both with downthrows south-west, as shown by sheets of lamprophyre and hornblende porphyrite. Indications of the continuance of the south branch occur 1 mile E.S.E. of the first "C" of "Cruach nan Capull," where a band of epidiorite is crossed by two closely parallel N.N.E. faults, each with a slight downthrow north-east, and the crushes are seen in the burn to the south-east. The more northerly branch can be traced to Lochgoilhead. Its course will be further described subsequently.

As already stated, p. 48, the north-west boundary of the garnetiferous schist is not well defined. We have adopted the most north-westerly zone of graphite schist as the boundary. There are crushes, accompanied with severe contortions, running with this graphite schist. Near the head of Eas Dubh there is an interesting section in this crushed schist, at a place about ⅝ mile north-east of the "h" of "Dubh." There is a well defined crush line hading north-west. The beds above it are mainly quartzose schists with the foliation planes approximately parallel to the crush line, contrasting in this respect strongly with the contorted graphite schist below the crush. As you look north-west, towards the section, the axes of anticlines in the contorted area are seen to have a tendency to point towards the north-east, as if there had been a lateral movement along the crush line of the upper schists towards the north-east over the lower ones. This direction is the opposite to that indicated by the crush in the graphite schist near Leanach (151), mentioned on p. 210. The section is also of interest as indicating that some of the crush movements are earlier than some of the lamprophyres (see p. 109).

At the head of Eas Dubh there are three W.N.W. faults which shift the north-west graphite schist, the south-west side to the south-east. Owing to these a series of thin epidiorites or chlorite schists, which occur a little north-west of the graphite schist, are repeated again and again. The middle fault is seen in a burn, not marked in the 1-inch map, ¾ mile

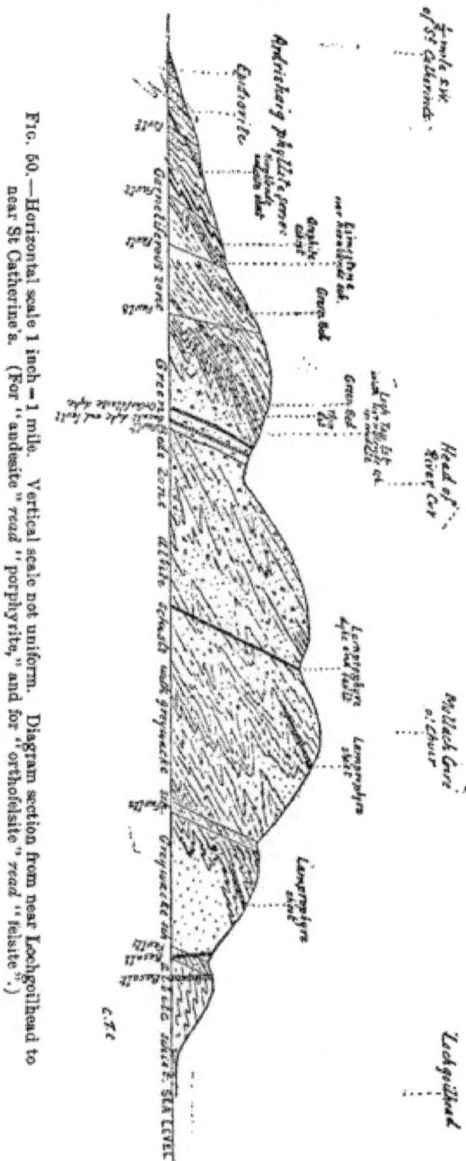

Fig. 50.—Horizontal scale 1 inch = 1 mile. Vertical scale not uniform. Diagram section from near Lochgoilhead to near St Catherine's. (For "andesite" read "porphyrite," and for "ortholfelsite" read "felsite.")

E.N.E. of the second "a" of "Creagan an Eich." It is in two parts, both throwing down to south-west. In the space between them there is a limestone, the same band as that in the burn ¼ mile south-east of Laglingarten.

On the north east of this limestone, at the other side of one of the faults, is the hornblende schist which comes below the limestone. The most southerly fault is seen in the burn ⅜ mile E.N.E. of the same letter.

S.W.   A   B   N.E.

C

Fig. 51.—1 inch = 4 feet. Vertical section ¼ mile N.E. of "h" of "Eas Dubh," observer looking N.W. [Line "AB" is a crush hading N.W. Above this line is quartzose schist. Below is contorted graphite schist. "C" is a lamprophyre sheet.

The E.N.E. fault a little north of the "h" of "Eas Dubh" is nowhere seen in section. It is inferred from the high level of the limestone and hornblende schist in the burn ½ mile east of the "h" compared to that in a burn, not marked in the 1-inch Map, ¾ mile slightly north of east of this letter. The hillsides adjacent are thickly covered with drift and it is not possible to determine exactly how the fault runs. At the foot of the east to west burn near the "h" the foliation of the phyllite dips north, probably owing to the proximity of the fault.

In the burn ¼ mile east of the "s" of "Creggans" there is a crush hading west and a lamprophyre dyke accompanying it. In the epidiorite ½ mile S.S.E. of Creagan an Eich the amount of surface displacement is about 70 yards, the west side to south. Graphite schist is seen in several places close on the west side of the fault, between the raised beach and rather more than ¼ mile north of this beach, but we noticed no clear exposure on the east side nearer than the north-east corner of the wood through which the burn runs. From the epidiorite the fault continues N.N.E., forming a bank feature facing west as far as the burn east of Ardnagowan Cottage : in this burn a crush and accompanying lamprophyre are again seen ; the foliation of the schist on the west side is vertical and parallel to the crush. Still farther north-east this fault probably joins the one in the burn ¼ mile N.N.W. of the "A" of "Ardchyline" (see p. 217).

The limestone ¼ mile south-east of the highest graphite schist is mentioned on page 53. As there stated, there is a little difficulty in regarding it as an outcrop of the Loch Tay limestone repeated by folding. This limestone is seen in the burn ¼ mile S.S.E. of Ardno, and is associated with hornblende schist as usual. Sixty to seventy yards south-west of this exposure a north-west fault seems required with downthrow north. There is no indication of the limestone in the burn west of the one mentioned, but in the next but one west there is, at a place a little over ¼ mile S.S.W. of Ardno. The outcrop in the last burn is striking west of that in the first mentioned burn. In its course north-west the fault may form the south-west ending of the felsite sheet a little more than ¼ mile south-west of Ardno. The shift in the felsite must be in the opposite direction to that in the limestone. The felsite is an intrusive sheet and not necessarily parallel to the bedding : in this locality the general hade may be different from that of the schists. A third north-west fault, with downthrow north-east, is needed to come between the limestone and hornblende schist outcrops in the two burns ½ mile east of Laglingarten.

One-sixth of a mile south-east of Laglingarten a north to south crush, hading west, intersects the top of the limestone. The base of the limestone does not appear west of the fault until we reach the south-west, or downthrow side, of a north-west fault. Fifty and 100 yards south-west of the north-west fault there are other parallel faults throwing in the same direction. The strike of the foliation between these faults is north to south, so that the displacements effected by the faults are soon compensated for by this abnormal direction. A little south of the most south-westerly fault the foliation strike gets normal again. South-west of the faults there is no indication of the limestone at the surface until we reach ⅜ mile E.S.E. of the "s" of St Catherine's, but the hornblende schist below it is seen ¼ mile north-east of this exposure.

In the burn ⅝ mile north-east of the first "C" of "Cruach na Capull" there are two N.N.E. faults, each throwing down to north-west. One is close to the base of the limestone, and the other near the top.

Five-eighths of a mile slightly north of west of the same "C" the strike seems twisted for about ¼ mile into a south-east direction, almost at right angles to the ordinary one. The change of strike is inferred, partly from the occurrence of the limestone and hornblende schist south-east of where the normal strike should take them, and partly from the section in a burn not marked in the 1-inch Map. The strike in the burn is in some places south-east and in one place sharply twisted along a nearly east to west axis. A felsite sheet occurs in the twisted beds and is itself twisted in the same way. We incline to connect these disturbances with the passage of a fault near their south-west side, probably the same as the east to west one north of Ardchyline (see p. 217). South of the supposed fault we see no good limestone until we reach a burn ⅞ mile slightly south of east of the "n" of "Ardchyline." Between this exposure and the disturbances there is also a N.N.E. crush which is seen in several of the burns south and south-east of Laglingarten : the hade is west and the adjacent schist foliation is sometimes quite vertical and parallel to it. This fault is the cause of a prominent feature, facing north-west, all the way from ⅝ mile N.N.W. of the "C" of "Creag Dhubh" to ¾ mile W.S.W. of this letter. On the south-east side of the feature there are sections of the limestone and hornblende schist, sometimes in repeated alternations, probably partly due to folding, in the different burns marked in the Map and a few others.

South-west of the burn ⅜ mile slightly north of west of the "C" of "Creag Dhubh" there is nothing seen either of limestone or hornblende schists until we reach a little burn 1 mile W.S.W. of the above-mentioned letter (see p. 53). It is possible that for a considerable part of this distance the two beds may be faulted out. As the fault is running nearly with the strike of the beds this would not necessarily imply a large throw. In the burn ⅝ mile east of the first "n" of "Creagan an Eich" there seem two faults, one running N.N.E. and the other north-east, and both throwing down north-west. The N.N.E. may be a continuation of that we have just been describing.

The Loch Tay limestone appears in Coire No burn, ¼ mile north-west of the "o" of "No." One hundred and twenty yards lower down the burn there is another limestone outcrop, much contorted and crushed. Between the two limestones, comes in, among other things, a band of green beds. A calcareous quartzite, with some limestone, is also seen a little above the band of green beds in the burn rather more than ¼ mile north-east of Cruach nan Capull. The Loch Tay limestone in the Coire No burn is no doubt considerably folded ; otherwise the base would come out of the burn more quickly than it does. A band of hornblende schist occurs within, but near the top of, the limestone in one place on the west bank of the burn, but seems to be folded out in a little distance. On the north-east side of the limestone there are several north-west faults. They are so near together that it is not possible to show the detail in the 1-inch Map. All the faults hade south-west. On the north-east side of the most south-westerly fault, the limestone is seen in the north-east bank of the burn. The most north-easterly fault must have the largest throw, the limestone not appearing on its north-east side until we reach the burn ½ mile S.S.E. of Ardno, more than ¼ mile away. Immediately north-east of the burn mentioned there is another subordinate fault which effects a shift in the same direction of 50 to 60 yards. One hundred yards east of the "o" of "No" the different faults appear to be together, and meet with an N.N.E. fault and an accompanying basalt dyke. The N.N.E. fault appears to shift the north-west fault, the east side to south, for perhaps 20 yards, but, as stated on page 229, we are by no means certain the N.N.E. fault is the older of the two.

The north-west fault in its continuation south-east crosses over into Hell's Glen. Various indications of this continuation are to be mentioned shortly.

The N.N.E. fault makes great confusion at the head of Coire No, north-east and east of Cruach nan Capull. In these localities it seems divided into three main branches. There are connections between them, and the beds are considerably twisted. A basalt dyke, not itself crushed, runs with the west branch. This branch must have a considerable downthrow east, for on the west side the hornblende schist below the limestone occurs against the schist overlying the green beds above the limestone on the east side. The middle branch has also a downthrow east in some places, but, owing to the changing dips of the beds on the sides, not always so. The east branch has a downthrow west usually, perhaps always: the limestone is never seen on the east side, though just on the west we may get either it or the overlying beds. On Cruach nan Capull there are several north-east or N.N.E. features, but it is uncertain whether they are accompanied with faults.

In the River Cur part of the Coire No N.N.E. faults appears 100 yards north-east of the north-east part of the "E" of "ARGYLLSHIRE." It is accompanied by two basalt dykes and a crushed lamprophyre dyke. There is probably another part 150 yards higher up the burn, where there is a feature running N.N.E., with no rock below for a little space.

Three-eighths of a mile S.S.E. of Cruach nan Capull a fault crosses the green bed above the Loch Tay limestone. The surface displacement is about 30 yards, the south-west side to the south-east. It is probably the fault we suppose to pass near the contortions ⅝ mile slightly north of west of the "C" of "Cruach nan Capull" (see p. 221). The effect on the limestone is not seen. In all the burns to the south-west there are sections in the limestone. The general character of it and of the beds near it are described elsewhere (p. 46). One-half mile south of the "D" of "Creag Dhubh" a north to south and an east to west fault meet and let down the beds in the north-west angle between them. The displacements caused by both are seen in the two green beds north-west of the limestone. Just before the north to south fault reaches the most north-westerly of these two green beds it appears to give off a branch in a north-west direction: the branch has a slight downthrow to south-west. One mile S.S.W. of the "D" of "Creag Dhubh" an east to west fault effects a displacement in the limestone and overlying beds, the south side to the west. The amount of shift in the hornblende schist is about 120 yards. A little more than ⅓ mile north-east of the "S" of

"Socach" a prominent north-east feature breaks through the schists. The hill on both sides is largely covered with drift. Probably the downthrow is north-west; massive quartzose schist, like that usually below the limestone, occurs on the south-east side on the same level as the hornblende schist in the limestone on the north-west side.

The character of the line adopted as the top of the main zone of green beds is described on p. 37. It is covered by drift for a considerable distance on either side of the Coire No north-west fault. In the burn ¾ mile north-east of Cruach nan Capull there are some thin green beds above the top of the main zone, but below the Loch Tay limestone. In a little burn ¾ mile E.N.E. of the second "n" of "Cruach nan Capull" there is also a green bed in this position. There is no exposure of the line south-west of this little burn until we reach the stream ½ mile south of the last "a" of "Cruach nan Capull." After this the line is frequently seen, keeping at a fairly uniform distance from the Loch Tay limestone, until it is obscured by drift south of the east to west basalt dykes ⅜ mile E.N.E. of the "h" of "Strachur."

The line taken for the base of the green bed zone comes down to the south-west side of the Hell's Glen fault ¼ mile W.S.W. of the "B" of "Gleann Beag." The surface displacement effected by the fault is not quite ½ mile, the south-west side to the south-east. The line is fairly defined all the way from Hell's Glen to Gleann Canachadan. There are several N.N.E. crushes going through it, but they seem to have no appreciable throw. The green bed above the line makes a rather wide outcrop. There are subordinate lower outcrops ¼ mile south-east and ¾ mile south of Cruach nam Mult. By the "l" of "Gleann Canachadan" there are three lower outcrops: the top outcrop is 6 feet below the base of the zone, below this is another 3 feet thick, and then a band 5 inches thick. A quartzose pebbly schist occurs immediately below the base of the zone.

A north-west fault runs up Gleann Canachadan at the base of the green beds, and displaces this base about 30 yards, the south-west side towards the north-west. On the south-west side of this fault there is a subordinate fault close to it, with a shift in the opposite direction. The chief fault is seen in the burn lower down the glen: it has a tendency to twist the schist into parallelism with itself, with a south-west dip.

One-quarter of a mile south-west of the "G" of "Gleann" the base of the green beds is again shifted, this time the south side to west. There are probably two faults converging towards one another as they go west, and continuing to join the one noticed (p. 222), ¾ mile S.S.E. of Cruach nan Capull. South of these faults there are several green beds outcrops, varying in thickness from 2 to 6 feet, below the base of the zone; in some places there are five such. Most of these can be traced more than ½ mile. The lowest band is probably more than 100 feet below the main base. The width of outcrop of the band, the base of which we have here taken as the main base, varies from 150 to 300 yards, but there are thin quartzose schists within this outcrop in some places. Above this band comes a wider outcrop of quartzose schist, which we have separated from the green beds zone for a considerable distance south-west. Above this quartzose schist green beds come on again, and extend far down the hill on a dip slope towards the River Cur—at least more than ½ mile.

The wide green bed at the base of the main zone can be followed pretty clearly along the south-east side of Socach Uachdarach. One-quarter of a mile south-east of the "S" of "Socach Uachdarach" the bed is thrown down north-west by the continuations of the Coire No N.N.E. faults. On the shoulder of the hill ½ mile north-east of Cruach na Cioba, on the east side of the faults, their is an ill-defined outlier of this bed. Below the outlier are one or two other outcrops of similar schist. Further south-west the bed gradually decreases in width. It is well seen about half way between the "C" and "H" of "STRACHUR," but is hardly 30 yards wide. One-half of a mile S.S.E. of the above "C" it is represented by two outcrops, each from 1 to 3 feet thick, and separated by 10 to 12 feet of quartzose schist. A little south-west of this it ceases to be traceable.

Near the "h" of "Leamhanin" the outcrop of green beds above the one we have been describing is hardly 50 yards wide, and is separated from the next overlying green bed by an interval of quartzose and phyllitic mica schist. The interval has been mapped to near the sides of the alluvium ⅜ mile east of Ballimore. The last clear section of it is in the burn ¾ mile slightly north of east of Ballimore : here the width of green beds below the interval is less than in Leamhanin and less than that of the interval.

The top of a greywacke schist horizon occurs on the south-west side of the Hell's Glen fault, between the first "a" and "d" of "Monovechadan." On the south-east side of Stob Liath, this line, together with the lamprophyre sheets a little above it, is thrown by a series of north-west faults, all of which shift the south-west side towards the south-east, the same direction as the Hell's Glen fault. One-eighth of a mile south of the first "i" of "Drimsyniebeg" the fault in Gleann Canachadan breaks through the line. The structure is complicated by other N.N.E. faults, some of which are the continuations of those described in Glen Kinglas (p. 233), and the ground is also largely obscured by drift. The line taken for the top of the horizon on the east side of the fault crosses Drimsynie burn 200 yards west of the sharp bend to the west. It is not well defined. In the burn at this place there are many crushes and small faults : one runs east to west along the burn, and there are several others almost at right angles to this. One-half of a mile north-west of the "y" of "Drimsynie" the line is better defined : there is a large proportion of quartzose schist below the line down to the fault at the west side of Drimsynie burn. The east to west basalt dyke, ½ mile north-east of "s" of "Drimsynie," is crushed and intermixed with ferriferous carbonate strings. The quartzose schist top seems shifted somewhat, the south side to the west, either by it or some line of fault between it and the north basalt dyke. The east to west basalt, ½ mile north-east of the second "n" of "Beinn Bheula," also shifts the line in the same direction.

The fault on the west side of the lower part of Drimsynie burn is seen in the little east to west burn in the wood. It is accompanied by subordinate crushes in the burn, some of them several yards wide. The hade is west. The strike of the schists is made parallel to the faults. Taken together the faults seem to have a large throw, for on the west side massive quartzose schists extend down to the bottom of the wood ; while on the east side there is almost unmixed albite schist up to near the 400 foot contour. A little east of the burn there is a parallel fault with a small downthrow in the same direction : the amount of surface displacement at the top of the albite schist seems 60 or 70 yards. The main fault crushes two basalt dykes at points ⅜ mile and a little over ½ mile north of the "m" of "Drimsynie." In both there is an apparent shift of 10 to 16 yards, the east side to north. The other dyke north of these is also crushed near the fault, but the direction of shift is not certain : it seems in the opposite direction to that of the other two. We suppose the fault to continue to those in Gleann Canachadan (see p. 223), and thence to the fault ⅜ mile north of the "a" of "Ardchyline" (see p. 217).

On the hilltop $\frac{3}{4}$ mile north of the "e" of "Drimsynie," and other places near, there are good exposures of early strain bands (see p. 24) folded sharply by folds of "anticline" type with axes hading north-west. These strains bands hade north-west less steeply than the bedding, or if the bedding be flat, they have a tendency to hade slightly south-east (see p. 28). The folds of anticline age are prominent and acute, and we see repeatedly that it is the under or south-east limbs of anticlines that are the most thinned.

There are good exposures of albite schist on the east side of the lower part of Drimsynie burn and between here and Drimsyniebeg. In various places, e.g., in the band 1 mile south-east of the "n" of "Drimsyniebeg," there are many quartzose portions which have perhaps been formed from quartz veins. These portions are often finely foliated parallel to the axes of "the anticline" folds, but where no such folds are present, they also usually show a fissile structure parallel to that in the schists near. The quartzose portions usually consist essentially of white quartz, and there are gradual passages between the most fissile, with splitting planes at intervals of about $\frac{1}{8}$ inch, and the thicker veins without fissile structure. As we should expect in veins, they vary in thickness and frequency. As a rule they run along the early folia-tion planes, but sometimes are only a few inches in length. At other times they cross the foliation markedly. "The anticline" folds have led to, or been accompanied by, the production of marked differences in the composition of some portions of the schist. The under limbs of anticlines are generally more quartzose than the upper ones. This seems opposed to the general rule in other places (see p. 22).

District of Lochgoilhead.

This district is inclosed by Allt Glinne Mohr, the Lochgoilhead valley, Loch Goil, part of Loch Long, and part of the east margin of 1-inch Map 37.

The schists throughout this area belong to the alternating series of albite or phyllitic mica schists and quartzose, more or less pebbly, schists. The north part of the area is on the north-west side of "the anticline," the south part on the middle, between Lochgoil-head and Stuckbeg, is an area, perhaps 2 miles in width across the strike of foliation, in which the foliation is nearly flat, the alternating dips, now north-west, and now south-east, about neutralising one another.

The promontory between Loch Goil and Loch Long, south of Clach Bheinn and Mark, is, with the exception of two bands of albite schist which have been separated out, in the main composed of massive quartzose, often pebbly, schist. This belongs to the greywacke schist horizon the top of which has been traced, with some interruptions, for a long distance south-east. At the extreme south end of the promontory, just opposite the Carraig nan Ròn, there are good sections of pre-anticline folds with their axes slightly waved. One-quarter of a mile north-east of Mark, a vein in the pebbly schist was found (slide 4198) to consist of quartz, felspar, and a little chlorite: Mr Teall states that the felspar is albite: it is pink macroscopically, like the albite in the pegmatites of the Dunoon shore section (p. 18).

Near the head of the west of the two burns on the south side of Cnoc Coinnich there is a large proportion of sharply jointed quartzite-like schist, but this is mixed with distinctly pebbly bands.

The marginal lines of various albite schists have been traced. Often only one of the margins of any particular band was found definite enough

P

to admit of satisfactory mapping.   We feel it unwise to make the structure
appear clearer than it is on the ground.   Some of the margins run parallel
to the foliation planes in the
neighbourhood, but in other
places there are broad bands of
albite schist which die or finger
out as we proceed along the
foliation.   These phenomena
we explain by supposing that
the beds are often folded along
axes which lie parallel to the
foliation, the apparent dying
out being really folding out.
A good instance of this on a
large scale occurs on the east
side of the valley between
Lochgoilhead and Polchorkan.
On the west side of the slack
which marks the course of the
N.N.E. fault near Inveronich,
there are good exposures of
albite schist extending from the
alluvium to above the 400-foot
contour.   Close to the slack,
practically the whole of the
exposure is albite schist.   After
going north ½ mile, we see, in
the most northerly of the two
east to west burns marked in
the map, a little quartzose
schist from 60 to 100 yards
from the edge of the alluvium.
The quartzose schist broadens
to the north : in the burn 200
yards further north it is 100
yards wide.   The top of it runs
north, not far off the 200-foot
contour, and between this and
the alluvium little else but
quartzose schist is seen.   Above
the 200-foot contour there still
continues a wide outcrop of
albite schist until we come near
the burn ½ mile south-east of
the "r" of "Polchorkan."   Two
hundred yards south of this
burn we notice a thin quartzose
schist within the albite schist.
This band of quartzose schist
rapidly increases in width
going north.   In the burn just
mentioned, it is 70 yards wide.
North of the burn it widens
further, and joins ultimately

to the quartzose schists above and below the albite schist.   Thus gradually
the whole of the thick albite schist, apparently 400 feet thick, on which

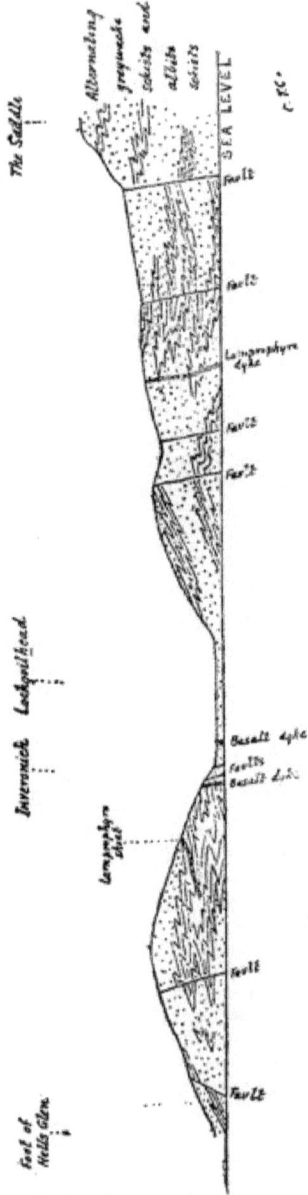

FIG. 52.—Horizontal scale 1 inch = 1 mile.   Vertical scale not uniform.   Diagram section from the Saddle through Lochgoilhead
to the foot of Hell's Glen.

we started near Lochgoilhead, has, by the time we get to Polchorkan, 1½ miles away, disappeared.

Another instance of apparent rapid dying out of albite schist occurs near Stuckbeg (153). In the burn running south-west there is a continuous exposure almost entirely of albite schist from 160 yards north-east of the house to the 700-foot contour. Albite schist is also seen on the shore 100 yards north of the burn foot. There may be a little quartzose schist between the exposures in the burn and on the shore, as a drift area 70 to 80 yards wide intervenes between them. The albite schist on the shore is underlain by a massive quartzose schist. The line between these schists is definite, and readily traced in the bare area by the loch side: it becomes drift covered ¼ mile north of the top of the "S" of "Stuckbeg," and north of this, in the two streams south of Inverlounin, is not so definite as to encourage us to trace it further. The top of the albite schist cannot be traced satisfactorily north of the burn, but the top of another similar, but thinner, outcrop above this, has been carried on as far as the stream ½ mile south-east of Tom nan Gambna. There is a good section in this stream for ½ mile below the albite schist, and very little else than quartzose schist is seen. What then has become of the wide outcrops near Stuckbeg? In the burn ¼ mile south of the last "u" of "Inverlounin" there is a width of about 170 yards of quartzose schist. In a south direction this gradually narrows, and forms a separation within the albite schist. But going north it widens, and the albite schists at the sides become subordinate, until in the burn ½ mile south-east of Tom nan Gambua we can hardly say there are any representatives of them.

A well-defined band of albite schist occurs on the side of Loch Long near Dail. This has been traced uphill to the north-east, often making outcrops about 140 yards wide, and it comes into a burn 1 mile W.S.W. of Coilessan. On the north side of the burn we found it impossible to trace the bed, and yet there are no indications of faulting. There is plenty of bare rock in the locality, and it seemed impossible that any great thickness of albite schist could escape notice.

These suggestions of folding prepare us for the difficulty experienced in matching on Ben Donich the greywacke horizon of which we have mapped the top on the west side of the valley between Lochgoilhead and Loch Restil, etc. From the height of Ben Donich, this greywacke should, after making allowance for the faults, circle all round it, if the beds continued uniformly parallel to the foliation. But no such thick greywacke can as a fact be traced round the hill. At the south-east and east sides there is indeed a thick greywacke, but on the west side this dwindles into insignificance. This is also the case on the north-west side. On the N.N.W. side there is a thick outcrop of greywacke, but this splits up into different bands: the lowest of them appears to run down into Allt Glinne Mhoir, coming into the stream ¼ mile east of the "i" of "Mhoir."

In Allt Criche (142) the top of the greywacke we suppose to match that traced on the west side of the Lochgoilhead valley, occurs about half way between the 1500-foot and 1750-foot contours. Fifty yards south-east of the burn it is thrown down to north by an east to west fault; the displacement at the surface is about 30 yards. The line north of the burn is tolerably well-defined for ½ mile, but on the south side the ground is obscure, so that we cannot see what are the effects of the N.N.E. and north to south faults that must cross it; probably, however, the displacements are not large. Seven-eighths of a mile slightly west of south of Ben Donich top, a spur of rock runs south into a drift area, and shows us the line again, close to the 1750-foot contour. In the burn that comes into Allt Coire Odhar ¼ mile below the first "l" of "Allt," the rock exposed is mostly massive quartzose schist from a little below the 1250-foot to the 1750-foot contour. Near the latter contour a north to south crush hading west appears, and the top of the quartzose schist is thrown down to west: the displacement at the surface is about 100 yards. In the burn to the east there is a N.N.W. fault which throws the top down to south-west; the displacement at the surface is about 40 yards. Between the last-mentioned burns and those that run into Allt Coire Odhar near the "O" of "Odhar," there is a drift area. The level of the line evidently is much lower on the west of this area than on the east. This we suppose is probably owing to the N.N.E. fault which breaks through the crags ½ mile E.N.E. of the "D" of "Ben Donich"; in the north to south burn ½ mile S.E. of the "D" of "Donich" a prominent face of rock, like those often marking crush margins, is seen hading east. Near the head of the east of the two burns that come into Allt Coire Odhair near the "O" of "Odhair," a north to south crush is seen. The direction of hade is liable to variation, but is generally west. The downthrow is also west. The top of the greywacke is not reached on the east side of the fault. The base, which is a well-defined line, is shifted at the surface about 200 yards.

The north-east spur of Ben Donich, from ⅛ mile north of the 2250- to a little north of the 1750-foot contour, is practically all albite schist. Below this comes a quartzose schist which we suppose matches the one the course of which we are describing, but its width of outcrop, where best defined, on the east side of the spur, is only about 100 yards.

The N.N.E. faults on the north side of Ben Donich do not seem to cause much alteration of level in the top of the quartzose schist, but the bottom seems lower on the west side. A want of correspondence in the thickness of the outcrops of the beds at the side of these faults is also noticed near Lochgoilhead. Near the latter place the top of the lowest albite schist is at about the same level at each side of the faults. But the base of it on the east side does not extend to within 160 yards of the alluvium, though on the west the base must be below the alluvium. The movements along the faults were perhaps more lateral than vertical, and outcrops affected by folding to a different extent may have been brought against one another. In two N.N.E. faults near Strachur there is also evidence of lateral movement (see pp. 210, 218), and in one in Garrachra Glen (p. 216).

A band of albite schist just north of Polchorkan has been traced east and north-east, and was found useful in disclosing faults which might otherwise have escaped observation. The fault nearest Polchorkan comes into the peat moss at the bottom of the valley near the "h" of "Polchorkan," and runs north-west in a straight line for the Hell's Glen fault (see p. 223). There is no rock in the River Goil in the place where the fault should cross. The outcrop of albite schist is displaced at the surface 200 yards, the south side to east, the same direction as, but much less in amount than, the green beds are displaced by the Hell's Glen fault. North-east of this fault there is another, again effecting a displacement, of about 60 yards, in the same direction. In its course south-east the last fault soon meets another coming from the N.N.E., and appears to be shifted by it 40 or 50 yards, the east side to south. This N.N.E. fault throws the albite schist, but the displacement, 60 to 70 yards, effected in this schist at the surface is in the opposite direction to the apparent displacement in the fault feature. The hade of the N.N.E. fault is east, but the downthrow in the albite schist is west. Though the north-west fault appears shifted by the N.N.E. one, it is not certain that the latter is the later of the two. It may be that the N.N.E. fault was in existence first, and that the north-west one subsequently proceeded along it for a little distance before resuming its normal direction. From the evidence elsewhere procured (see p. 109) we know that some of the N.N.E. faults are of very early date, and we should not like to conclude that the N.N.E. fault in this locality was the later of the two without clear evidence. The appearances are the same as those observed in Coire No (133) where the Coire No N.N.E. fault appears to shift the Hell's Glen north-west fault (see p. 232). We have continued the north-west fault to the south-east to two faults in Allt Criche, near the "h" of "Criche." These throw a lamprophyre sheet down to south-east. The effect on the top of the greywacke on the south side of the burn is uncertain.

A north to south crush, hading east, is seen on the shore of Loch Goil, 100 yards south of Inverlounin. This makes a conspicuous slack or depression at the top of the craggy ground south-east of Lochgoilhead. Part of it may probably be regarded as a continuation of the north to south crush seen in the burn $\frac{1}{4}$ mile north of the "G" of "LOCH-GOIL." Another part probably runs north-east to join an exposure of shattered schist, 20 to 30 yards wide, in Donich Burn, $\frac{1}{4}$ mile south-east of the "G." A little more than $\frac{1}{4}$ mile east of Inverlounin there is another north to south fault. This is not seen at the loch side, but it is probably the cause of the sudden change in direction of the burn $\frac{1}{4}$ mile S.S.E. of Inverlounin. The fault is traceable north with branchings in places as far as the burn $\frac{1}{10}$ mile north of the "d" of "Lochgoilhead."

The promontory between Loch Goil and Loch Long is intersected by a prominent feature running north to south which forms the east margin of a drift area immediately north of the loch. Between this drift area and Corran Lochan the feature crosses two bands of albite schist without making any appreciable shift.⋅ The length of the lochan lies along the continuation of this feature, and there are also slacks extending from the lochan in the same line. Between Tom Molach and Beinn Reithe the line is continued by a feature facing east. Soon after this the fault divides into two : both branches hade east and pass a little east of Cnoc Coinnich, keeping within 50 to 100 yards of one another.

Strong crushes occur in the burn south of Stuckbeg, running north-west and hading south. They twist the strike of the foliation into parallelism with themselves and cause it to dip south-west. On the north side of the Saddle they give rise to a conspicuous pass, and then run down the burn that enters Loch Long at Feoileann (1-inch Map 38). In the course of this burn a good many ferriferous carbonate veins occur in and near the crush breccia.

The outcrop of albite schist at the south end of Corran Lochan grows gradually wider to the east. At the lochan it is only 50 yards across : on the shore of Loch Long it is 300 yards. One-sixth of a mile west of Mark it is shifted by a fault with a general N.N.E. direction : the amount of displacement at the surface is about 130 yards, the west side to south. This fault makes a conspicuous feature, facing east, in its course north-east for about 2 miles. Not quite 1 mile south-west of the foot of Allt Guanan it crosses a lamprophyre sheet and throws it : the surface displacement is about 70 yards, the west side to south. On the south side of Allt Guanan it is marked by a slack. In the continuation of the direction of the slack a vertical crush crosses the burn, nearly $\frac{1}{4}$ mile above its foot.

On Meall Daraich, the rocky point between Loch Goil and Loch Long, there are a good many ferriferous carbonate veins running almost with the schist strike. One of the veins, $\frac{1}{3}$ mile N.N.E. of Carraig nan Ron, is as much as 6 inches wide.

District of Cairndow.

This district is inclosed by Glen Fine, part of Loch Fine, Hell's Glen, Allt Glinne Mohr, and part of the north and east margins of 1-inch Map 37.

The schists which occur in the district extend from the garnetiferous mica schist to the alternating series of albite and greywacke schists which come below the green beds. Much the largest part of the area is occupied by the latter series : there are good sections of it in Kinglas Water, Allt Glinne Mhoir, and the streams on the south side of Binnein an Fhidleir. There are dip slope exposures of the green beds in nearly all the streams that run north-east between Achadunan and the head of Hell's Glen. The fall of these streams is on the whole rather less steep than the dip of foliation, but for considerable distances the two may coincide.

There are no quite satisfactory exposures of the Loch Tay limestone. A section in the burn just south of the "o" of "Ardno" is crossed by dykes of lamprophyre and felsite running with the strike and accompanied by crushing, the effect of which is not exactly known. At the top of the limestone there is calcareous quartzite. Near the base of the limestone there is also calcareous quartzite, but the bottom bed is limestone again. There is no hornblende schist except a thin band near the top of the limestone. In the different burn sections further north-east there is always a thick hornblende schist in the middle of the bed : the part below this schist is largely mixed with calcareous quartzite.

In the north-east corner of the area a portion of the coarse granite of the Meall Breac and Garabal Hill complex (see p. 99), appears, and the schists near it are much altered. The usual strike of the schist is also lost near the granite. For a distance of $\frac{1}{4}$ mile or more the foliation dip is inwards towards the granite. Further away the dip is in some places away from the granite, towards the south-west: at other places it is almost nil, or towards the west, without any evident relation to the granite boundary. Near Achadunan, more than 2 miles off the granite, the common dip is still east, but south-west of this place, the usual strike and dip prevail. At the east side of Lochan Mill Bhig the foliation is often nearly flat.

The base of the green beds comes down to the alluvium of Glen Fine $\frac{1}{4}$ mile north-east of Achadunan. In the burn $\frac{1}{4}$ mile E.S.E. of Achadunan a N.N.E. crush is seen with a hade to west. The crush is accompanied with a throw, there being quartzose schist on the west side close against phyllitic mica schist on the east, but the amount of throw is uncertain. It's effect on the base of the green beds is not seen. In the burn $\frac{1}{4}$ mile south-east of Achadunan there is an E.N.E. fault, hading north, which shifts the base with normal throw. About $\frac{3}{4}$ mile south-east of Achadunan there are several faults running in the same direction as the last mentioned, but with opposite throws. One of these makes a conspicuous shift in a basic lamprophyre sheet.

Half-a-mile north-east of Cairndow there are two ochreous weathering lamprophyre dykes which are accompanied with crushes. These can be traced about $\frac{1}{2}$ mile, and along their east continuation the base of the green beds appears to be thrown down to south considerably : the lateral displacement effected at the surface is 330 yards. A thin dyke, apparently basaltic, is seen along the course of the fault nearly $\frac{1}{4}$ mile south of the "oi" of "Upper Clasheoin."

Between Achadunan and Binnein an Fhidleir the bottom bed of the green beds zone makes a wide outcrop, and there are no subordinate lower outcrops which have been noticed. On the hill-top west of Binnein an Fhidleir a thin, tough, green bed, only 1 to 2 feet thick, makes its appearance a little below the base of the zone. Half-a-mile south-west of the "B" of "Binnein" there is an 8-foot green bed about 40 feet below the base : the interval between is mainly massive quartzose schist. Near the last place, on the west side of a little N.N.E. fault, which has a downthrow west of 10 feet, there is a 6-foot green bed about 10 feet below the base : the lower green bed is still apparent.

In the burn which goes between the "G" and "l" of "Glen King-las" there is a good deal of faulting near the green beds base, and it is hard to show the detail in the 1-inch Map. The fault that runs up the burn has, where it crosses a lamprophyre sheet below the green beds, a throw of about 20 feet down to the south. But when it gets up into the green beds the throw is in the opposite direction. This seems owing to two faults which come in on the west side of the burn, each of them throwing the green beds base down to the north. After meeting the fault running with the burn, they keep along it, and more than neutralise its former effect. On the east side of the burn there are two massive pebbly quartzite-like bands about 8 feet thick. One of these is immediately under the green beds base and the other a little lower. A little below the lower one is a band of leaden-coloured phyllitic mica schist 30 feet thick.

It is a little uncertain where to take the base of the green beds in Kinglas Water. There is a 12-foot band about 120 yards up stream from the line drawn for the main base. On the hillside south of the burn, there is at the base a band of green beds, about 30 yards wide, separated by an approximately similar width of massive pebbly quartzose schist from the overlying green bed.

The section in the west of the two burns that run into Kinglas Water on the south side near the "G" of "Glen" shows an unusual proportion of massive pebbly quartzose schist. It gives the impression that there is as large a proportion of such schist as anywhere in Glen Kinglas, not excluding the horizon of which the top is mapped on the north side of Beinn an Lochain (134 east).

Near the course of the east to west basalt on the south side of Kinglas Water there is a sudden alteration of position in the base of the green beds as if it were thrown down south considerably. On the west side of the base there are various crushes hading south or south-west. Some of these are close by the basalt, and others by a thin lamprophyre dyke on the north of the basalt. On the east side of the base there are no certain indications of faulting in the supposed direction. The broad lamprophyre sheets $\frac{1}{2}$ mile east of the base are not appreciably shifted. This may be because these sheets happen to be nearly vertical. Or it may be that the appearance of faulting further west is really due to folding.

On the north-east side of Stob an Eas there are two bands of green beds below the line adopted as the base of the zone, and separated from it by massive quartzose schist. The lower of these makes an outcrop nearly 100 yards wide in places, but this is on a dip slope. Two outcrops in a corresponding position are seen again in the burns $\frac{1}{3}$ mile W.S.W. of the "S" of "Stob."

Owing to drift the base of the green beds is not clearly seen at the side of the road between Lochgoilhead and St Catherine's. In the burn south-west of the line drawn, there is a prominent joint hading south-west, and it is possible that this may be accompanied with a throw. Twenty yards west of the road there are many blocks of green beds just where we have taken the base line, but we cannot be certain we see the rock in place until we have proceeded more than $\frac{1}{8}$ mile from the road. Sixty yards south-east of the base line we see, on the south-west side of the road, a green bed, 5 to 6 feet thick, with massive quartzose pebbly schist on either side.

The big fault on the south-west side of Hell's Glen, and its effect on the green beds outcrop, is described on p. 223.

The line for the top of the green beds zone is entirely hidden, excepting in three burns. These burns are, taking them in order from north-east to south-west, by the "g" of "Bathaich ban Cottage," $\frac{1}{2}$ mile east of the "T" of "Tom Dubh," and nearly $\frac{1}{2}$ mile S.S.E. of the same letter. The Loch Tay limestone and the felsite sheet near it appear to be faulted with a large downthrow to west or south-west a little on the east side of Tom Dubh, but the effect of this fault on the green beds top is not seen. The limestone is not exposed in either burn at the sides of Tom Dubh, but it is certain that it is not south-east of the felsite sheet, for there is a good exposure of schist extending for some distance in this direction from the sheet, and without any limestone. On the east of the supposed fault, or faults, a hornblende schist is seen at the top of the wood $\frac{3}{8}$ mile east of the "T" of "Tom Dubh." We suppose the main part of the limestone occurs between this hornblende schist and the felsite, as it does in the burn a little north-east. If this is so, the felsite occurs, geologically, below the limestone on the south-west of the fault, but above it on the north-east.

Above the main zone of green beds there are various other thinner beds of a similar kind. One-third of a mile south-west of Rudha Bathaich Ban, a band of green beds, a little inland, and on the south-east side of a felsite, is shifted by a fault with downthrow to west. This fault may continue south, and help to separate the two outcrops of limestone. The throw between these outcrops is greater than that between the green beds mentioned: perhaps part of it takes a more north-west direction. A green bed still more to the north-west occurs on the shore $\frac{1}{4}$ mile south-west of Rudha Bathaich Ban. It is crossed by two north-west faults. Two green beds are also seen above the limestone in the burn that goes through the "n" of "Ardno." A band only a few feet thick also occurs in the wood $\frac{3}{8}$ mile E.N.E. of the "T" of "Tom Dubh": this cannot be much above the limestone. The schist below the felsite on Tom Dubh is garnetiferous mica schist. This is an instance of the occurrence of garnets S.E. of their usual province. Such instances are common in the north-east part of the area (see p. 48).

There are strong crushes a little within the high side of the wood between Ardkinglas and Ardno, running almost parallel to the side. These are seen in the burn that goes through the "ai" of "Bathaich ban Cottage," and the two burns to the south-west that are shown in the 1-inch Map. In the first-mentioned burn there are three parallel crushes and the hade is south-east. Judging from the hornblende schist, which occurs above the top of the green beds zone in the two south-west burns, there cannot be much throw with the crushes.

The top of the schistose greywacke that passes through the second "l" of "Hell's Glen" is a good line to adopt as the top of a greywacke horizon, but one cannot be certain that it corresponds with the line adopted on the other side of the Hell's Glen fault. The top of another, higher, greywacke schist, passes by the "e" of "Hell's." The line by the second "l" of "Hell's" continues well marked for some distance north-east. In and near the burn that passes east of the "n" of "Hell's Glen" there are several N.N.E. crushes close together that hade west, but they throw down the top of the greywacke to the east, and must be reversed faults. In its course north-east these crushes separate. The east branch makes a conspicuous feature on the south-east side of the long scar on the south side of Beinn an t-Seilich, and throws a felsite sheet down to the east. The west portion runs west of the long scar, and also throws the felsite down to the east.

The greywacke top in the burn that joins Allt Glinne Mhoir by the " r " of " Mhoir " is thrown down east by a N.N.W. fault. We suppose there is a more important fault with downthrow west a little on the east of this, and perhaps joining it as it goes south. We seem to need such a fault owing to the top of the greywacke on the east side being at a higher level than on the west. The greywacke top is tolerably well marked all along the south and east sides of Beinn an Lochain (134): it is shifted at least twice by north-west faults, once $\frac{1}{2}$ mile south of the "ai" of "Beinn an Lochain" and again $\frac{1}{12}$ mile S.S.E. of the same letters. On the north side of the hill the line is less well marked : in the burn that passes between the " W " and " a " of " Kinglas Water," there is a rather wide outcrop of massive pebbly quartzose schist a little above the line adopted in the Map, and separated from it by albite schist. Above this quartzose schist nearly all the exposures are albite schist right up to the hilltop. A little west of the burn mentioned, the greywacke top is affected by two large faults, which let down between them a strip of higher rocks, chiefly albite schist. The east fault has a tendency to twist the foliation into parallelism with itself. It is accompanied by a lamprophyre dyke in part of it's course. The west fault is inferred from the want of agreement in the schists in the burns that pass on either side of the " W " of "Water." The faults, if continued, should soon meet in a south direction, and probably neutralise one another in throw, but a crush line has been traced, and often makes a conspicuous feature in the landscape, for many miles south-west, far outside the particular area now being described. It makes a slack where it crosses the greywacke top $\frac{3}{4}$ mile slightly east of south of Beinn an t-Seilich, but has no appreciable throw.

On the north side of Kinglas Water, near Butterbridge, we are unable to see that there is any thick horizon in which greywacke schist is in distinct excess over albite schist. The only schist mapped in this locality is a massive quartzose schist near the top of the high crags a mile east of Binnein an Fhidleir, and running from these crags to the east margin of the 1-inch Map.

Various N.N.E. crushes have been already alluded to. There are a few others worthy of mention. The Coire No fault continues under drift to the burn $\frac{1}{2}$ mile E.S.E. of the "m" of "Tom Dubh." There is a section of the crush in this burn and the stream changes direction along it for a little distance. The next burn north-east also twists into the crush for a little way. From here we have prolonged it under drift in a straight line for the lochside at Cairndow. There is little doubt that this crush represents part of the disturbances that cross the west side of Newton Hill (126), and continue to Tyndrum, etc.

The crush that crosses the west shoulder of Ben Donich (142) perhaps crosses the head of Allt Glinne Mhoir a little below the 900 feet contour line. The schists near are discoloured with a yellow tint and some crushes are also visible, but are not so thick as we should have expected.

On the east side of Binnein an Fhidleir there are at least three N.N.E. crushes which make prominent features as they cross the hilltop between Glen Kinglas and the head waters of Eas Riachain. The two most westerly ones, about $\frac{1}{3}$ mile east of Binnein an Fhidleir, are accompanied with a good deal of contortion. In this respect they resemble the one already alluded to, that runs by the second " a " of " Kinglas Water," and they may be regarded with confidence as continuations of it. One of the crushes in the burn ⅓ mile slightly east of south of Binnein an Fhidleir is several yards wide.

South and south-east of Cairndow, a little above the green bed's base, the foliation dip is, in several places, south-east instead of north-west, Not far off there is a straight feature running nearly with the strike of the foliation. It is possible there is some disturbance along the feature, and that this is the cause of the unusual dip.

A N.N.E., fault crosses Hell's Glen near the "o" of "or." In the burn by the "o" the hade is west. About ¼ mile up the burn the fault branches. The west part throws a lamprophyre sheet down to west in the burn a little over ¼ mile south-west of Stob an Eas: the fault makes a prominent feature facing west on the west side of the Stob, and then comes into the east bank of the burn ¾ mile south-west of the "K" of "Kinglas Water"; the hade is still west. A little north of the branching a dark micaceous dyke gets into the east part of the fault and keeps along it for one or two hundred yards at least; the dyke is itself somewhat crushed. The next clear exposure of the east part is in the burn ¾ mile N.N.E. of Stob an Eas: here a reddish lamprophyre dyke accompanies the fault; the hade is still west; we could see nothing of the dyke on the north side of the east to west basalt. In the burn ⅝ mile north-west of Beinn an t-Seilich the crush is seen again. The sharp change in direction in the east to west basalt near this exposure is along a subordinate ferriferous carbonate vein hading west. On the north side of Kinglas Water there are several faults which may represent those on Stob an Eas, or branches of them.

The lamprophyre dyke running slightly north of west along the course of the burn near the "las" of "Kinglas Water" is accompanied by a fault. The dyke hades south and the throw is in the same direction. We lose sight of the dyke at the slight bend in the burn above the "l" of "Kinglas Water," but veins of barytes, as much as 1 foot thick, and sometimes hading north, run along the changed path of the burn. Below the "n" of "Glen" two lamprophyre dykes are met with, and are doubtless representatives of the dyke just alluded to. Both hade south and the south one is accompanied by faulting. The dyke in the burn south of the "G" of "Glen" also hades south and is probably connected to the others mentioned.

The nearly east to west fault that runs in the course of the burn below the Eagles' Fall has a downthrow south, probably of at least 60 feet. The crush is often several feet wide, and accompanied with strings of ferriferous carbonate. There are various other crushes which cross this burn at an angle; one of these, passing through the "F" of "Fall" and running north-west, makes a conspicuous nick in the steep crag on the south of the burn; on the north side of the burn it throws one of the porphyrite sheets down to the south.   C.T.C.

CHAPTER XVIII.

GLACIAL DEPOSITS.—GENERAL DESCRIPTION.

In the following description the illustrative details have been chiefly collected from that portion of Cowal which is contained within 1-inch Map 29, and lies on the east side of Glendaruel and Loch Riddon. The details in other portions are described subsequently.

Under the head of Glacial Deposits are included boulder clay, many shapeless widespread sands and gravels which exhibit no relation to the present streams but may occur scattered over the hill slopes up to considerable heights, more or less definitely shaped morainic mounds, and some beds of finely laminated clay or clayey sand which may contain glacial shells.

In the 1-inch Maps the localities where the most conspicuous moraines occur are indicated by the word "Moraines" engraved over them, but with this exception no attempt has yet been made to show any of the deposits mentioned. The approximate margins of the thicker and more widespread were, however, drawn on the field-maps,* and also the shapes of many of the more prominent moraines.

The boulder clay has the usual character of an apparently structureless tough clay wherein are distributed irregularly many rounded and striated boulders. The relative proportion of stones to the clayey matrix varies considerably, and in some cases is so large that the general character becomes more that of a gravel. In other places the matrix is so sandy that the word "clay" is hardly appropriate. This more sandy character of the matrix was, no doubt, original in many cases, and due to the more siliceous character of the rocks from which it was mainly derived, or to a washing out of clayey matter before it reached its present position. But in other cases it is secondary, being produced near the present ground surface by the slow action of water washing away the finer clayey particles : in these cases the colour also usually changes into a brown or ochreous tint owing to more complete oxidation of the ferruginous matter in the clay. In the bank of the raised beach nearly $\frac{1}{4}$ mile N.N.W. of Hunter's Quay ochreous streaks, about an inch thick and slightly projecting on the weathered face, descend nearly vertically into the boulder clay for 3 or 4 feet. They thin gradually away as they descend. The upper 1 foot of clay is of the same character as the streaks. Individual pebbles lie partly in the streaks and partly in the less ochreous clay at their sides, and so it would seem that the change of colour must be due to the percolation of water. There appear to be thin cracks down the middle of each streak.

In original colour the boulder clay varies greatly. In certain areas it is a deep hæmatite red, but, perhaps, a greenish-grey or brown colour is the most general. No attempt has been made to separate the red clay from the other kinds, but many places where it occurs have been noted in the field-maps. Among these are the following : the east shore of Loch Riddon, 170 yards north-west of Springfield House, 1 mile south of Springfield House, and $\frac{1}{4}$ mile south of Fearn' ach : many places along the shore of the Kyles of Bute between Fearna Bagh and Strone Point, e.g., 1 mile south-east of Fearn' ach, 240 yards N.N.E. of Altgaltraig Point, south-west of Newton, $\frac{1}{2}$ mile and $\frac{2}{3}$ mile south-east of South Hall : on the hillsides 1 mile north-east and east of Colintraive Pier : 250 yards north

* The sketch map of the superficial geology, near the end of the volume, has been compiled from the lines and notes on these field-maps.

and 200 yards north-east of South Hall ; in the burn 1 mile north-east of Meallan Riabbach (183 south-west), up to a height of over 1250 feet : on the west side of Loch Striven, ¼ mile north-west of Troustan (a little inland), and ⅓ mile S.S.E. of Braingortan ; on the east side of Loch Striven, ½ mile south-west of Cnoc Madaidh, a little more than ¼ mile north-west of the Craig, various places on either side of the foot of Invervegain burn, and 240 yards north-west of Inverchaolain Church : in Invervegain Glen by the " i " of " Invervegain," and again a little more than ¼ mile slightly west of south of this locality : in the burn nearly ¾ mile north-west of Dun Mor (172 south-east) : in the burn ¾ mile south-west of A Chruach (172 east) : in Gleann Laoigh a little below the " G," and ¼ mile south-east of Balliemore : ¼ mile west of Coraddie (182 north-east), at a height over 1350 feet : in Glen Tarsan in many places near the head of the glen and by the shepherd's house : in Glen Lean above the Powder Mills up to the head of the glen, and in particularly clear sections a little south-east of Glenlean Farm : in Coire Athaich (173 north-east), by the " ai " : in various places a little south-west and south-east of Ballochyle (175 south-west) : ¼ mile north-east of Clachaig Hill (173 east) : in Strath Echaig close by Uig : in Glen Kin in various places near the "G" : in the burn ¼ mile south-west of Ardentinny Church. It would thus appear from our notes that it is general in the district between Loch Riddon and Loch Striven ; and (on the east side of Loch Striven) on the north-west side of a line between Inverchaolain and Ardentinny, but we cannot be sure that it may not also occur in other areas which have not been noted. This red boulder clay may occur in the same areas as the clays of other colour without any marked physical separation, and is occasionally seen mixed with them in sections : in the burn nearly 1 mile west of Meall an Fharaidh (173 north-west) both red and brown clays occur, and the change of colour is sometimes gradual and sometimes sudden ; in a burn ⅓ mile E.N.E. of Bargehouse Point(194 north-west) red and brown boulder clay are seen within 20 yards of one another : on Strone Point (194 north-west) similar clays occur on the shore, at the same level and not far from one another.

It is not clear what is the cause of the red colour. The idea of derivation from rocks of the Upper Old Red Sandstone is suggested, but no pieces of these have been noticed in any of the exposures mentioned, and the boulder clays near to them, e.g., near Toward, Inellan, and Dunoon, are not red. It is unlikely that the boulder clay movement was ever from the red rock area in this district on to the schists. The boulders in the red clay do not, as far as has been observed, differ from those in the other clays, and consist either of schist, vein-quartz, or of various igneous rocks—granites, basalts, lamprophyres, hornblende schists, etc.— none of which seem likely by pounding down to have given rise to the peculiar colour. Thin hæmatite veins are not uncommon in parts of the schist area, but are not numerous enough to have coloured such widespread deposits.

Among the boulders in 1-inch Map 29 found in, or in all probability derived from, boulder clay, which have especially attracted our attention as being probably derived from without this Map, are many boulders of epidiorite, hornblende schist, granite with conspicuous porphyritic crystals of flesh coloured felspar, diorites with black mica, and some dark and occasionally much pyritized schists, which sometime contain albite of an unusual reddish brown colour. It is not possible to say with confidence whence the epidiorites and hornblende schists came, because these rocks are common both in this and adjoining maps ; but it is clear that many, in the north-east portion of this map, must have come from some distance outside it, as such rocks are locally scarce. The porphyritic granite with flesh coloured felspar can, on the other hand, be referred with certainty to the area between the heads of Loch Fine and Loch Lomond, which has been described by Messrs Dakyns and Teall (*Quart. Journ. Geol. Soc.*, vol. xlviii., p. 104, "On the Plutonic Rocks of Garabal Hill and Meall Breac)." A granite boulder of this character occurs by the roadside 130 yards slightly west of south of Trinity Church, Dunoon, and there are others on the shore of the West Bay of Dunoon, on the shore near Strone Point (194 north-east), on the shore near Altgaltraig Point (173 south-east), on the shore 70 yards S.S.E. of Toward Lighthouse (at least 4 or 5 cubic feet in contents), and on the east side of Loch Striven ½ mile south-east of Ardtarag. These granite boulders are much more numerous as we approach their source of origin, and together with the diorite, which has come from the same area, are conspicuous on the sides of Loch Long and Loch Eck, and in Glendaruel. The dark albite schists have only been noticed on the side of Loch Long. A boulder of this kind, measuring about 12 feet by 12 feet by 6 feet, occurs on the shore just south of the rocky point at Ardentinny. A similar rock is seen in place near the hyperite of Donich Burn, Lochgoilhead (1-inch Map 37), and seems to owe its peculiar character to contact alteration by this rock. A similar rock is also seen occasionally near some of the basalt dykes and lamprophyre intrusions. But in none of these cases does the altered rock extend over a large area, and it is probable that the boulders noticed have come from some other district where it is more widespread—from the area around the plutonic rocks of Garabal Hill and Meall Breac, etc.

In some comparatively rare cases the stones in the boulder clay have their longer axes arranged in evident parallelism. On the shore ¼ mile E.N.E. of Cluniter (195 north-west) a stiff grey boulder clay, close to high-water mark, has the stones in it arranged with their longer axes slightly south of east, and lying on their "keels," *i.e.*, with their flat sides approximately vertical. Some 30 yards further north the stones are still on their keels, but are arranged nearly parallel with the coast—slightly west of south: 20 yards still further north they lie with their axes E.S.E., and are partly on their keels. On the shore a little more than ½ mile N.N.E. of Tor Aluinn (184 south-west) a stiff brown clay shows the stones often on their keels and with their axes south-east. In the small stream ⅝ mile slightly north of west of Stronchullin (north 174) the flatter sides of the stones incline N.N.E., much with the slope of the hill. In a little burn ¼ mile slightly west of south of the "G" of "Gleann Laoigh" a red boulder clay shows a rough lamination sloping down with the hill towards the west. A stiff laminated clay with stones occurs in the boulder clay in the burn ⅝ mile slightly east of north of Cnoc à Mhadaidh (174). On the south-east side of Inverchaolain Burn, a little more than ½ mile above the church, the boulder clay is both mixed with gravel layers and also has a tendency to lamination in itself.

Some of the burns run through boulder clay for considerable distances, their pre-glacial courses being still deeply hidden. Brown boulder clay occurs in the bed of the burn ¼ mile S.S.W. of "G" of "Glen Kin" (183 north-east), and no rock is seen for ½ mile above or 130 yards below this point. In the burn ½ mile north-east of Meallan Riabhach (183) there is red boulder clay with a thickness of at least 20 to 30 feet, and for a distance of more than ¼ mile no rock is seen: this is at a height of above 1250 feet. In the burn above Corlarach (194 north-east), from a point due east of Corlarach for a distance of more than ¼ mile above this, no rock is seen, but only stiff grey boulder clay with a thickness of at least 20 feet, and up to a height of 950 feet. Brown boulder clay is seen in the bed of Tamhnich Burn (172 north-east) near the " c " of " Tamhnich," and also about ½ mile above the "n" of " Burn :" also in the stream bed ½ mile south of Cruach nam Mult (172 north-east). From a point ⅝ mile N.N.W. of Blairmore Hill (174 north-west), for about ½ mile up stream, the burn runs through a stiff yellowish-brown boulder clay, at least 40 feet thick : this clay is overlain by loose sandy and gravelly material which is at times developed into good morainic shapes. Near the junction of the burns ½ mile east of the Badd (183 north-east) we see at least 70 feet of boulder clay without reaching the rock below : the clay is generally brown near the top and greyer below : in one place, near the top, thin bands of red and brown alternate. The burn 1 mile W.S.W. of Kilbride Hill (183 south-west) shows, on its south side, a thickness of at least 60 feet of boulder clay, but there are rock exposures in parts of the burn. The burn ½ mile north-east of Kilmarnock Hill (194 and 183), for a distance of 200 yards, runs through reddish brown boulder clay overlain by a greyish-green more stony clay : the total thickness seen is 50 or 60 feet : the height above sea-level is 800 to 900 feet.

No rock is seen in Ardyne Burn (194) more than 50 yards below the main road, with the doubtful exception of a small exposure about ½ mile N.N.W. of Killellan, but exposures of boulder clay and other glacial beds are not uncommon. The shores on either side of the foot of this burn show a similar absence of rock for a long distance, and though now spread over by beach gravels, it is probable that glacial deposits underlie these gravels, and are indeed, from their readiness to yield to denudation, the cause of them making such a prominent show in this locality. The original rock floor of the mouth of the glen is clearly below the present sea-level, and is still covered by glacial deposits. Being under the sea no fluviatile or atmospheric denudation can now effectively work to free it from the overlying beds, and restore the old outlines : and during the times of the different raised beaches this protection from denudation must have been still more perfect. There are many other glens which are at their seawards ends entirely occupied by boulder clay or later deposits, e.g., Strath Echaig, Glendaruel, Gleann Laoigh, the valley of the burn which runs into the West Bay of Dunoon, and that of the burn between Ardhallow and Clyde View (184 south-west). There must be rock hollows under the present sea-level in each of these cases, for rock is seen at either side of their mouths, and if we suppose the hollows to have been formed by river denudation, the land must in preglacial times have stood at a higher level relatively to the sea than it does now.

It is probable that the sea lochs themselves, however modified by sub-sequent glacial action,* represent glens originally formed by river action. We do not know in what proportion the floors of these are of rock or boulder clay, but the shallower parts near the shore are often of the latter. In these places the clay is often partly obscured by the shifting shingle of the shore, and therefore the exposures at one time noted may not be observable subsequently, but may be represented by other similar ones. Very commonly, e.g., in the West Bay of Dunoon, north-east of Tor Aluinn (184), near Strone Point (194 north-west), and north-west of Inverchaolain Church, these boulder-clay exposures are covered, near high-water mark, by the gravels of the raised beaches, from which numerous fresh-water streamlets descend and trickle down across the partly shingle-hidden boulder clay.

---

* For the contours under sea-level see *Trans. Roy. Soc. Edin.*, vol. xxxvi., "The Clyde Sea Area," by Dr H. R. Mill. Also Admiralty Chart, Sheet $\overline{2131}^{2}$.

Sand and gravel beds are not uncommon within or under boulder clay. Examples occur on the east side of Inverchapel Burn (164 south-west), nearly ¼ mile above the bend, and ⅓ mile above the bend ; in the east to west burn 1 mile west of Meall an Fharaidh (173 north-west) ; by the "ai" of " Coire Athaich " (173 north-west) ; in a little burn (not marked in the Map) about ⅓ mile south-east of Ballochyle (173 east) ; ⅓ mile N.N.E. of the last " n " of " Inverchaolain Glen " ; on the east side of the burn a little more than ½ mile north-east of Inverchaolain Church ; ⅓ mile W.S.W. of Ardnadam Farm ; ₁⁄₁₂ mile south-west of " G " of " Glen Kin," in a section on the south side of the burn.

In the section ⅓ mile south-west of Ballochyle, only the 6 or 7 feet of coarse gravel at the top seem referable to the high-raised beach ; below this at the west end of the section, are sand layers, sandy gravel, and stiff laminated stony clays ; at the north-east end, just under the gravel at the top, there are 10 feet of deep red boulder clay, and below this indications of sand and gravel.

In the two burns west and south-west of Ardentinny there is a considerable thickness of sand and gravel lying over boulder clay. The sand is in part very fine and evenly laminated ; on the east bank of the burn, a little more than ⅓ mile south-west of the church, it is overlain by a stiff clayey sand with striated pebbles and boulders, and without lamination. Over this is a 4-foot bed of ochreous gravel and boulders, the base of which undulates but is rudely parallel to the present ground surface : it has a somewhat smooth flat top and may belong to an old burn deposit, but has not been mapped as such.

In the burn nearly 1 mile west of Meall an Fharaidh (173 north-west) the boulder clay contains irregular pockets of gravel, and a small fault runs vertically through it.

Much more commonly seen than the sands and gravels below boulder clay are those above it. These last are so abundant over the hill slopes that it is not necessary to cite particular examples. Should you notice on any burnside a line of springs and wet ground you will find this due, in nine cases out of ten, to a coarse gravel or sand overlying the boulder clay, the line between the two being sharp and usually straight. Even on the hillsides it is often possible to distinguish with confidence the lines between such beds, by the drier character of the ground where the sand and gravel occur ; but no general attempt has been made to separate them.

These deposits are generally very irregularly bedded : sometimes when lying on steep slopes they are seen to be bedded roughly parallel with the slope. They spread over the hills far away from any streams, and up to heights where there are now no streams of any magnitude. They form surfaces usually not distinct in the distance from those of the boulder clay, but in other cases well-defined mounds and morainic shapes of a recognisable character. It has been suggested that these well-defined mounds are of later age than the other gravels and sands, and that the great age of the last is the cause of their want of mound-shape, denudation having destroyed the original outlines. But in absence of clear sections it is not possible to say that one gravel is younger than another, or younger even than a certain boulder clay ; the sands and gravels without mound-space do not, as far as is known, differ in character from other beds seen under the boulder clay, and it is possible that some of them are overlain by boulder clay in some areas.

On the east bank of the Ruel, 300 yards north of Ardacheranbeg, a section shows a wedge-shaped exposure of fine sand, with occasional fine gravel bands, lying in coarser gravel. The sand shows distinct and repeated contortions. The folds are very sharp, and their axes, as seen in the section, which runs nearly north to south, are approximately horizontal. There are some lines which appear like horizontal " thrusts," but perhaps they are only lines of sharp twist. The sand layers that show the contortions are a little variable in thickness, their outcrops not being exactly parallel. Above the contorted beds comes a red sandy clay with stones; the matrix of this is stiff, but we did not see any striations on the stones; it is a good deal mixed with irregular patches of fine yellow sand. At the top of the section are 3 or 4 feet of evenly stratified gravel and sand which clearly belong to a river terrace. The underlying beds are probably all glacial, and the contortions may be due to the passage of ice over them.

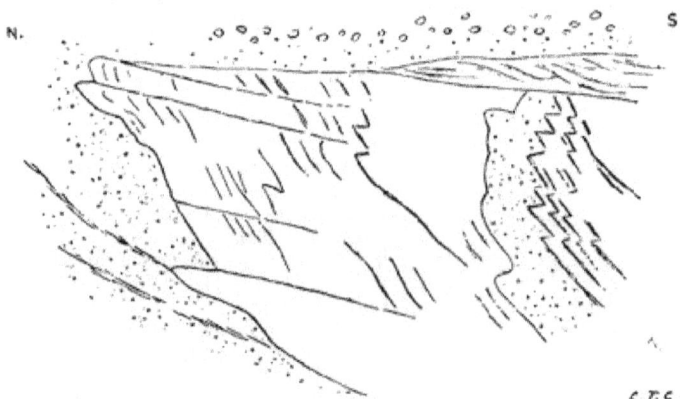

FIG. 53.—× ¹⁄₁₂. Section on E. bank of river Ruel 300 yards N. of Ardacheranbeg. Fine sand and gravel, with nearly horizontal folds and thrust-like lines, surrounded by gravel. Over the last is red sandy clay with stones.

A little further up the same river, ¼ mile E.N.E. of Glendaruel House, stratified clays, sands, and fine gravel, occur in a series of false-bedded sands and gravels. The clay layers are at times sharply folded with both limbs of the fold inclining in the same direction. A little below the folded layers is a 6-inch band of dark clay which is traversed by a reversed fault, of about 6 inches throw, the direction of which is the same as that of the axes of fold.

In a little burn 260 yards south-west of Ardnadam Pier 4 feet of gravel, presumably belonging to the raised beach, overlie 5 feet of finely laminated clayey sand with occasional stones, which is much contorted. This sand rests on brown boulder clay.

On the east shore of Loch Striven, ¼ mile N.N.W. of Finnart, a laminated stoneless clay shows occasional contortions.

It is possible that the contorted beds described belong to the same series as the beds containing glacial shells. Some of them closely resemble the last in character, excepting for the fact that no shells have yet been observed in them: and we know that the shell beds do not unfrequently exhibit similar contortion.

The localities for some of the best morainic shapes are indicated in the Map, but it was not possible to show all, and as moraines are not so common in this district as in the higher parts of the Highlands, we give below a list of them :—at the north part of Gleann Laoigh (163 south-west), on either side of the burn; the south-west side of the Massan (163 south-west), west of Glenmassan, and on the north-east side (rarer) ; the head of the glen $1\frac{5}{8}$ mile east of Glenmassan, and extending down from here to within about $\frac{1}{4}$ mile of the foot of the burn of this glen ; the hillside $\frac{5}{8}$ mile north-west of Cnocan Sgeir (172 north-east), at a height of 500 or 600 feet, and extending hence north-east ; the east side of Glendaruel just above the river terraces near Auchateggan ; the hillside south and south-east of Duillater (162), at a height of from 500 to 800 feet ; by the junction of the burns nearly $\frac{1}{2}$ mile slightly west of south of Corrachaive (173 south-west) ; a little north of Corrachaive, separated by alluvial deposits ; the north-east side of burn $\frac{5}{12}$ mile north-east of Ballochyle (173 south-east), and hence extending up stream for more than $\frac{1}{2}$ mile ; about $\frac{1}{2}$ mile slightly east of north of Leacann nan Gall (173 and 183) ; near the "c" of "Coire Athaich" (173 north-east) ; near the junction of the burns $\frac{5}{8}$ mile N.N.W. of Blairmore Hill ; $\frac{1}{2}$ mile west of Ardbeg (183 north-west) ; north-east side of Glen Kin, $\frac{7}{8}$ mile north-west of the Bishop's Seat (183 north-east) ; hillside between $\frac{1}{4}$ and $\frac{1}{3}$ mile E.N.E. of Blar Buidhe (183 south-west) ; east side of the north to south burn 1 mile north-east of Blar Buidhe, and on the west side (rarer).

The semi-circular shape is rare in the moraines of this district. Commonly they consist of mounds rather longer in one direction than another, and at times slightly curved. The longer axes are usually roughly parallel to what appears to have been the course of the glaciers which produced them : this is seen at the head of the glen $1\frac{5}{8}$ mile east of Glenmassan, and on the east side of the north to south burn 1 mile north-east of Blar Buidhe : in the last locality two of them extend north and south for a distance of $\frac{1}{4}$ mile, though they do not often exceed 30 yards in breadth. Individual mound shapes are less common than more or less confluent banks with rather indefinite margins.

In the typical moraines the material consists of an irregular mixture of huge boulders, sand, and gravel, the fine and coarse matters lying together without any sorting, and their tops are strewn over with angular blocks of schist. But there are also other mounds which do not contain such big blocks, and the sides of which are smoother. About $\frac{1}{4}$ mile north-east of Corlarach such a mound occurs, about 150 yards long and 80 broad, at a height of about 600 feet : a section near the top shows stones varying from 1 to 3 inches in diameter, and not very rounded in outline.

On either side of the burn which runs into the West Bay of Dunoon there are moraines running much in the direction of the burn, and these descend to a level of 280 feet. The morainic material lies over grey boulder clay, and about $\frac{1}{3}$ mile east of the last "d" of "Badd" the surface is strewn with huge blocks of hornblende schist.

In the more hilly areas moraines are seen at a lower level even than this. Near the foot of the Massan, just before it joins the Echaig, there are four morainic mounds which stand out like islands in a wide gravel flat. The gravel probably belongs to one of the old raised beaches more or less modified by river action. The bottoms of the mounds are at a height of about 40 feet. This low elevation was long ago called attention to by Mr Charles MacLaren (*Edin. New Phil. Journal*, 1855, "Notices of Ancient Moraines, etc."). That some moraines are of later date than the raised beaches is clearly made out in some areas of Scotland (L. W. Hinxman, *Trans. Geol. Soc. Edin.*, vol. vi. p. 249, "On the Occurrence of Moraines later than the 50-foot beach in the North-West Highlands"), but in the Glenmassan locality we think it doubtful whether this relation obtains. We should prefer to suppose that the gravel flat is of later date, and lying on the sides of or over morainic material. At all events, we are not aware of any section that shows the reverse relation.

In 1-inch Map 37 there is an absence of the highest raised beaches near the heads of the sea lochs in the more hilly districts, and it is suggested that these areas were occupied by ice while the terraces were being formed in the less mountainous parts, but we have not seen any instance of contortion of the beach deposits, as if by the advance of ice, and there is too a general absence of coarse boulders within or on them.

The glacial sands and gravels, when near sea-level and not showing morainic shape, may be hard to distinguish from the deposits belonging to the 100-foot beach, these latter being occasionally so worn by denudation that they no longer possess a rudely flat or evenly-sloping surface. East and north-east of Altgaltraig Point (182 south-east), stratified sand and gravel, sometimes showing a reddish tint, cover the slopes above the 20-foot beach. These beds ascend considerably above the 100-foot contour line, and we have not mapped any as beach deposits, but it is quite possible that they are partly so. On the east side of Loch Striven, between Brackleymore and Tighnuilt, and a little on the south of Brackleymore, there is a considerable spread of sand and gravel, the character of which is doubtful. Some of it has been taken as beach deposit, and some as glacial.

The steeply-dipping sands and gravels below the gravel of the top of the 100-foot and lower beaches may possibly be glacial. Their edges are often seen to be cut across by the gravel at the top, and there may be a great difference in age between the two. The sands and gravels alluded to are seen in the relation described in the following places,—in the burn bank $\frac{1}{6}$ mile north-west of Colintraive Pier ; on the west side of Loch Striven, 60 yards south-west of Troustan (the lower gravels dipping steeply down the burn, and overlain unconformably by coarser gravel) ; on the east side of Loch Striven, on the east side of the foot of Invervegain Burn (the lower sands and gravels with a regular dip seawards of about 25°, and overlain by a coarser and less regular gravel dipping seawards at about 11°), and at Tighnuilt (the lower gravels dipping seawards at 24°, and cut across at the top by more nearly level bands of gravel). One-sixth of a mile north-west of Brackleymore, gravel layers incline seawards at 20°, and are cut off by the more nearly level gravel of the 20-foot beach.

The drift gravels are not uncommonly stained, and partly cemented, by some black material—probably the black oxide of manganese. On the hillside, 1 mile S.S.W. of Coraddie (182 east), the cement binds the little stones into small lumps.*

In various sections of drift, particularly the looser and more permeable varieties, we are struck by the very soft, decomposed condition of some of the contained blocks, such as basalt, epidiorite, green beds, and granite. These may even be quite friable to the fingers. It is clear the decomposition, and resulting softness, must have been brought about since the formation of the beds in which the boulders occur, and since these beds were in their present position : the boulders in their present condition could not keep their shape while they were being dragged or rolled about in rapid streams. Boulders of basalt sometimes 6 to 8 inches in diameter may now be only represented by concentric rings of soft brown crumbly rock. The schistose laminæ of the green beds sometimes flake off from one another on the slightest touch. The bank on the east side of the river Cur about $\frac{1}{2}$ mile east of Glensluan (141) shows at the top layers of sand and gravel almost flat. These perhaps represent an old

---

* Similar occurrences in the north part of Cowal are mentioned on p. 258. Reference may also be made to the black manganese staining found in the recent freshwater alluvia of the district (p. 271).

river terrace. Below these come in order, coarser gravel, which contains tongues and lenticles of stiff greenish-grey laminated clay with rare small stones, then stiff grey stony boulder clay, then laminated sand and gravel. In one part of the boulder clay we noticed a boulder of the green beds which was cavernous all through, with bits of vein quartz and chlorite projecting into irregular hollows. The boulder seemed to represent part of a green bed which had been crossed by veins of ferriferous carbonate, quartz and chlorite, such as are so common in these rocks. The ferriferous carbonate is now all dissolved away. The boulder broke up at a slight tap of the hammer, but on incorporation in the bed was probably quite sound, and with carbonate still connecting the other constituents of the veins. The boulder clay is very stony, and allows of the percolation of water to some extent. Porphyritic granite boulders, in a much decomposed condition, occur in the same bed. Granite boulders occur also in the gravel below the clay. The section is also of interest in some other respects, and is alluded to again (p. 259).

Three-quarters of a mile E.N.E. of Ballemeanoch (141) the morainic drift at the side of a little stream contains various granite boulders about 8 inches long, which crumble readily under the fingers; the porphyritic felspars break up also, so that their shapes cannot be isolated.

The widest spreads of drift, including in this term both boulder clay and glacial sands and gravels, occur in Inverchaolain Glen, Glen Fyne (194 and 183), on the hill-slopes near Toward, in the glen leading into the West Bay of Dunoon, Glen Kin, between this glen and the head of the Holy Loch, and in the glens that open into Loch Long by Gairletter Point and Blairmore. If we were speaking of the boulder clay apart from sands and gravels, it is not probable that any of the above localities would have to be omitted. The arrangement of the boulder clay on the different sides of the glens not unfrequently bears witness to the usual rule, that the greatest development is on that side from which the "carry" has been. The head of Glen Kin, in which the ice carry as shown by glacial striations was from north-west to south-east, shows this well. But the carry was down rather than across the more important glens and sea lochs, and therefore this feature is not brought out strongly.

It is certain that all this district, up to the highest hilltops, was overridden by ice, for, even in cases where no striations have been found, we may still see boulders which have been derived from, or which have travelled across lower areas, e.g., on the top of the hill 1 mile south of Glenlean, at a height of 1988 feet (only 7 feet below the highest point in the 1-inch Map 29) there are a good number of basalt pebbles; these have not necessarily come far, but they must have come from a lower position, as there is no basalt on this hilltop.

In some areas the hillsides have been polished so smooth that a first glance suggests that they are covered by drift, but closer examination shows that the covering may be merely a few inches of turf, or at most a foot or two of drift, lying on a polished rock surface. Examples of this occur on the east side of Kilmun Hill; on the slope 200 yards slightly west of south of Trinity Church, Dunoon; and ⅔ mile slightly east of north of South Hall (194 north-west).

The glacial striations are marked in the Map, and so need not be mentioned in detail. The general direction along the bottoms of the more important glens is nearly with these glens,* the depth having been enough to determine the flow locally; thus, in parts of Glen Lean, it is nearly east to west; in the lower areas of Glen Massan, south-east; in

* For an early reference to the glaciation of Glendaruel, see Sir A. Geikie, *Scenery of Scotland* (1865), p. 157.

Glen Tarsan, S.S.W. On the higher grounds, away from the influence of the valleys, the most general direction is perhaps from north-west or N.N.W. This is so on the hill between Bishop's Seat and Strone Saul; on the south side of Cruach nan Capull (183); on the south-east side of Sgarach Mor (173 north-east); and on the hills east of Loch Eck.

Different sets of striations crossing one another are not uncommon, but it is not generally clear which set are the earlier. In the Burnt Islands of the Kyles of Bute there is one set trending south or a little east of south, and another set trending east or a little south of east. On the south-west side of Glen Massan, above Stronlonag, several occur which point slightly east of north, about at right angles to the general direction of the glen and most of the striations in it. Perhaps we may regard these as later than the others, and due to local ice descending directly down the steep side of the glen. The striations on the still higher ground to the south-west of these run south-east again, like those in the glen generally. On the hill slope nearly 1 mile slightly south of west of Ardentinny there is also a set pointing straight down the slope, and almost at right angles to the general direction in the neighbourhood.

Perhaps the most distinct striations occur between tidemarks, e.g., on the Gantcocks, near Dunoon, on the Bull Rock, $1\frac{1}{2}$ mile south of Dunoon, and in many places on the west side of Loch Striven. But in higher areas they are also often wonderfully fresh, and we are repeatedly coming across little vertical scars which are deeply scored, and may be existing now in essentially the same state as when the ice left them. Numerous small hollows can be proved to be of glacial or preglacial origin, the rocky walls being striated on both sides: e.g., Inverchapel Burn (164 south-west) is running along such a hollow, which is still partly covered by boulder clay.

Few freshwater lochs occur in this area, but it is certain various others formerly existed which are now filled by alluvium or peat. Some sites of this character are referred to in chapter xx. There is no rock seen in the bottom of Strath Echaig between Loch Eck and the Holy Loch. The barrier at the foot of the former loch consists, as far as seen, of gravel and sand which show a fairly flat surface, and may be referred to a raised beach deposit. The barrier reaches to a height of about 90 feet. On the east side of the outflow the surface is, for an acre or two, uneven and moundy, and reminds us of morainic outlines partly washed down; and it is very probable that the more even surfaced area also represents an original morainic surface, which has had its inequalities levelled up by rearranged material.

Loch Loskin, between Dunoon and Sandbank, is perhaps a rock basin. The barrier is formed, on the east side, by a massive hornblende schist, which displays striations pointing slightly east of south. West of the outflow only alluvium, and the gravel and sand of the 100-foot beach, are seen, but it is possible that a rock barrier may occur on this side also, hidden under the beach deposits.

Lochan na Leirg, on the hill between Loch Riddon and Loch Striven at a height of over 1500 feet, is surrounded by rock on all sides.

Parts of the areas of the Clyde and of the neighbouring sea-lochs would, if raised sufficiently, form land-locked water basins, the bottoms of some of the more inland parts being known to reach a greater depth than those more seaward. Particulars of these basins are given by Dr H. R. Mill in his work on "The Clyde Sea Area" (Trans. Roy. Soc. Edin., vol. xxxvi. p. 641). From this we gather that a basin, called the Dunoon basin, extends from the north end of Great Cumbrae up to the junction of Loch Goil and Loch Long. On the south side of this is another basin, the Arran basin, an extension of which runs up Loch Fine as far as Otter. In upper Loch Fine there is another basin extending from Minard to Inveraray. At Minard Narrows the depth is 18 fathoms : near Creggans it is 83 fathoms. In Loch Goil there is also a well-marked basin. The deepest part, opposite Stuckbeg, is 45 fathoms : the depth at the mouth of the loch is only 7 fathoms.

Contortions of schist planes due apparently to the drag of the ice sheet are occasionally observed. On the shore $\frac{1}{4}$ mile north of Blairmore Farm, $\frac{1}{4}$ mile south-east of Kilmun Established Church, and north-east of Glenacre (195 north-west), the schists seem to dip north-west instead of south-east, and have probably had their edges twisted back by the ice coming from the north and north-west. On the shore $\frac{1}{2}$ mile east of Hafton House (174 south-east) thin yellow schists show a reversed dip, and are churned up and mixed with boulder clay. In a small burn (not marked in Map) $\frac{1}{2}$ mile slightly north of east of Corlarach (194 north-east), and in Corlarach Burn itself, $\frac{6}{12}$ mile N.N.E. of Beinn Ruadh, the schist is much twisted near the surface.

Beds containing glacial shells have long been known in the district. Such were first described by Mr Jas. Smith (of Jordanhill) in his paper "On the Phenomena of the Elevated Marine Beds of the Basin of the Clyde," *Mem. Wern. Soc.*, vol. viii. More recently Messrs Crosskey and Robertson have examined them in great detail, and given lists of the fossils procured from them ("The Post Tertiary Fossiliferous Beds of Scotland," *Geol. Soc. Glasgow*, vol. v., p. 29, 1875).

In Cowal these beds have only been noticed up to a height of a few feet above high-water mark ; frequent exposures occur on the shore by this mark, or between it and low-water. They generally consist of very finely-laminated clays or sandy clays, of a reddish-brown or yellow colour, alternating layers differing slightly in tint. At other times the clay is grey and shows no signs of lamination. The proportion of stones in the clay varies considerably, and in some places is so large that probably half the bulk of the bed is of stones, while in others stones are almost absent. The stones rarely exceed 3 or 4 inches in diameter, and are generally somewhat angular in shape : we have not noticed any with striated surfaces.

We are not aware that any of the beds in this district lie within boulder clay, though in other areas shell-bearing beds can be seen in this position, *e.g.*, in Tangy Glen, near Campbeltown, described by Messrs Robertson and Crosskey (*Trans. Geol. Soc. Glasgow*, vol. iv.), and in the south of Arran, described by the Reverend R. Boog Watson (*Trans. Roy. Soc. Edin.*, vol. xvi. p. 523).

In the bay at Port an Eilein, east side of Loch Riddon, a reddish shelly clay occurs which contains many small Tellinas. Mr J. Bennie, who collected from this and other localities, informs us that many of the small shells had a bright epidermis, and had evidently been well preserved, but that others were much weathered.

In the bay $\frac{1}{4}$ mile W.S.W. of Fearn'ach (182 south-east), close under the raised beach, there is a stiff shelly clay which contains Pecten Islandicus. Mr Bennie found here a great number of Littorinas, many in a broken condition, suggesting that they had had a "squeeze," but none showed marks of abrasion.

On the shore $\frac{1}{2}$ mile or so south of Fearn'ach, at various places for a distance of 80 yards, a stiff chocolate-brown clay, somewhat sandy in parts, is seen 3 or 4 feet above high water mark. The clay is full of rootlets and vegetable fibres descending from the ground surface. It is overlain by rather fine gravel belonging to the raised beach, and there is a wet line between the two deposits, the clay being evidently much less permeable to water than the gravel. The clay shows no lamination on the weathered faces, and is sticky, moulding well to the trowel and fingers. It contains many shells, but they are somewhat brittle and readily broken : among them Pecten Islandicus is abundant, and attains a length of 4 or 5 inches : the two valves of one individual sometimes occur in close apposition. The large arctic barnacle is also common, sometimes fixed on stones and sometimes on the larger shells. A sea urchin was also found, the plates fitting exactly together but only kept in position by the clay. The same bed forms the shore below high water mark, and seems here to contain shells in a less brittle state.

On the shore ½ mile north-west of Colintraive Pier the shell bed again occurs a few feet above high water mark. It is here much harder to dig than south of Fearn'ach, as it is very full of stones. Mr Bennie found Myas frequent and most of them with both valves in contact, but they were soft and easily broken : Littorinas frequent : Pecten Islandicus small, only ¾ inch, but perfect, not rubbed or worn.

The north bank of the Ruel, ¼ mile slightly east of north of Ormidale House, shows a section as below :

Grass.

Three to four feet gravel becoming coarser downwards : stones of 8-inch diameter not uncommon near the base.

One foot sandy clay with nests of shells and big barnacles.

Six feet finely laminated greenish-grey sandy clay, apparently stoneless and shell-less, and conformably under the bed above.

The sand and gravel at the top probably belong to a river terrace. No unconformity is seen here between them and the underlying deposits, but, at a point ½ mile south-east of Ormidale House, in the west bank of the Ruel, a laminated blue sandy clay dips down stream rather steeply, and the edges are cut off by gravel of the overlying river terrace. In the 1 foot of sandy clay Mya truncata was found in abundance, and often in the habit of life with the mouth uppermost. The shells are brittle and easily broken near the surface, but stout and well preserved further in : they are generally filled with coarser sand ; the epidermis is in a good state in all, except where weathered. The Cyprinas are large and stout, occasionally with the valves together, but many single : about half the size of those in the Loch Long localities south of Ardentinny : the epidermis is preserved. The barnacles are 1½ inch long, often in clusters adhering together, and generally free from attachment to shells or stones. One stone had apparently serpulæ on it. The clay at the bottom reminds Mr Bennie very much of the white mud found under the shell beds of the Clyde, in which, after much search, Mr David Robertson found some foraminifera : near Ormidale the mud was sometimes dyed black from decaying matter, from the shells in the bed above.

One hundred and twenty yards E.N.E. of the cottage at the foot of Ardyne burn (194, south-east), a section on the west bank of the stream shows (March, 1893) the section below :

Six to seven feet sand and gravel, horizontal and evenly stratified, belonging to the 20-foot beach.

Three feet of dark grey boulder clay.

Similar boulder clay is also seen under the water of the burn. The line between the gravel and the clay is perfectly sharp and nearly horizontal. A few yards south of this section a change takes place, and, just at the edge of the water, a finely laminated stoneless clay, in alternating yellow and reddish-brown colours, overlies the boulder clay, the line between being quite sharp and distinct. The laminæ of the clay are not conformable to its junction with the boulder clay, and, further under the water, more of it comes on, until the thickness reaches 2 or 3 feet. The laminæ dip 10° to 12° in a direction about 20° west of south. It is clear that the 20-foot beach gravel is unconformable to the laminated clay, and must cut across its inclined edges, so as to rest on boulder clay below it. About 20 yards above the section the same clay is seen in the burn bed, from a point about 10 yards below the new suspension bridge to 6 or 7 yards above. We could see no shells in the laminated clay either in the bank section or the burn bed, but 6 or 7 yards above the bridge, on the west side of the burn, this clay dips east at a gentle angle, and is overlain by a dull brown or dark grey clay, which is full of shells : perhaps there is one shell for every 16 to 18 square inches of surface. This clay does not show any lamination : it is only seen under the ordinary water level of the stream, and no specimens have been obtained from it. It is stated by Sir Arch. Geikie ("On the Phenomena of the Glacial Drift of Scotland," *Trans. Geol. Soc., Glasgow*, vol. i. p. 133) that a laminated brown or reddish clay, without stones and shells, is almost always found between the shelly clay and the boulder clay in Bute and the Clyde district. Further up the burn, ⅛ mile south-west of Killellan, a yellow shelly, nearly stoneless, clay was seen under the river terrace on the east side of the burn : 220 yards north-west of Killellan a similar shelly clay was seen in the burn bed, the laminæ dipping down stream in a southerly direction : and in several places a little further up the burn, on either bank, similar clay occurs, but without shells as far as noticed.

A laminated shelly clay is exposed on the shore just north of the rocky point at Toward Quay. It is very stiff, and coloured yellow and reddish-brown in thin alternating layers. The layers often dip steeply, as much as 45°, and make curved outcrops on the shore, as if the strike was contorted. Pecten Islandicus, sometimes 3 to 4 inches long, is abundant, stout, and well preserved. Astarte, a big barnacle, and a stony seaweed, growing attached to boulders, are also common. The deposit is largely obscured by shingle, and is overlain at high water mark by the gravel of the raised beach, from which numerous fresh-water streamlets descend.

On the west side of Loch Long several exposures of shelly laminated clay occur. About ¾ mile south of Ardentinny church this clay is seen to overlie a reddish-brown boulder clay. About 1¼ mile south of the church the Myas were good, the Pecten Islandicus, Cyprinas, and Astartes, large, and Mr Bennie considers these localities very promising for further search.

Laminated clays, but as far as known without shells, occur on the shore ⅛ mile E.S.E. of Altgaltraig Point (182 south-east) ; on the shore ₁⁵⁄₁₂ mile north of Port an Eilein ; and on the west side of Loch Striven, nearly ⅔ mile N. of Strone Point. In the first mentioned locality the clay clearly overlies boulder clay.

Lists of the shells collected in the different localities are given in the Appendix, but we reproduce here the substance of some observations Mr Bennie made on concluding his researches. Nothing was observed differing much from what is generally seen in the Clyde glacial shell beds, excepting for the great number and large size of the barnacles near Ormidale House, and also the entire absence here of the stones and shells to which they must once have been attached : this suggested that they may have been transported, but the place from which they came could not be far off, as there are no marks of travel-wear upon them. In most of the beds there is a mixture of shore and shallow water shells with those that inhabit deep water : this also is a common feature in the Clyde beds, where Littorinas and mussels are often found in clay which must have been deposited in water of some depth.

It has been already mentioned that the Toward Quay bed shows indications of contortions. It is perhaps reasonable to suppose this due to the drag of ice. In this locality a block of calcareous freestone, measuring about 5 feet by 4 by 2, lies on the shore within 4 or 5 yards of a clear exposure of the bed, and is in all probability lying on it; it is somewhat angular in shape. Beside this, blocks of schist, attaining a length of 5 feet and a breadth of 2 to 3, occur within 2 or 3 yards of some of the exposures. From 260 yards north-west of the west gates of Toward policies the shore to the westwards, at mean tides, is covered by gravel, and shows no boulders so big that they may not have been derived from the raised beaches, or brought as ballast, or derived from a wrecked sea-wall. This applies to the shore as far west as Ardyne Point, and for at least a little distance north of the point. With the exception of the blocks from the old sea-wall, there are none much more than 1½ foot in length, and very few as large as this. Blocks of the size of those on the shelly clay do not occur among the gravels of either the present or the old beaches in this locality.

Along the shore between Ardyne Point and Toward Quay, small blocks derived apparently from areas lying south or south-east occur, e.g., Upper Old Red sandstones, magnesian limestones, crush-rocks like those near the schist boundary, and fine-grained purple andesite-like rocks.

Between Toward Quay and Toward Lighthouse, there is an abundance of large boulders on the rock exposures, and on the tidal flats which may be formed out of boulder clay, but they do not occur on the raised beach gravel, excepting when this is within a few yards of the margins of the other formations: the boulders in this situation seem in some cases to be resting firmly on beach gravel.

The shore just east of Colintraive Pier shows an abundance of boulders, as much as 3 feet in length, some of which seem to rest firmly on the beach gravel. Good sections of the old beach occur near, and show no stones so large as these: the largest are about 9 inches long.

In various other localities we find stones on the shore which are not noticed in the boulder clays, and which seem confined to the shores, and to have been derived from areas lying south or south-east. In Fearna Bagh there are many rather soft, coarse-grained freestone boulders, containing small quartz pebbles and few or none of schist: we did not notice any exceeding 1 foot in length: they are not very rounded. There are also small boulders of fine-grained hard white sandstones, coarse chocolate-coloured freestone with clayey lenticles, conglomerate of quartz pebbles, with the pebbles as much as 2 inches in length, and purple andesite-like rocks. To the north also, as far as Springfield House, red freestone blocks have been noticed.

In the West Bay of Dunoon there are a good number of boulders, about 6 or 8 inches long, of green serpentinous rock like that at the schist boundary, and of purple andesite-like rock.

The garden walls about Blairmore contain so many blocks of reddish freestone, often several feet long, that we examined the shore to see if they did not occur as boulders on it. We found a good number both of red and yellow blocks, sometimes quite coarse and not the kind that would be chosen for good building stones: in one place, about ⅓ mile south of the pier, there is a boulder of calcareous false-bedded free-stone measuring 4 feet by 2 by 2: in another, ⅓ mile north of the pier, there is a grit boulder with a length of 3 feet and a breadth of 2: but it is rare for them to exceed 1 foot in length.   C.T.C.

CHAPTER XIX.

GLACIAL DEPOSITS (continued).—DESCRIPTION OF PARTICULAR DISTRICTS.

District of Tighnabruaich.

This district includes that part of Cowal which lies west of the lower part of Loch Riddon, and the west Kyle of Bute, and south of a line that runs east to west between Eilean Dearg and Drum Point.

There are no wide spread areas of glacial drift in the peninsula south of Kilfinan. The boulder clay or till, always of a bright red colour, generally occurs in small irregular patches in the hollows and slopes between the numerous rock exposures, and seems nowhere to be thick except in some of the burns and valleys. Perhaps the drift areas are the largest in the Acharosson valley. In Allt Dubh to the north-east of Beinn Capuill at a height of 950 feet above the sea, the till is 15 to 20 feet thick. It is very thick in the lower part of the Acharosson burn : a thickness of 20 to 30 feet may be seen in the north bank near Auchanaskioch, and it is probably much thicker than this. A similar thickness of drift may be seen in the most westerly branch of the stream that runs into Loch na Melldalloch. Thick red clay was observed at Tighnabruaich reservoir, and it may be seen on the shore in many places from the east side of Tighnabruaich Pier, round to south of Kames Pier. There is a good thickness of red clay at Rudhaban, Port Driseach, underlying the gravel of the 100 feet beach, and also in the Craignafeich Burn, 200 to 400 yards below the lowest dam. Red till may be seen about Stillaig Farm and in the burn to the north-east, and there is a good section of it in the small stream near Asgog Farm. It may also be noticed up the valley of Allt Osda in several places, under the high beach gravels and sand, and about Millhouse. In addition to these localities may be mentioned the neighbourhood of Ardlamont and Corra, especially a small stream west of the bend of the road on the west of Corra.

There is a patch of gravel, probably glacial, to the west of Allt Mor, and about ½ mile north-west of the Established Church, Tighnabruaich. The gravel, in which there are small pits, is dirty-looking and subangular, and covers an area about 100 yards long by 50 yards broad. It appears to overlie schist.

Glacial striæ have been noted on the 6-inch Map in this district about 50 times, but all of these are not placed on the 1-inch Map. They vary in direction from north-west round to N.N.E., the most numerous directions being about due north, and north 10° west, in nearly equal proportions. The northerly direction prevails along Loch Riddon, but over the greater part of the area the general direction is west of north. The places where the direction is to the east of north are nearly all on the coast, e.g., at Callow and Rudhaban, Loch Riddon, on the shore opposite Tighnabruaich, and to the south of Kames Pier, and on Loch Fine, south of Black Harbour. Some of the striæ most to the west of north must be regarded as abnormal, e.g., one which is about north-west, by the roadside on the way to Kilfinan to the south-west of the lower dam, is on a nearly vertical face of rock, which has evidently influenced the direction of the ice at that point. The same explanation must be given of one to the north of Callow, which points about north 40 west, nearly parallel to the direction of the cliff there.　Sometimes we meet with more than one set of striæ on the same surface, as, for instance, by the side of the upper road, 300 yards south of Tighnabruaich Established Church, where they vary in direction 20° to 40° west of north ; and north-west of Stillaig Farm, where they vary from 10° to 30° west of north.

Not many noticeable boulders were observed in this area.　A large block of white granite, 8 feet long by 4 or 5 feet broad, was seen on the shore on the west side of Ardlamont Bay near Rudha na Peilige, and boulders of granite are built into the wall bounding the plantation 1000 yards north-east of Ardmarnock House.　A boulder of serpentine, 6 feet by 5 feet by 3 feet 6 inches high, occurs on the moor, 1000 yards north-east of Loch na Melldalloch, and 800 yards north of the upper dam.　About ½ mile N.N.E. of Ardlamont House, there is a group of large schist boulders, one of which is 15 feet long by 7 feet broad.

W.G.

District of Kilfinan and Stralachlan.

This district is bounded on the east side by Newton Bay, Caol-Ghleann, Glendaruel, and the upper part of Loch Riddon.　On the south side the boundary is the Tighnabruaich District.

This area is comparatively poor in glacial deposits.　Boulder clay and drift are only met with in isolated patches among rock hollows, and are generally only a few feet in thickness.　In the high ground but few striated surfaces of rocks have been observed.　As the lower ground is approached towards Loch Fine on one side and Glendaruel on the other, such markings become more numerous, and when the coast is reached they are discovered in abundance.　Those on the coast have the same general direction as those found on the hilltops and slopes, and indicate a movement of the ice-sheet from a north-easterly direction.

The area that lies between Stralachlan Glen and Loch Fine is well glaciated, the rock surfaces preserving the rounded outlines that have been impressed on them by the passage of the ice-sheet.　In this district most of the rocks consist of massive epidiorites, which, by their power of resisting the processes of weathering more than the rock masses in which they have been intruded, have been the means of indicating the degree of glaciation to which the whole district has been subjected.　Besides the general smoothness of the surface, good striations are met with up to the summits of these epidiorite ridges, at heights between 500 and 600 feet.

The direction of movement of the ice-sheet corresponded nearly to the general trend of Loch Fine, and to the upper portions of Glendaruel and Caol Ghleann.　Where, however, the valley of Glendaruel deflects from this course, as at Ballochandrain (172), the glacial striæ at the lowest parts of the valley likewise deflect.　At Ballochandrain (172) striæ by the roadside have been noted in a direction nearly north and south.　At the head of Loch Riddon on the western side of the Ruel estuary (172) north and south striæ have been observed.　On the published Map (sheet 29) striæ have been shown on the coast at Lochead on the seaward margin of a raised beach ; the striæ are preserved on mica schist.　About 200 yards further south along the shore another set of striæ occurs with a similar trend, the scorings being well preserved on the surfaces of a massive quartzose schist.　North and south striæ have also been observed on the west shore of Loch Riddon opposite Eilean Dearg.　On reference to the Map, it will be noticed that the north and south direction of these glacial scorings between Lochead and Eilean Dearg corresponds almost exactly to the trend of Loch Riddon and the Ruel estuary near these localities.　It is evident, therefore, that these striæ were produced by the advance of an ice-sheet from Glendaruel into Loch Riddon, the course of which was determined by the trend of the valley of Glendaruel and its extension into Loch Riddon. While the bulk of the striæ along Loch Riddon conform to this general direction, a few have been observed somewhat oblique, but in these cases they follow the unevenness of the coast-line.

As stated previously, the general trend of the ice-sheet that once covered the area was in a direction from north-east to south-west. Some instances, however, occur, in which the rock surfaces have been striated from a direction nearly due east. These are well shown on some of the epidiorite masses between Stralachlan Glen and Loch Fine at elevations of 500 to 600 feet.

The nature of the boulder clay has already been treated of. Two distinct types of boulder clay occur differing from one another in colour ; the prevalent colour of one type is greyish to blue when fresh, and yellow to yellowish-brown when weathered ; the colour of the second type is red to chocolate. In this district the red boulder clays do not occur north of Otter Ferry (171) on Loch Fine ; the boulder clays in the northern part of the area are of the grey-blue type.

It is quite clear that the difference in colour of the boulder clay cannot be ascribed to difference of composition of the rocks on which it now lies. The rocks underlying the red clay are similar to those underlying the blue clay. Neither can the difference be accounted for by ascribing one variety to the more elevated regions and the other to the lower country. In the areas peculiar to each clay these deposits are found from sea-level to the most elevated situation.

In this district sandy and gravelly beds in the boulder clay are rare. The boulder clay is sometimes laminated : more usually, however, it is devoid of structure. The majority of the contained boulders are semi-rounded, and a small proportion show striations.

The bulk of the area consists of an irregular hill slope dipping away from the high ground that rises precipitously out of Glendaruel down to Loch Fine. Boulder clay is seldom met with on this hill slope, and when it does occur the deposit is usually thin and covered by peat. It is common, however, for the slope to be thickly strewn with a coating of blocks and rubbish, which appear to have been left behind on the hill slope after the drift deposits in which they were embedded had been denuded away. This is a common feature of the slopes in the Stralachlan area.

The raised beaches on the coast of Loch Fine are generally found resting on a boulder clay floor. At Barnacarry (151) the gravel of the 20-foot beach rests on a stony stiff boulder clay. A mile further south two small raised beaches consisting of sand and gravel are found resting on blue boulder clay. In fact, without further enumeration in detail, it may be said that in nearly the whole of the raised beaches within this area sections are exposed, revealing boulder clay below the beach deposit.

At the roadside leading south-west from Lephinmore (162) just beyond the upper limit of the 50 foot beach there is a section of blue boulder clay showing contortions. The section is 40 feet above sea-level, and the contortions were evidently produced before the formation of the 50 foot beach. In this area sections of contorted boulder clay are rare, but it must be borne in mind that such contortions can only be detected when the boulder clay happens to be laminated. From its position the contortion may very well have been brought about by floating ice.

Where the raised beaches rest upon the boulder clay, the beaches are unconformable to the boulder clays. The gravel frequently overlaps the boulder clay and rests on the rock.

Erratics and perched blocks are common over the whole district : the latter are often found on the summits of the ridges. As a rule the boulders have not come from any great distance. Where the original locality of the boulders can be made out, they invariably indicate their transport from a general north-easterly direction. The boulders comprise every type of rock which the district contains, and in addition some rocks which are not found in the area. Boulders of the Glen Fine granite are often met with as well as fragments of coarse garnet schist, which have probably been derived from the same area. No boulders have been found which correspond to rocks peculiar to the north-western side of Loch Fine, nor any of the Lorne Andesites, which lie about 11 miles to the north-west of this area. Although the Lorne Andesites are not found on the eastern side of Loch Awe, they are nearer this district than the Glen Fine granite.

The glaciation must be referred to a great south-westerly movement of the ice-sheet, which spread over a large area of which this district forms but a part, the area extending across Loch Fine to Loch Awe. The highest points of this area have been covered by the ice-sheet, which must have had a thickness of at least 2000 feet.

Moraines are scarcely ever met with in this area, except near the extreme north-east boundary where numerous moraines occur in the upper part of Caol Ghleann, resting on a deep deposit of boulder clay. This is the most extensive deposit of boulder clay in the area. As these moraines are described in the description of another area (see page 257), they will not be further alluded to. They occur here and there the whole way down the glen as far as Dunans, and along the sides of the Kilbridemore Glen (162).   J.B.H.

District of Strachur and Lochgoilhead.

This district includes most of the north part of Cowal. It includes that portion of the area being described which is contained in 1-inch Maps 37 and 38, with the exception of the part which lies west of Newton Bay, Caol Ghleann, and Glendaruel.

Near the tops of the higher hills in the northern part of Cowal, e.g., on Sgor Coinnich, Beinn Bheula, Cnoc Coinnich, Beinn Reithe, Carnach Mor, the two Beinn Lochains, Ben Donich, Binnein an Fhidleir, etc., the glacial striations point south-east or S.S.E., though on the lower ground they are either parallel to the glens, or to Loch Fine. The south-east direction is not found at heights much under 1500 feet, and in some places, e.g., on the north-west side of Stob an Eas, just over the 2000 feet contour, and $\frac{1}{2}$ mile east of Binnein an Fhidleir, over the 2250 feet contour, a movement in a south-west direction is indicated at higher levels than this. There is no ground so high as the hills mentioned in exactly the direction whence some of the striæ point, anywhere between Loch Fine and the Firth of Lorne, though Ben Cruachan (3689 feet) is not much off the direction. The south-east striæ are not abundant, and to some of them exception might be taken, but of the most of them there can be no doubt : they are usually on the surfaces of quartz veins in the albite schists. From the carry of the boulders we suppose the movement was from the north-west and not the south-east. Sometimes boulders are found not far off the striations, which do not occur anywhere on the south-east side of the striations, e.g., $\frac{1}{2}$ mile north of Cruach nam Miseag (153), there are many angular bits of garnetiferous schist mixed with others of quartzite schist and lamprophyre, at a height of about 1900 feet. The garnet schist is like the schist which occurs on the north-west side of the Loch Tay limestone. There is no rock like it in the area south-east. Also, there is a total absence of boulders of the less altered beds which are found near the south-west boundary of the Highlands. The boulders of garnet schist occur at an elevation greater than the supposed parent rock. In the north-east corner of 1-inch Map 37, the same rock occurs at higher elevations : this is in a direction north or slightly east of north of the boulders. But as the nearest striæ indicate a movement in a north-west direction, it is perhaps unnecessary to go so far for the source of the boulders.

On the lower areas we meet again and again with boulders, such as those from the green beds on the north-west side of "the anticline," which have come from a north-west direction, but this is along the course of glens, such as Hell's Glen and Gleann Canachadan.

In spite of the striations running north-west on the higher hills, we cannot point to any boulders that have been brought from the north-west side of Loch Fine on to the south-east side. The absence on the south-east side of any boulders of the andesitic traps of Lorne, and the comparative rarity of quartz felsite boulders, seems to negative the supposition of any great ice movement across the loch from the north-west side. It may be that at the same time as the ice-sheet was moving down Loch Fine, and the lower areas near it, from the north-east, a portion of the sheet near the high hills mentioned was shedding itself off in a south-east direction.

On the south side of Drynain Glen, about ⅛ mile west of the "T" of "Tom nan Con" (164), and ¼ mile south-east of the "n" of "Con," there are striations pointing north-east, almost at right angles to the general direction in the neighbourhood. We suppose these due to a movement from the high ground on the south-west, at a time subsequent to the general glaciation. They are clearly of the same type as those a little south of here, west of Ardentinny (see p. 244).

The boulders of porphyritic granite derived from the Garabal Hill and Meall Breac complex (p. 96) are much more abundant in the north part of Cowal than in the south part. This is, of course, but natural, as the north part is so much nearer their source. It is needless to detail where these boulders may be seen : suffice it to say, that both on the shores of Loch Fine, Loch Goil, Loch Long, Loch Eck, and on the hillsides near Loch Restil, Socach, and Caol Ghleann, they cannot but attract attention. The clean-washed boulders on the shores of Loch Fine are, quite apart from their glacial interest, worthy the attention of any who is studying the rock masses from which they came. We see, mixed with the boulders of granite, others of diorite, of granite with diorite inclusions, junction specimens of diorite and schist, altered albite schist, and schist with opaque spots of white felspar about the size of peas. We are informed by Mr Dakyns that a rock like this last mentioned forms nearly the whole of Ben Damhain and the south face of Troisgeach. This assemblage of boulders can only have come from a north-east direction from the complex already alluded to. At the point ½ mile north-east of Rudha Bathaich Bhain, we estimated these boulders to form quite one-quarter of the boulders on the shore. The following list of places, nearly the highest in their respective localities, at which the granite boulders have been seen, will indicate to what an extent the granite-bearing drift has overridden this country :—240 yards north-east of the Ordnance Station on Beinn an Fhidleir, at a height of 2600 feet ; ¼ mile E.N.E. of the same station, at a height of about 2450 feet ; 170 yards north-west of the same station, at a height of over 2500 feet ; close to the top of the water-shed $\frac{1}{4}$ mile and $\frac{5}{12}$ mile E.N.E. of Mullach Coire a' Chuir, at heights of over 1750 feet ; by the top of the "h" of "Beinn Tharsuinn," at a height of 1750 feet ; a little more than $\frac{1}{4}$ mile N.N.E. of the "B" of the same words, at a height of over 1500 feet ; 260 yards south-west of the top of Cruach nan Capull, at a height of 1800 feet ; nearly $\frac{1}{8}$ mile slightly north of west of Sith an t-Sluain, at a height of a little more than 1000 feet—this boulder is unusually large, 8 feet by 6 feet by 5 feet. The highest point at which this granite occurs in place is on Meall Breac, in the north-west corner of 1-inch Map 38. The height is 2115 feet. It is clear, therefore, that all the boulders mentioned on Beinn an Fhidleir must have been derived from areas lower than their present positions.

Some boulders of felsite, like the sheets of this rock, which occur in abundance on the sides of Loch Fine, particularly the north-east side, have been noticed within the Clyde watershed, but we cannot be certain that these have come from the Loch Fine watershed, as a band of this rock also occurs on the south side of Stob an Eas and Beinn an t' Seilich (134), and along the course of the N.N.E. disturbances which pass on the east side of Cruach nan Capull (133). One such boulder was noticed $\frac{1}{2}$ mile north of An Cruachan (162) ; another on the watershed nearly $\frac{1}{2}$ mile north-west of Meall Dubh (163), at a height of over 1700 feet ; another by the road a little more than $\frac{1}{4}$ mile south-east of Drim-syniebeg (142) ; and others in Donich Burn below Donich Lodge, and on the shore of Loch Goil near Ardnahien (153).

A boulder over 5 feet long, seemingly nearly all composed of large crystals of hornblende, was found $\frac{1}{2}$ mile east of the second "n" of "Beinn Tharsuinn" (142), at a height of about 1250 feet. Other boulders rather like this, but not quite so basic, were noticed $\frac{3}{4}$ mile south-west of Mark (143) and $\frac{1}{8}$ mile below the foot of Criogan (141). Mr J. R. Dakyns thinks the boulder near Mark is like the rock of some diorite bosses which occur east of Glen Falloch, and Mr J. B. Hill says the Criogan boulder is like some of the deformed basic rock on Beinn Buidhe (1-inch Map 45).

There are several boulders of the coarse dolerite of Sith an t-Sluain about $\frac{1}{4}$ mile south-west of this hill. There is no other rock like this in the neighbourhood with which it could be confused. We noticed no boulders carried in any other direction. The glacial striations in the district are all running north-east to south-west.

At various places north-west of the large albite schist areas, boulders of this rock were found :—e.g., on the shore near Tighe Claddich (133), in the burn $\frac{1}{8}$ mile S.S.E. of Laglingarten (133), by the road a little more than $\frac{1}{4}$ mile S.S.W. of Arduo, and 80 yards above the road a little more than $\frac{1}{2}$ mile east of Socach (141). In the last place, the height is about 400 feet, and the boulder is 5 feet long. It is suggested that there was a carry in a north-west direction from the albite schist hills down to the lower ground by Loch Fine. But it is not necessary to suppose this, for the same movement which carried the granite boulders south-east, would carry albite schist, and, as the strike of the schists is rather nearer north and south than the "carry," might, after carrying it a long distance, leave it on green beds or other rocks which lie on the north-west side of the strike of the albite schist. There are some exposures of albite schist which lie on the north-west side of the main outcrop of green beds, but these are only small, and not likely to be the source of the boulders mentioned.

A similar explanation may be given for the boulders of garnetiferous mica schist that are sometimes seen resting on rocks which lie north-west of the strike of this schist.

The deep red boulder clay so common in the south part of Cowal (p. 235) extends also into 1-inch Map 37. It is seen near the head of Glen Finart (164) ; in sections by the roadside on Larach Hill there are sharp junctions between it and overlying ochreous morainic drift. In Strath nan Lub (163) the boulder clay is sometimes brown and sometimes red ; in the river bank near the "a" of "nan" both colours are seen in the same section ; near the top of the clay there is a close wavy lamination. Red boulder clay also occurs in a burn at the south margin of 1-inch Map 37, ⅜ mile south-east of the last "n" of "An Cruachan" (162) ; on the side of Loch Eck, by the foot of the little stream south of Rudha Garbh ; in a drift area west and south-west of Tom Soilleir (163), generally overlain by loose morainic drift ; on the south side of Drynain Glen (164) between the "Drynain" and "Glen" ; ½ mile S.S.W. of the "C" of "Tom nan Con" (164). We have not noticed any red clay within the Loch Fine watershed. The general colour of the clay both on the shores and in the glens that lead down to this loch, is pale greenish-grey or pale buff. These colours seem to pass gradually into one another, also at times into more decided brown.

Signs of lamination in the boulder clay are not uncommon. One-half mile south-east of the "r" of "Laglingarten" (133), a stiff grey boulder clay shows a lamination inclining down hill. In the glen west of Creag Dhubh (141), a thickness of over 30 feet of drift is sometimes seen : the under part is boulder clay with laminæ sloping south-east much with the hill. In a burn $\frac{5}{12}$ mile south of St Catherine's Pier stiff buff boulder clay, containing granite boulders, shows lamination dipping north-west, again down hill. A greenish-grey clay in the burn ½ mile south of the "h" of "Bathaich ban Cottage" (134) also shows lamination in a north-west direction.

A section on the east side of "Coire No" (133), by the last "o" of "Coire No," shows laminated sand and gravel, at least 30 feet thick, under stiff boulder clay, probably about 60 feet thick, under morainic drift.

In the burn 120 yards below the basalt dyke in Leamhanin (141) there is at least 50 to 60 feet of drift. The lower part is stiff grey or buff boulder clay, the upper part looser morainic. There are good sections for some distance above the basalt, and in more than one place the till is seen below the water of the burn : the stones in the till are sometimes lying with their broader sides approximately horizontal.

In the burn south of Corrow (142), there are good drift sections ; at the top is loose morainic stuff ; under this comes a tough grey or brown, rather sandy till ; below this again is looser sand and gravel.

In Sugach Aodainn (152), ½ mile slightly east of north of Inverneaden, 40 feet of sand and gravel is seen under loose bouldery drift. The sand and gravel contain some more clayey pieces, which are very irregularly distributed, almost as if they were pebbles, and with a lamination sometimes at right angles to that in the gravel. Lower down the burn there are sections of stiff grey till under loose sand and gravel.

In the burn ½ mile S.S.W. of Meall an-T (163), there is 10 feet of sandy or gravelly clay and reddish-brown almost stoneless clay, with a lamination dipping S.S.E. at 24°, under coarse morainic drift. A little further down the burn, 8 feet of sand and gravel, with boulders and roughly horizontal bedding, is seen to overlie stiff brown till.

The areas of thick and wide-spread drift have been mapped, but it is impossible to give a satisfactory idea of them by verbal description. We

must, until the margins are engraved in the 1-inch Maps, be content with
enumerating a few of them, as follows : the west side of Caol Ghleann
(152), and the tract to the north passing over into the Loch Fine watershed ; both sides of Strath nan Lub (163), and the tracts extending,
south-west to the large peat moss on the south-east side of An Cruachan,
and north-east over into Glen Branter ; the west side of Garrachra Glen,
above the second "a" of "Garrachra," and extensions thence in scattered
patches to join wide deposits in Glen Shellish ; both sides of Eas Dubh
(141), and the area crossing over the watershed into Coire No ; the Coire
No, and the ground on the east side of it over which the road passes
between Lochgoilhead and St Catherine's ; the south-east side of the River
Cur (142), and the area extending up Glen Leamhanin to a little east
of Curra Lochain, and also far up Liogan and Criogan ; the south-east
side of Lettermay Burn (142), and the tract extending over the watershed
into Coire Ealt ; both sides of Donich Burn and Coire Odhair, and the
sides of the streams 1 mile east of Lochgoilhead ; the north side of Glen
Kinglas (134), ¾ mile below Butterbridge ; and the south side 2 miles
below the same place ; the sides of Eas Riachain (126) and its tributaries—
this area is connected by a narrow strip with another wide area on either
side of the glen north of Meall Beag.

In most of these areas the surface is occupied by loose morainic drift,
and well-defined banks and heaps of coarse bouldery material are common.
In some places there are definite boundaries to the areas occupied by the
better-shaped moraines. The moraines west and south of Loch Restil
pass over the watershed into Glinne Mhor ; the upper limit of good
moraine shapes is a fairly regular line gradually descending as it goes
west from 1200 to 900 feet ; it never gets more than about 200 yards
from the roadside. The ground north and west of this line is for some
distance occupied by smoother surfaced drift, with isolated rock exposures
in places.

The moraines on the top of the watershed between Hell's Glen and
Loch Fine cannot but attract the traveller's attention as he passes on the
coach road. Hell's Glen itself is comparatively free of drift, but just on
the watershed moraines appear, and extend for a long distance by the
roadside. They spread up the west slope of Stob an Eas to a little above
the 1250 feet contour. The shapes on the south-west side of the road
usually take the form of banks running north-east to south-west, and
with their steep sides facing north-west. Near the sides of Coire No, the
south-west margin of the morainic area is well defined : the boundary
crosses the burn by the " C " of " Coire " : on the east of the burn it runs
almost due east for a little over ½ mile : on the west side it forms a
curved line which can be traced about ¼ mile in a north direction.
There is drift further up the Coire, almost to the head, but there
are no well-defined moraine shapes.

Moraine shapes are prominent on the sides of Glen Branter (152) ; some
of them on the east of the glen have a long axis almost at right angles to
the course of the glen. The best shapes end near the head of the glen,
along a line running south-west from the "S" of "Sron Criche" ; the
ground beyond this line is for some distance much smoother, but ill-
defined hummocks appear again on the higher ground to the east and south-
east, between the 1250 feet and 1500 feet contours, and the surface
drift at the south end of the moss on the watershed between Glen Branter
and Strath nan Lub is morainic in character. The best moraines along the
lower part of this strath have the form of banks with a direction roughly
parallel to that of the strath. Such are seen near the "W" of
"COWAL," and ½ mile south and S.S.E. of Maol Odhar : the last run

W.S.W., bending as it were down the glen to Garvie. The valley which continues in the direction of Strath nan Lub, S.S.W., from the junction of this strath with Garvie Burn, is occupied by moraines on both sides until the peat moss, ½ mile E.S.E. of An Cruachan, is reached. Moraines are particularly prominent near the moss, and there is one long bank, running in a rough semi-circle with convex side facing south, which has clearly caused the formation of the moss by damming the drainage outlet. This bank is smoother than typical moraines, and a section at the south side shows fine yellow sand layers, inclining in places at a considerable angle. Another section in the burn ¼ mile E.S.E. of the last "n" of "An Cruachan" shows coarse bouldery drift intermixed with sand and false-bedded laminated clay, and it seems impossible to separate this bank from the others a little further north, which also show occasional sand layers mixed with coarse angular gravel and boulders. South of the bank which margins the moss there are no good moraine shapes, though loose morainic drift is still seen in various sections, at times overlying deep red boulder clay.

Dunans (162) probably takes its name from the multitude of small moraine heaps (dun = a heap or castle), which occur in the glen near this place and to the north-east. Such are prominent all the way from about 1 mile above Dunans to the head of Caol Ghleann, and, on the east side of the valley, extend over the watershed to a little beyond the boundary between the parishes of Stralachlan and Kilmodan. North-east of this the ground is still drift-covered for a considerable distance, but the surface is much smoother, and stiff grey boulder clay is seen close under the turf in several places.

On the west side of Gleann Dubh (152 and 141) the boundary of the more lumpy drift surface is fairly definite, between the head of the glen and ½ mile of the hamlet of Glensluan ; below this the moraines rest directly on rock, and this continues to be the case at intervals until we get north of the hamlet. From the hamlet to within a few hundred yards of Tombuidhe the boundary of the moraines runs evenly, a little higher than the 200 feet contour ; here and there on the east side of this line stiff grey boulder clay is exposed to day. On the east side of Gleann Dubh moraines extend to higher levels than the west, but they become gradually smaller and less definite on the higher grounds. Mr C. MacLaren (*Edin. New. Phil. Jour.*, vol. i. p. 189, 1855) mentions the moraines near Glensluan ; he supposes them to be terminal moraines belonging to the glen.

The watershed between Eas Dubh, which flows into Loch Fine, and the Cur, which flows into Loch Eck and thence into the Clyde, is very low, and is occupied partly by a peat moss and partly by drift, with one or two small rock exposures near the north end of the moss. The drift generally exposed is stiff greenish-grey boulder clay, but there are also some mounds of sandy material with rabbit burrows. Mounds are much more prominent on the east side of the Cur near Ballemeanoch, and extend south to Invernoaden. Many of these are quite smooth, and have been ploughed over. One, at the edge of the alluvium $\frac{5}{12}$ mile south of Ballemeanoch has been artificially levelled at the top, and the sides evenly graded, to make an old burying ground. The slacks and low parts between the mounds are often occupied by alluvial wash which passes gradually down into the alluvium of the Cur, so that this alluvium has the appearance of fingering off between the mounds, or of quite isolating them. We suppose the mounds to approximately represent original shapes, and that the side streams have, as a rule, merely taken advantage of the original hollows between them, without modifying them largely. There

R

are moraines a little up the west side of the Cur ½ mile north-west of Bridgend, running almost at right angles to the course of the river; stiff grey boulder clay underlies the morainic material.

Mounds occur on the north-west side of the Cur ½ mile above Ballemeanoch, with a north-east to south-west direction. About ⅓ mile further up the river there is a large one which has a direction almost at right angles to the adjacent Cur. On the north-west side of the river 1 mile above Ballemeanoch, a section on the roadside shows loose morainic drift lying irregularly on, or vertically at the sides of, boulder clay. There seem also to be some patches of the latter enclosed in the morainic drift. It is clear that there has been some erosion between the deposition of the two beds. In the same locality, in a little burn ¼ mile south-west of Socach, 66 yards from the foot of the burn, there is an interesting section of a somewhat similar nature. A coarse bouldery gravel cuts through a buff, finely-laminated, almost stoneless clayey sand. The laminæ dip north-west, nearly with the stream and hillslope; they are cut sharply through on the south side of the section by the coarse gravel, and are also overlain by a thin band of it. The junction line between the two beds is in places overhanging. Twenty yards above this section, a similar gravel is seen, lying with an irregular junction on a stiff laminated sand with occasional stones. This sand is no doubt a continuation of the laminated sand above mentioned, but with more stones.

There are two mounds a few hundred years east of Glensbellish farm, which are probably morainic in origin, though some people regard them as artificial.

On the south-east side of the River Fine the deltas of the side streams have spread out irregularly among moraines, and there are moraine shapes quite isolated by alluvial gravel. The old mound fort ½ mile N.N.E. of Achadunan may possibly be a modified moraine.

In the angle between the Cur and the Cab there is a well-marked boundary to the good moraine shapes. This boundary makes a curve 300 yards north of the "C" of "Cab," corresponding roughly to the angle between the two rivers. From the curve it runs E.S.E. until it gets to the "i" of "Liogan," and N.N.E. for nearly ½ mile. Loose sand and gravel beds often cover the hillslopes outside of, and at higher levels than, this boundary. Good sections of these beds occur in a little stream ½ mile and ¼ mile north-east of the bottom of the "L" of "Liogan." In the first place we see layers of fine and coarse gravel, inclining south-east nearly with the stream and hillslope; the layers are occasionally cemented with some black cement, probably in part the black oxide of manganese (see p. 242). In the second place there is no clear stratification. The north-east side of Leambanin is ridgy and banky, and usually shows a sandy drift at the surface. There are comparatively few boulders on the ridges. A pronounced mound runs from near the sharp bend in the burn ¾ mile north-west of Carnach Mor in a south-west direction, and this too has a smooth surface.

There are a few good moraines close at the sides of Loch Eck: on the east side, ½ mile north-east of Dornoch Point, and on either side of Rudha Croise: and on the west side, from the north side of the Bernice delta for ½ mile north, and for a little distance on the south side of the same delta. All these have directions parallel with the loch. Mr C. MacLaren mentions them (op. cit.) and supposes them to be remnants of lateral moraines. Other moraines occur in the coires on the east side of Beinn Mhòr, but these are generally small.

The moraines are particularly rough, with boulder-strewn tops and sides, on the north side of Kinglas Water near Butterbridge, on the sides of the different tributaries of Eas Riachain, in the glen north of Meall Beag, in the glens north and south of Carrick Castle, and the glen south of Cruach nam Miseag (153). In these localities they sometimes consist of great heaps of schist boulders with little or no fine soil between. There are heaps of large boulders too on the east side of Glen Shellish, about ½ mile and ⅔ mile below the burn fork near the head of the glen. There are several island-like knolls in the alluvium ₁⁶₂ mile east of Butterbridge : some are rock, but others are heaps of boulders, probably remnants of moraines. On the high delta on the south side of the foot of the glen south of Cruach nam Miseag, there are some large isolated schist boulders ; these are also probably remnants of moraines which the stream has not had power to clear away. In the alluvium ¼ mile south-east of Drimsyniebeg (142), and in the river near this place, there are also great boulders of probably the same character.

A very large block occurs on the south slope of Ben Donich at a point ½ mile north-east of the "h" of "Inveronich." It forms a prominent object in the view of the mountain from near Lettermay, and goes by the name of "Clach a' Bhreatunnaich" (the Briton's stone). The slope above the stone is comparatively gentle, and we saw no indications of landslips near. Probably the stone must be regarded as ice-borne. The height of it is about 20 feet, and the sides 18 and 30 feet respectively. It is composed of albite schist, and a little on the south side of it, there is a bank which is thickly strewn with smaller blocks of the same rock.

Some glacial gravels cemented in part by black cement have been already mentioned (p. 258). Such have also been noted ½ mile slightly north of east of "h" of "Strath nan Lub," in a stream ¼ mile slightly west of south of "S" of the same words, and in a stream a little more than ¾ mile E.S.E. of Socach ; in a specimen from the last place the presence of manganese was proved. In the east bank of the stream ½ mile E.N.E. of the first "n" of "Cruach nan Capull (133), there are small nodules in the top drift soil with a black binding material.

A section on the roadside 30 yards south of the burn south of Cruach nam Miseag (153) shows 3 to 4 feet of fine sand lying between coarse gravel. The top gravel is probably part of the raised beach. The sand varies in fineness in different layers, and in places the bedding is sharply contorted and folded, with both limbs dipping the same direction.

The section in the bank of the Cur a little more than ½ mile east of Glensluan has already been alluded to (p. 242) for some points of interest. It has to be further mentioned that the lenticles of laminated clay enclosed within the coarse gravel sometimes contain laminæ which are sharply twisted, though the boundaries of the lenticles are not so. The ends of the twisted laminæ sometimes end suddenly at lines which rather resemble thrust lines. The axes of the folds always incline up stream, towards the N.N.E. . . The lowest gravel may include or overlie stiff laminated clay, like that of the lenticles ; sometimes it cuts across the laminæ slightly. Small faults or slips occur in this gravel, and they are filled, for a breadth of an inch or two, with fine sand.

Lochan Mill Bhig (126) is surrounded by moraines, except where there are two small rock exposures, near the north-east end. The other smaller lochans near this are also mostly surrounded by morainic drift. Curra Lochain (141) has morainic drift all round it ; the burn that runs from the moss at its west end shows no rock till you get more than ½ mile west of the loch. Lochan nan Cnaimh, south-east of Beinn Bheula, is surrounded by a narrow margin of drift, except in one place near the alluvium at the head. There is a drift-covered strip extending from the lochside, south of the outlet, to the east side of Lettermay burn. The various peat mosses a little more than 1 mile south of Carrick Castle, east of Tom Molach (153) and ⅓ mile north-west of Cruach nam Miseag, probably represent old lochs. Many of the sudden expansions of alluvium at the sides of the streams may also do so.

In two places on the side of Loch Fine the boulder clay contains broken shells. One of these, 90 yards south-west of Ard na Gailich (133), shows a stiff greenish-grey boulder clay overlain by the raised beach gravel. Close under the gravel the colour of the clay gets brown. The line between the two colours may abut against the middle of an enclosed boulder, and probably represents the advance of water and oxidising agents. Shell fragments occur in both the grey and brown clay; they are in an extremely friable condition, crumbling readily under the finger, and are in pieces which rarely attain ½ inch in length. They are in fair abundance, there being perhaps one fragment visible on the average for every square foot of exposure. One bit, a little over ½ inch long, showed clear glacial striations. The clay is full of well-smoothed rounded boulders of Glen Fine granite; there is an obscure lamination in it for a few inches near the top.

Two hundred yards south-west of the sharp bend in the shore at Airidh a' Ghobhainn (133 and 141), a stiff grey boulder clay, with lamination dipping north, contains broken shell bits.

In contrast with the above, some glacial shell-bearing beds occur in various places on the coast, in which the shells are probably in approximately the position of growth. None of these have yet been searched by the Survey collectors. In the little burn by the lodge ₆ₓ mile E.N.E. of Rudha No (133) 1 to 2 feet of greenish-grey sandy laminated clay is seen, with a gentle dip down stream. This is only a few feet above high-water mark. About 30 yards above the lodge the clay is exposed again, rather more distinctly; it is clearly laminated, but contains stones as much as 7 inches in length, in the arrangement of which we could discern no order. The colour of the weathered clay is buff or greenish, but internally the clay is dark-grey, and contains many shiny spangles, probably of minute mica flakes. Shells occur scattered through it near the water-level of the burn, sometimes singly and sometimes in discontinuous streaks as much as ½ inch broad. The shells are often much broken: we noticed several Astartes among them.

A greenish-grey laminated clay with shells occurs on the foreshore ⅓ mile and nearly ½ mile north-east of Creag nam Faoileann (141). In one place the laminæ dip south-east at 30°, in another north-west, and we thought there were signs of contortion. A stiff buff clay is also seen on the shore in several places to the south-west, nearer to the Creag. Old exposures are no doubt often covered up and new ones exposed, owing to the shifting of the shore gravel. At the time of the Survey, the clay was seen within a few yards of some of the big boulders near the Clach Dubh na Criche (a little more than ¼ mile north-east of Creag nam Faoilean), and we have little doubt that both they, and the Clach Dubh itself, are resting on it. If this is so, the case is parallel to that mentioned near Toward Quay (p. 248). The Clach Dubh and the other boulders are referred to again (p. 261). This shell-bed is described by Mr W. Ivison Macadam, ("Preliminary Notice of a Clay Shell-bed between Newton and Strachur, Loch Fine, Argyllshire," *Trans. Edin. Geol. Soc.*, vol iv., 1883, p. 94, and "Further Notice of the Tigh-na-criche Shell-bed, Loch Fine, Argyllshire," *op. cit.*, p. 232). He gives lists of the shells collected by him from the bed, and also an analysis of the clay in which they are found. The analysis is as follows :—

Ferric oxide, Fe_2O_3,	.	.	.	16·08
Aluminic oxide, Al_2O_3,	.	.	.	22·83
Calcic oxide, CaO,	.	.	.	2·55
Magnesic oxide, MgO,	.	.	.	1·96
Silica, etc., SiO_2,	.	.	.	56·58

100·00

It is stated that the clay is employed by the people of the locality to whiten the hearthstones before the fire.

One hundred and thirty yards south-west of Corriesyke (142, close to Douglas Pier, which is not shown on the 1-inch Map) a laminated stoneless clay dips steeply seawards. Similar clay also occurs on the shore 1 mile and $\frac{1}{4}$ mile north-west of this place, and $\frac{9}{12}$ mile south-east. On the side of Loch Long, 100 yards south of the foot of Allt Guanan (142) a laminated clay is seen near high-water mark. All the clays mentioned in this paragraph look like glacial shell clay, but no shells were actually observed in any of them. At the locality $\frac{1}{3}$ mile south-west of Corriesyke the edges of the clay laminæ are distinctly cut off by the raised beach gravel.

The Clach Dubh na Criche (Black Stone of the boundary, so called because it marks one end of the boundary between the parishes of Stralachan and Strachur) is said to have been slightly shifted by shore ice during a severe frost 50 or 60 years ago. Upper Loch Fine is distinctly less salt than the open sea (Dr H. R. Mill, "The Clyde Sea Area," *Trans. Roy. Soc. Edin.*, vol. xxxvi. p. 697), owing to the number of large burns which empty into it, and freezes with comparative readiness. The Clach Dubh is a boulder of epidiorite, about 10 feet long. The height and breadth we have no note of, but we think they are each about 5 feet. The rock type is very like that of the epidiorite bands near Creggan's Point. The sides are rather rough, but the top is smoothed and striated in a direction 4° east of south. There are some other rather large epidiorite boulders on the shore a little south-west of the Clach Dubh ; one of them is striated at the top in a direction 30° west of south, but it is rough at the sides ; on another one there are striations both on the top and the sides, but the directions vary. If the Clach Dubh, one of the largest boulders on the coast, was moved by shore ice so recently, it is still more possible for the smaller ones, so abundantly scattered on the shore, to have been so. But perhaps most of the large ones occupy much the same positions as they did ages ago when the climate was much more severe than now. On the shores of Loch Fine there is generally a marked absence of large boulders at the mouths of the more gently-flowing burns, both on the fore-shore, on the deltas now in process of formation, and on the deltas of raised beach times. So that it would seem that there have been no boulders left in these places by ice since the deltas mentioned were formed. At the foot of steep running mountain torrents, on the contrary, large boulders may be found, but these have probably come down in flood time, and are no larger than others in the courses of these torrents.

In Strachur Bay there is neither rock nor boulder clay, and the raised beach deposits extend for some distance inland. Here we saw no boulders on the shore which much exceeded 8 inches in length. But on either side of the bay, where we get on rock or the laminated shelly clay. boulders are abundant. Near Rudha No (133) there are no big boulders on the shore for a distance of about $\frac{1}{2}$ mile, but at either side of this tract. where the raised beach is represented in the main by a bare shelf of erosion, in rock or boulder clay, boulders are abundant.

The above instances certainly follow the usual rule, but there are some exceptions which deserve notice, e.g., on the shore south-west of Airidh na Gobhainn, there are boulders of granite 3 to 4 feet long which are surrounded by, and seem resting on gravel; boulder clay is seen on the bank of the raised beach within a little distance.

We examined the shore of Newton Bay, and found two pieces of red sandstone like that of the Upper Old Red sandstone, but both these were small. The dimensions of the larger were 1 foot by 6 inches by 4 inches; this was a rather pebbly sandstone, with some well rounded quartz pebbles as large as a blackbird's egg, and a greenish-grey matrix. The other was 7 to 8 inches long, and contained pieces of greenish-grey shale nearly 1 inch long. Both might well have been brought as ballast to the bay. They are not the kind of rocks likely to be used for building stone. They are too small to compare with the boulder mentioned (p. 248) near Toward Quay. The largest and most abundant boulders in the bay consist of Glen Fine granite, epidiorite and hornblende schist. Pieces of quartz porphyry with rare quartz crystals, like the rock of the quarries at Furnace on the opposite side of the loch, are common, but they are generally angular, and with fresh fractured sides, and most of them have probably been brought across the loch for building purposes. This is still a common practice.   C. T. C.

CHAPTER XX.

MARINE AND FRESHWATER ALLUVIA.*

Marine Alluvia.

There is in many cases no satisfactory distinction between the high gravel and sand deposits taken as marine and freshwater. Shell remains are rarely observed in either, and we may be left to judge of the character of a deposit merely from the distance from the sea at which it now occurs. In the same way at the present day the gravel in the burns passes quite gradually as they approach the sea-level into the gravel under the sea, below the burn mouths.

In cases, however, in which the deposits can be traced running approximately level along the shores, far away from any burn mouth, it seems clear that the terrace shape must be due to sea action, and the gravel and sand on such terraces are no doubt marine in most cases. It is not certain that all of them are, for the terrace may originally have been one of erosion merely, unaccompanied by deposition, and, long after its formation, little burns coming down the steep hillsides may partly cover the terrace with alluvial material, the more gentle slope of the terrace largely helping to this result. There is, too, a doubt whether some of the deposits now seen below the terrace tops may not be of earlier age than the terrace itself,—may be glacial sand and gravel, the edges of which have been abraded by later sea action. The way in which the lower beds in terrace sections may have their edges cut across unconformably by a more gentle sloping, and generally coarser bed, the top of which approximately represents the slope of the terrace, certainly suggests that the lower beds

* A sketch map of the superficial geology is inserted near the end of the volume.

may be separated by a long interval from the upper, and have been formed under different conditions. The best instances of these apparent unconformities are given elsewhere, p. 242.

The greatest breadths of these deposits generally occur near the mouths of glens, a fact which indicates that the material in them has not been produced on a large scale by wave action along the shore, but rather brought down ready formed into gravel, etc., by rivers, though by the sea it has been subsequently rearranged and spread out. For examples of this, we may refer especially to the deposits at the foot of Ardyne Burn (194), at the head of the Holy Loch, at the foot of Glendaruel, and to the smaller beds at the mouths of the different burns that flow into Loch Striven.

Shell remains are very rare in beds which we can with certainty refer to the terraces. It is said that such are occasionally ploughed up in the field near Colintraive Pier, within the area of the lower raised beach, and also in the higher terrace near Ardentraive farm, but we do not know the particulars. In some of the small wave worn hollows at the back of the old beach in the garden of Buthkollidar, 2 miles south of Dunoon, sea shells are also said to have been found mixed with hazel-nuts.

The highest terrace shape usually seen is that known as the 100-foot beach. This title passes current from its brevity, and perhaps expresses sufficiently well the general height of the surface of the bed, but there are many places, even apart from the neighbourhood of burn courses along which the level might naturally be supposed to rise as we go inland, where the height of 100 feet is considerably exceeded. On the southwest side of Dunoon the height is about 120 feet. Close to Dunloskin farm (180) it is about 130 feet. And near Creag na Cailliche (194) it is even higher still. The surface of it, and also of all the lower beaches, including the present beach among these, has generally a distinct slope towards the adjacent sea margin. It has often been so carved out and denuded by the burns that the terrace shape is no longer clear, and we may doubt whether a certain bed belongs to it, or to an earlier series of glacial deposits. The landward margins near Hafton House (174) and Brackleymore (183) are very uncertain.

The deposits on the different terraces on the east side of Loch Riddon and the Kyles of Bute possess generally an even stratification. We have seen no sharp contortions in them such as the drag over of ice might be supposed to cause, nor inclusion of boulders enormous in size in comparison to the pieces among which they occur, such as might have been dropped by floating ice. No glacial shells have been seen at a height of more than a few feet above the present high water mark, though many of the varieties found in the glacial beds belong to littoral forms. Hence it seems probable that these beds belong, not to the period of the 100-foot beach, but to an earlier and probably colder age.* It should be stated, however, that, as mentioned elsewhere, p. 242, nothing of the 100-foot beach is seen at the heads of Loch Goil or Loch Fine, in the more mountainous district lying in 1-inch Map 37, and it has been suggested that this may be due to the heads of these lochs having been still filled with ice at the time the sea stood at the 100-foot level. The terrace at this level is not known for certain further up Loch Goil than ½ mile north of Gairletter Point: this is about 10 miles south of the head of the loch. The highest distinct flat in Glen Finart does not much exceed 40 feet in height. That at Ardnahein is about the same height at its landward margin. Those at Corrow and Lochgoilhead are rather lower than this.

* Mr J. B. Hill is inclined to dissent from this opinion. There is certainly room for different conclusions about the matter.

At Stuckbeg (153) the terrace rises over 40 feet, but this is close to the mouth of a large stream, and the terrace slopes steeply down towards the sea : just east of the terrace there is a section showing evenly stratified sand and gravel between coarser unstratified gravel, but there is no even top to the ground surface, and we cannot safely class these beds with the 100-foot beach.   C. T. C.

In Loch Fine the 100-foot terrace does not appear above Largiemore (171), about 20 miles below the head of the loch. The high flat at Strachur reaches to 100 feet or even more inland along the burn side, and its height near the sea averages perhaps 20 to 40 feet. The high flats at Ardno (133), and Ardkinglas (134), increase in height inland in much the same way.   C. T. C., J. B. H.

At the foot of the burn ½ mile north-west of Carrick Castle, well-shaped boulder strewn moraines come down to a level of about 50 feet, and are in close contact with the high flat. There is no section to show the relation in age of the one to the other. We suppose the moraines are earlier than the flat : on the south side of the stream the flat extends up it to form the apex of a " v " shape, and there are some huge boulders on the top of it which may be remnants of morainic material which the stream has not had the power to carry away. On the north side, the landward margin of the flat tapers off between moraines to some extent in the same way as the high river terrace in Glen Fine, and south of Ballemeanoch (see p. 257).

At the foot of Loch Eck the surface of the beach at the east of the outflow is uneven and moundy, and may represent an old morainic surface which has had its inequalities partly washed down. The surface on the east of the loch is also moundy in places : north of the north end of it there are some smaller patches of gravel which may also belong to it, but the stratification in these is very irregular, and they are not shown in the Map. On the west side of the loch, close to the north margin of 1-inch Map 29, the bed has a somewhat flat surface, but resting on it are two large schist boulders : the adjoining hill slope is not very steep, and it hardly seems that the boulders can be ordinary tumblers. Perhaps in this area the ice was still mantling the hills while the 100-foot beach was being formed.

A section on the north of the Little Echaig ⅓ mile south-west of Ballochyle shows 6 to 7 feet of coarse gravel at the top, overlying, at the south-west end of the section, sand, sandy gravel, and tough laminated reddish clay. This laminated clay may represent some of those seen within the boulder clay occasionally (see p. 237). We are inclined to refer the upper bed only to the 100-foot beach. At the north-east end of the section there comes under the gravel some thickness, at least 10 feet, of tough red boulder clay : this boulder clay is at a higher level than the sand and clay in the south-west part, and very likely is really of later age than them, for just under it undoubted signs of sand and gravel occur.

                                                                   C. T. C.

On the east shore of Loch Fine from Newton Bay to Kilfinan Bay, marine terraces are fairly abundant wherever the nature of the ground has favoured their formation and subsequent preservation. They are most numerous at the mouths of the old river valleys where they have escaped the destructive influences of marine erosion. In these cases they have to a large extent been modified by river action, and their partial destruction has furnished material for building up the river terraces with which they are associated. The raised beaches at Kilfinan offer a good illustration. They occupy an area of about ½ square mile at the mouth of Kilfinan Burn. On the south side of the burn their upper boundary between Otter House and Kilfinan attains a height of rather over 100 feet, from which there is a gradual slope towards the river. The corresponding terrace on the north side of the burn has been cut into at two or three places by tributary streams and its surface broken up. The ground between these two exposures of the 100-foot beach has been occupied at different times by Kilfinan Burn, and the marine beaches are entirely replaced by subsequent river terraces. The 100-foot beach on the north side of the burn is continuous east of Lindsaig with a deposit that at the side of the river reaches a height of 200 feet. It is possible, however, that the upper portions of this deposit may be fluviatile, and that we have here an instance of a freshwater terrace passing insensibly into one of marine origin. The deposits are made up of stratified sand and gravel. At Otter Ferry, a few miles further north, a set of terraces occurs, the upper limits of which reach a height above sea-level varying from 170 to 200 feet. It is possible these higher portions may be of glacial origin, but we were unable to find sufficient grounds for separating them from the deposits at lower levels into which they merged.   J. B. H.

Near Ardyne Point (194) there are besides the 100-foot beach no less than six different flats at the following approximate levels : 10 feet, 18 feet, 23 to 30 feet, 36 feet, 50 to 64 feet. In cases like this where there are many different levels it is impossible to show them all by different tints, and several have to be coloured the same. The lines between them are, however, still engraved.

Near South Hall (194) four terraces can be made out. At Ardtarag (173), Ardbeg (183), Invervegain (183), Couston (183), Colintraive, Dunoon, and many other places three. In all these places the 100-foot beach is present.

The lowest raised beach that is generally found is called the 20-foot beach. This forms an almost continuous fringe around the shore, though in places, especially in the more rocky areas, it is represented by a terrace of erosion merely. This is so, for instance, for considerable distances on both sides of Loch Striven, for a little space near Strone Point (Holy Loch), on the south side of Gairletter Point near the basalt dyke, ½ mile and ½ mile south of Ardentinny, between Ardentinny and the entrance to Loch Goil, on the east of Loch Goil, from ½ mile north of Woodlands to near Stuckbeg, on Loch Fine side between Leak and Stucreach, and between Aird Cottage and Creggans. In these cases the landward margin of the terrace is alone intended to be coloured burnt sienna, but the terraces are sometimes so narrow that it is hard to leave an area free from this colour between these margins and the sea line. A similar style of colouring is also used to distinguish the terraces of erosion without deposition on the higher levels, but there are not many instances of these known. On the south-east side of Castle Toward there is a terrace, in part higher than 20 feet, which is bare for some distance. Near Toward farm, on the east of the burn, there is a feature apparently referable to a high terrace, reaching to 70 feet or more, and there may be no gravel on the flatter area below.

Even when the terrace is of boulder clay there may be little or no deposit on it. This is seen well at Strone Point (Kyles of Bute).

As an illustration of the common fact that the terraces slope downwards to the sea, even when quite away from any large burn, we may mention the low terrace at Toward Point. The landward margin of this is in one

place 45 feet above sea-level, but the surface near the sea is only 11 or
12 feet.   Probably every terrace shows this to some extent, just as the
present beaches do.

It is readily noticed that the same physical causes which give rise to
features, indents, or points, on the present sea margin, have acted in a similar
way in the times of the raised beaches.   The same dykes that project
prominently on the shores of to-day did so also on the shores of long ago.
The same crush lines which form lines of weakness for the waves to
advance along and form gulleys did so also in old beach times.   Often the
old beach margins are partially obscured by the slipping of soil over
them, so that the smaller gulleys and caves are hidden, but in some of
the gardens near Dunoon, e.g., at Ardfillayne and Buthkollidar, the old
sea caves and wave-worn hollows have been carefully cleared out and the
old features restored.

It is probable that the lowest flat at Ardyne Point is to some extent a
storm bank.   The third beach, too, is of a very uneven surface, and may
be due to older storm banks.   The raised beach on the south side of the
Echaig near Kilmun Cottage is also very uneven, partly perhaps owing
to storm banks, and partly to old river channels.   The lowest flat on the
east side of the burn in the West Bay of Dunoon is banked up close to
the sea-margin, so that in very high tides pools of sea-water form in the
gardens, between the outer bank and the edge of the next higher flat.

C. T. C.

Near Drum Point south of Kilfinan are two terraces of erosion with
well-defined margins landward, but rocky and rough toward the sea.   At
a higher level are two curved gravel ridges meeting at a point—storm
beaches of the olden time—and still higher, rising to above 100 feet is an
extensive flat which is probably an old erosion terrace.   At Port Leathan
to the southward are gravel beaches at three different levels ; these are
rocky in places toward the sea and not everywhere distinct from one
another.   There is a gravel pit in the middle one.   In Auchalick Bay
there are two old beaches, well defined, but probably somewhat modified
by the stream.   There is a section of the higher one in the bank ¼ mile
from the mouth, and on the east side of the valley at Melldalloch and
Auchanaskioch there are patches of the 100-foot beach.   At Ardmarnock
the second beach is pretty extensive, and there is a patch of the highest
beach also.   The second beach north of Bagh Buic is covered with peat.
Patches of the two lower beaches occur in several places to the south of
this, and between Portavaidue and Derybruich there is a fine example of
the 100-foot with a good section in the banks of the burn.   To the
east of Low Stillaig the 50-foot forms an extensive peat-covered flat, and
there are traces of higher terraces both to the north-west and to the south-
east of this,—in the latter case often small gravelly patches interspersed
among rocky knolls covered with wood.   There are others east of Asgog
Bay, and gravel patches of the highest beach about Stillaig Farm, and on
either side the little stream to the north-east.   The lower beaches up Allt
Osda, north of Kilbride Bay, are largely composed of sand, and there is a
fine section in the banks of the stream west of Kilbride Farm.   Higher
up the valley the gravel of the higher raised beaches occurs interspersed
with the modern alluvium of the stream, with patches of glacial drift, and
some outcrops of rock.   The 100-foot beach runs up beyond the powder
works at Millhouse, and there is a section of it in a pit near the road-side
opposite Auchagoyl Cottages.   We heard that shells had been found in this
pit, but could not be satisfied of the truth of the rumour.   There is a pit
in sand and gravel on the west side of the valley near the old mill, and

another at the north end of the plantation in which are the powder mills,
which gives the following section :—

Gravel rather fine, in beds approximately horizontal, 15 feet or more.
Layers of sand, dipping steeply, seen below in places.

Where the road to Tighnabruaich passes through the plantation, the old
beach rises to nearly 120 feet above the sea, and south of the road seems
made up mainly of sand, sections of which may be seen in the banks of
the stream and in a pit near it, where was observed the following :—

Soil, 1 foot.

Sandy laminated clay, 1 foot 6 inches.

Brown sand with irregular layers of gravel, 5 to 10 feet.

South of the plantation the sand is seen to rest on red till.

The greatest extent of the highest beach in the south-west part of
Cowal is shown at Ardlamont where it reaches over 100 feet above the
sea, though the gravel is not very thick in places, and there are peaty
hollows in it. North of Carry Point for some distance the beaches are
inconspicuous being only small patches on the solid rock here and there.
From Blair's Ferry the lower beach is nearly continuous, but narrow,
round to Port Driseach, and patches of the higher beaches are seen
occasionally as at Over Inens in Tighnabruaich, and near Rudha Ban. At
the latter place they mount up to 130 feet above the sea and the gravel
was seen resting on 10 feet of boulder clay. In Loch Riddon beaches are
found at West Glen, Callow and Upper Callow. They are most conspicuous at the latter place where there are pits in the highest one. W. G.

In a good many places on the coast we find rock knolls surrounded by
old beach deposits, or by them and the present beach. These must have
formed islands in olden times. Examples occur in many places on the
east side of Loch Riddon : between Loch Riddon and Colintraive : at
Ardyne Point : at Dunoon (including the conspicuous Castle Hill) : in
Loch Long, at the foot of Knap Burn, $\frac{1}{8}$ mile south-west of the " P " of
" Port an Lochan," $\frac{1}{8}$ mile east of Feoileann, 1 mile S.S.W. of Coilessan :
in Loch Goil, the Carrick Castle Hill, 1 mile south-east of Cruach nan
Miseag : in Loch Fine, at Leak, Stucreach, a little north-east of Creag a'
Phuill, and St Catherine's.   C. T. C.

At the period of greatest submergence the coast line on the west side of
Loch Riddon and the west Kyle of Bute, presented much the same
appearance as now, but in Loch Fine all the adjacent islands disappeared
beneath the waters, except perhaps the east part of Eilean Aoidhe, and
new islands were formed south of Black Harbour, north of Bagh Buic,
west of Low Stillaig, East of Low Stillaig, east of Asgog Bay and north of
Ardlamont Point (two) : and the sea ran up Allt Osda to near Craignafeich
making a long narrow peninsula between it and the Kyles. At the 50-foot
level of the present islands, would still remain, but much reduced in size,
Eilean Buic, Eilean na Beithe, and Eilean Dubh (Loch Riddon), while
Eilean Aoidhe would have been two islands, and new islands would have
appeared west of Kilbride Bay and north of Port a' Mhadaidh. W. G.

Where the large burns discharge into the lochs we can generally see
that delta-shaped deposits are being formed below the sea line and
between tide marks. These marine deposits are in direct continuation of
those formed by the burn at higher levels. The tides in the lochs are
not strong usually, and so the material deposited under the water is not
spread out to great distances from the burn mouth. In upper Loch
Fine we have noticed that the deltas at the mouths of streams are

generally composed of finer material than the raised beach deposits on either side of them ; this is seen, e.g., at Rudha No. and $\frac{1}{8}$ north-west of Ardkinglas House.

In most parts of the area the present beaches are free from large boulders where the entire thickness of the shore bank is composed of raised beach material or of the delta deposits of the present streams. But if the terraces are represented by shelves of erosion merely, or by only a thin covering of gravel overlying rock or boulder clay, large boulders may be abundant, in size far exceeding any commonly seen in the sections of the raised beach beds. The deltas of very steeply-flowing burns may in places include boulders as large as these, but, as already intimated, many of the delta deposits are finer than the raised beach deposits near them. There are places, however, and at some distance from the burn mouths, where large boulders are .abundant on the shore quite surrounded by, and apparently resting on, gravel.

In the bay south of Creag a' Phuill (133), and for a little distance south-west of it, their are many boulders, chiefly of Glen Fine granite, sometimes 3 or 4 feet long, and these seem to be lying on gravel; in various places in the shore bank, however, boulder clay is seen underlying the gravel of the raised beach, and at a higher level than the boulders. In some districts where the foreshore is very free from boulders it is possible that this is due to artificial clearance to make sea walls, etc. There are many big boulders on the foreshore south-west of Strachur Bay. The raised beach here is represented by an erosion shelf merely, and so this is in accordance with the usual rule. Exposures of stiff finely laminated, and somewhat contorted, clay with glacial shells, occur close to several of them. The largest boulder called " Clach Dubh na Criche," is an epidiorite block about 10 feet long and about 5 or 6 feet high ; it is striated at the top in a direction 4° west of south. This big boulder is said according to information kindly supplied by the late Mr Rhynd of Strachur, to have been slightly shifted by shore ice at the break up of a severe frost some fifty-two years ago.

A big landslip, the slipped material of which descends to the shore line 1 mile south-west of Kilmarnock Hill (194) breaks the line of the 20-foot beach, and must be of later date than it.   C. T. C.

Freshwater Alluvia.

There are no broad spreads of freshwater alluvium in this district, the glens being all somewhat narrow, or else filled at their broader seaward ends with deposits which, in part, are marine. There is no sharp line in nature between the old marine and the old freshwater desposits, any more than between the gravels of the present beaches and of the streams which come down to the beaches. In the maps we are obliged to start the conventional colours of these deposits at some line, but lines already in the maps, such as road sides, have been utilised for this purpose, and they are not supposed to indicate more than the approximate positions of change.

It is probably that part, perhaps a large part, of the gravels spread over the different sea-margins is not marine, but freshwater, and of later date than the flats on which the gravel rests, having been deposited by the various little side streams which come laden with material from the steep sides of the lochs, but have their rapidity checked as soon as they get to the old sea flats, and so lose part of their load on them. Instances of this occur at the foot of the burn $\frac{3}{4}$ mile E.S.E. of Meallan Glaic (194 northwest), at Inverlounin (west side of Loch Goil), and on either side of Rudha nan Eoin (entrance to Loch Goil), etc.

Strath Echaig contains a breadth of freshwater alluvial deposits which reaches to ½ mile, and shows, by adding to these the probably marine alluvia, a spread from side to side of nearly 1 mile of flat land composed of sand or gravel. The river flat in Glendaruel has, above the bridge near Ballochandrain, a very uniform breadth of about ⅛ mile. In Glen Massan the greatest width is about ¼ mile, but this continues for hardly the length of a mile, and in the lower part the river runs for some distance through rock, or between rock and moraines without any alluvium. In the lower part of Glen Finart (163), there is sometimes a breadth of ½ mile ; outside this breadth the soil is also sandy or gravelly in character, but we refer it to glacial deposits. At the heads of Loch Eck and Loch Goil, and along the river Cur between Ballimore and Ballemeanoch (141), the breadths average about ¼ mile. In the valley of Stralachlan (151), the freshwater deposits are over 1 mile in length and ¼ mile in breadth. In parts of Glen Tarsan and Glen Lean there is a breadth of 200 to 300 yards. In Glen Kin and Glen Fyne (183 and 194), the breadths are less than that amount, and so much divided among terraces of different elevations, with occasional separations of drift banks, that a first sight fails to give an adequate impression of their extent.   C. T. C., J. B. H.

It is probable that in some of the glens the greater breadths occupy the sites of old loch basins which are now filled up. The lower part of the river Massan above Corrarsk is a rapid-flowing stream with but little alluvium, but above the rapids ⅛ mile south-east of Stonefield there is a flattish area extending for about ⅔ mile, and no more rock in the river for this distance, except in places where it is running just at the side of the flat area.

This absence of rock in the river beds, excepting where they are just at the sides of the flat areas is not at all uncommon. It occurs, e.g., in the higher part of Glen Massan above Stronlonag, in Glendaruel above Ballochandrain bridge, in Glen Tarsan " between the " t " and the " n " of "Tarsan," and in Glen Lean from a little above Glenlean Farm to ½ mile north-west of Corrachaive.

In the south-western part of Cowal there is nowhere any great extent of alluvium. There is some near the head of the Acharosson Burn at a height of over 1000 feet above the sea, and lower down the same burn near Acharosson in several places there is a high alluvial terrace. The alluvium which forms a narrow strip below this expands considerably in width near the sea, where it merges into the raised beaches. The only other strips worth noticing are those along the streams running into Asgog Bay and Kilbride Bay. The latter stream is called Allt Osda, and its continuation the Creag-an-fhithich burn. In the lower part of this valley it is not always easy to distinguish between the more modern freshwater alluvium and that of the raised beaches which run up the valley to beyond the Kames Powder Works. There is a pretty extensive alluvial flat partly overgrown with moss opposite Craignafeich, and there is another a good distance further up, above the upper dam, which may have been the site of an old lake like that above Loch na Melldalloch.

W. G.

The extension of the flat on the east side of Polchorkan, above Lochgoilhead, is occupied by a peat moss, and we suppose this to be of earlier date than the alluvium of the adjoining river, and that this alluvium is in large part spreading over the old extension of the moss.

The extension of the flat on the north-east side of Ballemeanoch towards Strachur is also occupied by a peat moss. The boundary between the peat and the alluvium is not well defined, but it is not far from the "mor" of "Strachurmore." The alluvium would not, in all probability, have extended so far as this, had there not been a flat area lying ready for it to spread over. The rise between the north-east end of the moss and the boundary of the adjacent flat in the Loch Fine watershed by Strachur is very slight, and mostly drift covered. Perhaps at one time the river Cur may have discharged into Loch Fine, instead of the Clyde as it does now. The west edge of the flat for ½ mile south of Ballemeanoch spreads out along peaty slacks between moraine-like mounds; in all probability the whole flat was once occupied by ground of this character —peaty areas here and there divided by drift mounds. One of the mounds ⅝ mile slightly west of south of Ballemeanoch has been artificially levelled at the top, and smoothed at the sides, but we regard it as natural in origin. Five-eighths of a mile south-east of Ballimore there are many small moraines surrounded by alluvial material. This material is probably more due to small side streams and rainwash than to the river itself. We suppose the moraine shapes are approximately as they were originally, and that the side streams have taken advantage of the low ground between them.

The east margin of the flat in Glen Fine between the "g" of "Glen," and Eas Riachain tapers out in an indefinite way between moraines, and one moraine shape is quite isolated by alluvium. There is no rock seen in the river between the head of Loch Fine and ¾ mile south-west of the "G" of "Glen."

In Glen Leau the stream running through the broad alluvial area is an extremely small one, and the alluvial deposits have clearly come in the main from the numerous side streams which descend the glen sides. This is no doubt the case to a large extent in some of the other glens also. Where each side stream begins to lose its steepness it forms a more or less prominent delta, the outer margins of which sometimes merge insensibly into the alluvium of the main stream, e.g., on the north-east side of the Massan near Stonefield. In other cases the deltas of the side streams are cut through by the main stream, and have their outer margins passing gradually into one of the higher flats of the valley.

The deltas or cones on the west side of the river Fine ½ mile S.S.W. of the "R" of "River," and above the "v" of "River" are very steep, and composed of unusually coarse torrential material; their upper margins are not easy to distinguish from screes and drift.

The alluvium in Hell's Glen consists in the main of coarse delta deposits brought down by the steep streams on the south-west side of the glen.

For a great part of Glen Kinglas the higher margins of the cones of material brought down by the numerous side streams are very prominent, but the detrital material does not extend down as far as the Kinglas Water, excepting in the case of the larger streams.

The different side streams flowing into Loch Eck show good delta deposits. The head of this loch is also rapidly silting up with the materials coming in with the River Cur.

Not unfrequently steeply-sloping deltas lie at the foot of crags high among the hills, their material being brought down in wet seasons along little gulleys which are too small to be named or marked in the map, e.g., at the foot of the crags ⅓ mile north-east of the top of Stronchullin Hill (164 south-west), on the south-west side of the burn rather more than ½ mile south-west of Stronchullin Hill, and ⅓ mile W.S.W. of Cnoc Breamanach (182 north-east). On the top of the watershed between Caol Ghleann and Stralachlan River there are delta deposits brought down by the streams from Tom a' Bhiorain. There are sand and gravel deposits filling up part of the old peaty slacks on the north-east side of Leanach ; 1 mile north-west of Leanach these are not less than 9 feet thick.

The deposits along some of the steeper streams consist of coarse and irregularly bedded gravel and boulders, but such as these are usually narrow, and have not often been mapped. The broader deposits all lie beside less steeply-flowing streams, and consist of less coarse material,—gravel, sand, or laminated sandy clay. The upper parts of the flats at present liable to floods usually consist of this last material.

The alluvium on the south-east side of Glen Lean (172 north-east) is often a fine clay of a deep-red colour. The hillsides are partly covered with a deep red boulder clay, and it is clear that the red alluvium has been derived from it : on wet days we have often seen the road, in places where it passes close under the steep glen sides, traversed by little streams carrying red clayey matter in suspension. A similar red alluvial clay occurs also ½ mile east of Rashfield (173 north-east).

The high terrace on the east side of Tamhnich burn, ½ mile west of A' Chruach (172 north-east), consists of gravel which is partly cemented by some black material. A similar black cement is not uncommon elsewhere, and, judging from examination of a sample obtained from ⅓ mile east of Ballimore (141 south) consists largely of the black oxide of manganese,* mixed with iron ores.

We have been often obliged to group several terraces under one sign and tint, owing to the impossibility of clearly distinguishing more than three tints of one colour, while the number of flats in any area may exceed this number. In such cases we have still had lines engraved to separate the different flats, where the scale of the Map allows it.

In some places the separating bank between two flats becomes quite indefinite, and is replaced by a gradual slope. In these cases we cease to draw a line between them, but keep the colour difference in a straight line, until we come to the next locality where there is a distinct bank between the two, provided this is not too far away.

Of the cultivated areas Glendaruel has a specially bad reputation for flooding, the coincidence of a high tide and a heavy spate leading almost inevitably to this result. We have also seen a large part of Glen Lean below the farm under water, also portions of the valleys above Loch Eck and Loch Goil, in spite of the earthen mounds built along the river sides to keep the water within bounds.

The flat by Ballimore(141) is in places only 3 feet above the ordinary water level, and for ¾ mile to the north-west it averages only about 5 feet above. Up the River Cur between Strachurmore and Ballemeanoch the slopes of the terraces increase, and their tops rise to a greater

---

* Dr John Murray and Mr Robert Irvine state ("On the Manganese Oxides and Manganese Nodules in Marine Deposits," *Trans. Roy. Soc. Edin.*, vol. xxxvii. p. 721), that manganese dioxide has been very commonly observed by them as a coating on the pebbles and sand in the channels of the river Clyde and its tributaries, and that traces of manganese were found in Glen Morag burn, both in a state of solution and also in an insoluble condition. In the different streams examined it was a rule that the manganese in solution was more abundant towards the head of the stream, especially near deposits of peat. They attribute its less abundance near the mouths. of the streams to the oxidation of the carbonate of manganese in the solution, and the consequent deposition of the black manganese dioxide on the stones of the bed of the stream, etc. For the original source of the manganese, whether in freshwater or marine deposits, they look to the rocks composing the earth's crust, and show how widely spread manganese is in these rocks. They give (*op. cit.*, p. 723) a list of rocks in which they have found manganese, including among these a mica schist from Glenmorag. This schist is stated to contain from 0·1 to 0·5 per cent. of manganous oxide (MnO). See also analysis of chlorite from St Catherine's, p. 64.

height above the present stream surface. It seems therefore that the stream must once have had a steeper fall than it has now; its course within the hilly area has been eroded and deepened while that in the more level parts has been less eroded, or even filled up in part by the material brought down from the hills.

The terraces between Glenmassan and Garrachra (163), slope more steeply than the course of the present burn, so that what is the lowest flat at Glenmassan becomes the second at Garrachra.

The surface of the flat between Loch Eck and the foot of Glen Branter burn is only 4 or 5 feet above the ordinary river level. The lower parts of the banks of the river are generally composed of fine even layers of sand, while the parts just above these, and the shoals and travelling material in the burn bed, are of gravel.

The alluvium above Loch Goil is in most places only a few feet above the ordinary river level. The materials in the flats always become coarser, as we should expect, when we approach the entrances of the more steeply flowing side streams.

A slight horizontal feature of erosion, about 2 feet above the ordinary surface of the loch, may be seen in places at the side of Loch Eck, e.g., between Dornoch Point and the foot of Coire Ealt.

Between 100 and 250 yards south-east of Inverchapel (163, south east) a flattish area extends in a south-west direction from Inverchapel burn. It is partly covered by gravel, and seems to have been used in times of flood as a channel into the river Echaig.

At the foot of Kinglas Water there are excellent examples of old river channels cutting through the higher flat. In the gardens a little more than ¼ mile south-west of Cairndow Church ship's anchors are said to have been found, thus testifying to the amount of silting up there has been.

Small outliers of higher flats surrounded by a lower one are not uncommon: an example occurs ½ mile above Garrachra (163).   C. T. C.

In the Glendaruel valley between Achanelid and Glendaruel House (162) some curious gravel mounds occur; they are noted on the published 1-inch Map 29 by the engraving of the words "gravel mounds." They vary in size from long serpentinous mounds to isolated hillocks. The River Ruel before reaching Achanelid flows in a south-westerly direction. Between Achanelid and Glendaruel it sweeps round and takes an approximate southerly course, and it is in the vicinity of the river bend that these mounds are met with. They have an average breadth of about 15 to 20 yards, and some of them attain a length of 150 yards. They attain a height of about 12 feet above the alluvial terrace. At present they are all planted with timber, and good sections cannot be observed. Their surface is however free from blocks, and where small sections can be seen they are seen to consist of rounded gravel similar to that of the terrace.

It has been suggested that these mounds may be glacial, but at present there is not sufficient evidence to decide the point.   J. B. H.

CHAPTER XXI.

PEAT, LANDSLIPS, BLOWN SAND, PREHISTORIC REMAINS.

Peat.

The amount of peat, especially in the north-east part of Cowal, is less than in many Highland areas, the sides of the hills being generally too steep and too rapidly drained to be favourable to its growth, and their tops often narrow and somewhat ridgy.

Peat areas are not coloured in the Maps except where they occupy basin-shaped or fairly well defined hollows. Among the wider, gently sloping hill tops which are covered with some thickness of peat, we may mention the following,—those between the head of Glen Tarsan and Gleann Laoigh; by the north margin of the 1-inch Map 29 near the "Creachan" of "Creachan Mor" (163); on the north-west side of Stronchullin Hill; on the south-west of the hornblende schists on Cruach Mhor (162); between Beinn Bhreac and Meallan Riabhach (183); on the top of Meallan Sithean (183); on the south side of Kilbride Hill (183); between Dunans burn (162) and Caol Ghleann on one side and Strath nan Lub, Eas Davain, and the heads of Glen Branter and Gleann Dubh on the other; north of the head of Eas Dubh (141). A peaty area about 5 miles in length extends from Garbhallt Lochain (151) to Lochan Chuilceachan (162), with a slope towards Loch Fine: it is ½ mile wide at its north extremity and widens out to 1 mile on the south side, and sends offshoots in various places along the dip slope towards Loch Fine. A mass of thick peat occurs between Cruach nan Gearran, and the main road between Otter Ferry and Glendaruel, which extends for about ¾ mile on the slope towards Loch Fine. In the part of Cowal between Loch Fine and the Kyles of Bute the hill peat is thicker and more extensive than usual on the west side of Beinn Capuill, the north side of Cnoc Dubh, and between this hill and the Acharosson Burn.

Some of the peat areas contain stems, etc., of birch at elevations considerably higher than any trees now growing in the neighbourhood, e.g., the areas ¼ mile north-west of Stronchullin Hill (163) at a height of 1500 feet or more, near the Horse Seat (183) at a height of about 1250 feet, ½ mile north-east of Maol Odhar (162) at a height of over 1000 feet.

Basin-shaped areas which are well-defined all round are not common. Many of the mosses mapped are well-defined on two or three sides, but thin away gradually in the other directions along gentle slopes. At the sides of most of them there are, in places, small delta deposits of alluvium brought down by side streams and partly spread over the peat. It is not usually possible to separate these deltas from the peat, the slope from the head of the delta down to the moss top being quite gradual.

Numerous peaty hollows have been mapped in the south west part of Cowal. Some of these have not well-defined margins on all sides, and may consist partly of ordinary hill peat overlying flat spreads of drift, especially those at or near the heads of streams, e.g., both north and south of Cnoc na Carraige (Kilbride); at the head of Loch na Melldalloch stream; south of Creag Mhor, by the side of the stream running into Glenan Bay; both north and south of Achanlochan (Tighnabruaich), etc. Some again consist probably of ordinary stream alluvium with a growth

of peat above.   There is a well-defined basin of peat to the east of Black
Harbour and S.S.W. of Ardmarnock House, which possibly overlies a
raised beach.   There are several fairly well-defined peaty hollows between
Low Stillaig and Loch Asgog, and north-west of Achanlochan.   In the
big moss between Kames and Millhouse stumps of trees were to be seen.
Other noticeable mosses lie east and south-east of Kilbride Church.

W. G., C. T. C., J. B. H.

There are cases where large streams have cut through mosses and formed
distinct gravel terraces in them at a lower level than the moss surface.
In other cases there is quite a gradual slope between the moss and the
terrace surface, the alluvium having spread over the moss to some
extent, or having been formed cotemporaneously with it.   The mosses
at the flat sides of streams have probably in great part grown on old
alluvia, the flatness of the surface prepared by the stream having furthered
the growth of moss.   In cases where we feel some confidence that this
has been the process of events, we omit any special indication of the peat
and map the whole area as alluvium.

The mosses earlier than, or cotemporaneous with, the river deposits at
their sides, have grown since the commencement of post-glacial times in
the hollows of uneven drift deposits or in rock basins.   For example, the
moss $\frac{1}{4}$ mile east of Polchorkan (142) makes an angle to the north, so
opposed to the general direction of the river adjacent, that it does not seem
likely that the flat which it occupies has been formed by this river's action.
The sudden expansion of the alluvium near Polchorkan seems to indicate
that there was here some area specially easy for the river to wind about
in and overflow, i.e., some low area of soft material.   We may either
regard the moss as a remnant of a once larger moss which has escaped
destruction by the river, because it lies so far off its natural path towards
the sea, or the moss and the alluvium may have been gradually formed
together, the moss in the part of the old hollow that lay out of the river
path and the alluvium in the part nearer that path.

The low watershed between Glen Lean and Glen Tarsan is partly
occupied by a peat moss.   The west side of the moss is separated from
the alluvium of Glen Tarsan burn by a well-defined margin, caused
apparently by the cutting action of the burn.   Part of the east margin is
also well defined from the deltas, etc., of some small side streams.   The
position of the moss, so near the watershed between the two glens, makes
it unlikely that since glacial times any large stream has traversed the area
occupied by it so as to denude a previously uneven site down to the flatness
favourable to peat growth.   More probably the flat area was there, much
as it is now, at the close of glacial times.

The moss on the watershed between the valley of Loch Eck and
Strachur, on the north side of the alluvium at Strachurmore, is another
instance of much the same kind, except that the boundary between
the alluvium of the Cur and the moss is less definite.   The shape of the
moss, the narrow constriction at the middle, together with the absence of
any signs of river deposits at its sides, excepting in a direction towards
the Cur, make it unlikely that any large stream has passed over the
watershed and prepared the site for the moss since glacial times.   We
would say rather that the surface features which led to the growth of the
moss are of glacial origin.   This is the more likely since there are a considerable number of peaty slacks or depressions to be noticed at the east
edge of the alluvium of the Cur south of Strachurmore.   These slacks are
not due to denudation by the river, but are hollows essentially due to the
uneven surface of drift deposits, moraines, etc., though some of them have

probably been subsequently modified by small side streams, or covered with alluvium by these streams.

The mosses a little more than 1 mile south of Carrick Castle, east of Tom Molach (153) and ¼ mile north-west of Cruach nam Miseag, probably represent sites of old lochs. The large moss on the watershed between Glen Laoigh and Strath nan Lub is also perhaps of this character. The north end of this is a prominent moraine-like mound (see p. 257).

The mosses on the hill tops ½ mile S.S.W. of Dun Mor (173), ¼ mile slightly west of north of Kilmarnock Hill (183), at the head of Curra Lochain, on the watershed between Strath nan Lub and Glen Branter, are none of them surrounded by higher ground on all sides.

At the head of Loch Restil there is a peat moss extending for about 100 yards from the loch end. Probably this does not represent an in-filled old bay for there is no stream near to have filled in such a bay. The peat has probably grown in an old hollow in the moraine drift edging the loch. There are peaty hollows of much the same character at the side of Lochan nan Cnaimh (153). The peaty flat at the head of Loch Loskin (184) is probably due in part to the in-filling of part of the loch. This case differs from the last mentioned in that there is a considerable burn near, bringing down alluvium into the loch, and still continuing to fill it up. But it is not likely that the loch ever extended as far as the north-west margin of the peaty area shown in the map, for the rise from the loch in a north-west direction is considerable, and the height of the surface becomes greater than that of the barrier at the south end of the loch.

C. T. C.

Landslips.

The position of the more important landslips have been indicated by writing on the maps.* In the 6-inch maps the boundaries of the slipped material were roughly drawn. The sites of these are formed by great hummocks and irregular ridges of slipped material covered by scattered blocks, and they have a very wild appearance in the landscape. It is perhaps most usual for the displaced schist masses to dip into the hill from whence they have come. This is seen at the head of the slip west of Carnach Mor (141). In the slip north-east of Strone Dearg (183), a little lochan, about 150 yards long and 40 broad, has been formed, and in another, in Hell's Glen, there is a narrow pool 50 yards long; but generally there is a very ready drainage through the slipped material, so that slip sites afford better feeding for the sheep than the adjacent hill slopes, and give rise near their bases to good springs of water. Such springs occur, e.g., in the slips on the north-west side of Beinn Mhòr, the south-east side of Caol Ghleann, the west side of Carnach Mor. They are also well fitted for rabbit warrens, the banks in some of them seeming to be quite alive with these animals.

The lower parts of the slipped material are elevated in a rubbish bank above the surrounding hill slopes. At the top there is usually a somewhat coombe shaped hollow in the crag, often itself in a somewhat shaken condition, and with deep narrow cracks at the back which the pedestrian has to be on his guard against. These cracks and also those in the lower material may extend underground for some distance, and form caves, some of which have been used in old time for dwellings, e.g., Paul's cave in Caol Ghleann, or as places of concealment for treasure and valuable documents, e.g., on Clach Beinn (163). Some of the caves in the parish of Lochgoilhead are described in the "New Statistical Account of Scotland," Argyle section, p. 703.

* See also one of the sketch maps at the end of the volume.

The following list gives the more important slip localities. The west side of Creachan Beag (163), sloping down to Glen Massan. The north side of Am Binnein (173). One-third of a mile north-west of Clachaig (173). The east side of Ballochyle Hill (173). The south-west side of Meallan Sithean, with its head extending for $\frac{3}{8}$ mile along the 1000 foot contour. The east side of Sgian Dubh (183). The north-east side of Strone Dearg (183), with the head extending $\frac{7}{12}$ mile at the level of 1500 feet or more, and the slipped stuff reaching down to Inverchaolain burn. The south-west and south-east sides of Kilmarnock Hill (194). The north side of Clach Beinn (163), extending down to the Coire an T burn. North-west, $\frac{1}{4}$ mile north, and $\frac{1}{2}$ mile north-east of Beinn Mhòr ; the head of the first is about 220 yards north-west of the hill top and the slipped material descends to the burn in Garrachra Glen. South-east side of Caol Ghleann from 1 mile north of Maol Odhar for $\frac{1}{2}$ mile north-east. North-west side of Clach Bheinn (153). One-third of a mile south-west of Cruach a' Bhuic. South side of Sugach Aodainn (152) between the "h" and "A." South-east side of Cnoc Coinnich (142), from between the 2250 and 2000 foot contours down to Loch Long, a distance of $1\frac{1}{3}$ mile. North side of Mullach Coire a' Chuir, the head extending nearly a mile from east to west, and up to the top of the ridge west of the Mullach. Five-eighths of a mile S.S.W. of Mullach Coire a' Chuir. South side of Allt Coire Odhair, near the "Allt." At the sides of Allt Criche (142). South side of Liogan, $\frac{1}{2}$ mile west, and close at the side of the "L" of "Liogan." West and north sides of Carnach Mor. The south-west side of Hell's Glen from near the top of Stob Liath. North and west sides of Beinn an Lochain (134). Three-quarters of a mile east and $\frac{5}{8}$ mile slightly south of west of Beinn Fhidleir. North side of Beinn an t-Seilich. One-half mile W.S.W. of the "1" of "Glen Kinglas." One-quarter mile north-east of Upper Clasheoin (126). $\frac{5}{8}$ mile S.S.E. and 1 mile north-east of Achadunan.

Many of the slips e.g. those on the north-east side of Strone Dearg, the south-east side of Kilmarnock Hill, north-west side of Beinn Mhor, south-east side of Caol Ghleann, south-east side of Cnoc Coinnich, south side of Allt Coire Odhar, and near Upper Clasheoin and Achadunan, are on hillsides the directions of slope of which agree with that of the foliation, a circumstance which must have helped towards the occurrence of the slip. But there seem at least an equal number which have descended down hill slopes at right angles to the foliation dip, e.g., the one in Hell's Glen, and that on Mullach Coire a' Chuir. And in some cases the slip is in a direction opposed to that of the dip, e.g., $\frac{3}{4}$ mile east of Binnein an Fhidleir, and $\frac{1}{2}$ mile north-east of Beinn Mhor. The margins of some of them are along basalt dykes from which the slipped material has broken away in a somewhat clean cut, e.g., the south margin of the slip on Strone Dearg ; or along lines of vein or crush, e.g., the head of the slip $\frac{1}{2}$ mile south-east of Lochgoilhead, portions of the north-east side of the slip $\frac{1}{4}$ mile north-east of Upper Clasheoin, and, apparently, the north-east side of the west slip on Caruach Mor.

The coombe shaped crags at the top remind us of those we may frequently see elsewhere, where now there is no trace of slipped material —crags high on the hills with conspicuously smoother areas below and at the sides, of which no satisfactory explanation can be given. Is it possible that these localities are sites of older slips, glacial or pre-glacial, the rubbish of which has now all been cleared away by glacial action?

The slipped material on the south-west side of Kilmarnock Hill (194) breaks the line of the 20 foot beach, so that the slip must be later than it. There is a tradition that the slip on the south side of the Kinglas, $\frac{1}{2}$ mile W.S.W. of the "1" of "Glen Kinglas," took place a little more than 200 years ago. From the slipped masses $\frac{1}{2}$ mile north-east of Beinn Mhor large masses still occasionally slip. A large block, about 11 yards by 7 yards by 7 yards, by the side of the stream $\frac{3}{4}$ mile E.N.E. of Beinn Mhor, was said by Mr Proudfoot, the late shepherd at Bernice, to have descended with others in 1885. It is very probable that portions of many of the other slips are also still liable to break away now and then, being in a state of equilibrium which is easily disturbed.

Small slips along the drift banks at the sides of streams and steeper hill slopes are very common. These generally take place after a time of unusually heavy rain. At the time the Survey was in progress, the coach road in Hell's Glen was blocked for several days by a small slip of this kind, and we noticed in more remote places various other slips which appeared to have taken place during the same wet season.   C. T. C.

Blown Sand.

The only area of blown sand noticed in the Cowal district occurs in Kilbride Bay or Bàgh Osde (203), west of Ardlamont Point, on the east of the mouth of the stream called Allt Osda. It extends along the bay for nearly half a mile from the stream, and its greatest breadth is about 150 yards. The sand, as is usually the case, forms a series of irregular shaped mounds, but none of these attain to any great height, probably not more than 10 or 12 feet. It has been accumulated on the lowest raised beach, part of the 20-foot.   W. G.

Prehistoric Remains.

Mr MacNaughton, the gamekeeper at Craigbrac (152), informed us that some years ago he found in the delta a little below Craigbrac House, a polished wedge-shaped instrument of stone, perhaps an old axe-head. The length was about $2\frac{1}{2}$ inches, and the extreme breadth about $1\frac{1}{2}$ inches. The stone was fine in grain and pale grey in colour, and according to our informant, not unlike some of the finer grained lamprophyre sheets of the neighbourhood.

In the National Museum of Antiquities of Scotland there is preserved (AF 138 in catalogue of 1892) a polished stone axe-head, $5\frac{1}{4}$ inches long by $2\frac{1}{2}$ inches broad, which is said to have been found at Strachur. The stone of which it is composed is a fine grained buff whinstone with a good number of amygdules, some of which are distinctly elongated. The sharp edge of the instrument is not symmetrical. Perhaps it has been worn back more at one side than another, and subsequently re-edged. In the same Museum there is also a stone knife (AA 12) from Strachur. The dimensions are $2\frac{3}{4}$ inches by $\frac{3}{4}$ inch. It is ground to a sharp cutting edge from both faces.

In the policies of Strachur House, there is a large standing stone of epidiorite. This is marked on the 6-inch Map. A little north-west of this there is another smaller standing stone, and it seems as if there may originally have been a group of such. At the head of Garrachra Glen (163), there is an old cairn about 100 yards east of the " n " of " Glen." Five-twelfths of a mile north-east of Strachur House there is a large mound by the side of the River Cur, which is supposed to be the site of an old fort. Another mound of very similar appearance occurs on the south-east side of the River Fine, $\frac{1}{3}$ mile N.N.E. of Achadunan. There seemed to be no mason work now remaining in either of these mounds. It is possible that both are natural in origin, though they have been subsequently modified and used as forts.

There is a round fort near the top of the wood, 1½ mile slightly east of
north of Castle Toward, at a height of over 1100 feet.   The fort is called
Buachailean in the 6-inch Map 194.   We are informed by Mr D. Whyte,
of Glasgow, that a number of stones from this fort were removed and
used in erecting the farmstead at Killellan.   Another round fort occurs
on a knoll, called Cnoc nam Fiantain in the 6-inch Map, 174, on the
shoulder of the hill ₁⁵₂ mile N.N.W. of Strone Point.   Mr Whyte states
that a flagstaff once stood on this knoll, and that at the time of its erection
a large quantity of wood cinders was discovered at a depth of 2 feet.   It
seemed as if the hillock had formerly been the site of an old signal station.
The position commands extensive views, particularly over the different
sea lochs adjacent,—both up and down the Clyde, and up Loch Long and
the Holy Loch.   C. T. C.

There is an unusual number of old forts and cairns in that part of
Cowal which lies within 1-inch Map 29 and borders the east side of
Loch Fine.   Those that occur on the north side of Kilfinan are detailed
in the following list :—
>   One-third of a mile north-west of Carn al Tilgidh (161).   A cairn con-
>   sisting of a big heap of stones at a height of 634 feet overlooking
>   Loch Fine.
>   One-third of a mile north of Largiemore.   Carn Evanachan.   A circular
>   heap of stones on the beach below high water mark.
>   Two-thirds of a mile E.S.E. of Largiemore.   Cairn on hill slope flanked
>   by two gorges.
>   One-quarter mile east of Otter Ferry.   Site of cairn.
>   One and a half mile E.S.E. of Otter Ferry.   Cairn Bàn.   A small cairn.
>   One hundred and fifty yards south-west of Ballimore House.   Fort on
>   round gravel mound.   Cnoc na Eoghainn.   Since used as a private
>   burial ground.
>   One mile east of Ballimore.   Fort on hill, Bàrr Iola.   The best pre-
>   served fort in the district ; part of the circular enclosing walls being
>   still standing.
>   Two miles east of Ballimore House.   Cairn at Meall Reamhar.   This
>   and the two preceding forts are in a line.
>   One and a quarter mile south of Ballimore.   Two forts within ¼ mile
>   of one another in a north and south line, on a ridge of limestone.
>   Both within ¼ mile of Loch Fine.
>   One and one-sixth mile south-east of Ballimore.   Cairn and standing
>   stone.
>   One and one-third mile south-east of Ballimore.   Fort.
>   One-half mile north-west of Kilfinan.   Cairn and standing stone on the
>   50 foot raised beach.
>   Nearly ½ mile N.N.E. of Glendaruel House.   Dùn an Oir.   Now used
>   as a family burial ground.   J. B. H.

There is a group of standing stones south of Polphail (192), one of them
10 feet high, another only a little over 2 feet.   There is another standing
stone among the woods east of Low Stillaig, and several to the west of
Kames.   Cairns are to be found to the S.S.E. of Ardmarnock House and
on the raised beach north of Eilean Aoidhe.
There is a vitrified fort on one of the Burnt Islands in the Kyles of
Bute, and another on an island in Loch Fine called Caisteal Aoidhe, near
Rudha Preasach.   The wall of the first-mentioned fort consists of pieces
of the mica schist of the district fused together.   A black or dark
coloured glass cements the schist pieces into a firm mass and also runs

into these pieces in thin layers, about $\frac{1}{20}$ inch thick. A specimen of the partially fused rock was sent to Mr Teall for microscopical examination (slides 5083 and 5084), and he reported in it as follows :—"The specimen consists of agglutinated fragments of the granulitic biotite-schists of the neighbourhood. The layers of biotite have fused to a dark coloured glass containing vesicles. Under the microscope the two principal constituents are seen to be glass and quartz. The glass varies from deep brown to colourless. It contains minute octohedra (spinelle). The quartz occurs in grains which have been strongly corroded by the glass." The vesicles in the specimen sometimes attain the size of peas, and are generally quite spherical in shape. Ordinary forts without vitrified walls are to be found—one in a plantation east of the road nearly a mile to the north of Ardmarnock House, and another, unmapped, over 20 yards in diameter, on the west side of Allt Osda, opposite Kilbride, at the western edge of a wood, and 200 yards from the stream. There is a fort called Creag a' Chasteil, at the edge of a crag 600 yards to the north-west of Low Stillaig, and another fort at Camp Cottage, Ardlamont.

There are no less than three crannogs or lake-dwellings in Asgog Loch. That nearly opposite the ruins of the old castle and about 70 yards from the shore is shortly described by Dr Robert Munro, in a paper in the *Proc. Soc. Antiquaries of Scotland*, vol. xv. pp. 205–222 (Notes on Crannogs or Lake-dwellings in Argyllshire, 1892–93). He states that this is known by the natives as *crannaig*, and that bars and huge beams of timber appeared in 1836, when the powder factory was started, the water being then at a lower level than ever before. He seems to have heard of the two others, but not to have seen them. They occur near the west end of the loch and were noticed by us in 1888, when the loch was unusually low. One of them is near the south-west end of the loch, and is 70 yards from the shore, nearly opposite the centre of that narrow side of the loch. We were able to walk out to it and found it 35 yards in diameter, nearly circular, and the piles distinctly visible, as was also a double row of piles leading for some yards from it towards the shore, but no piles were seen close to the loch side. Along the south-east side of the loch 200 yards from its southern outlet occurs another—about 90 yards apparently from the shore, but we were unable to reach it though we could see it distinctly, and also the double row of piles leading to it.

W. G.

CHAPTER XXII.

GEOLOGICAL ASPECTS OF THE SCENERY.

The "anticline" schist area even when of low elevation, as between Colintraive and the head of Loch Riddon, makes a much rougher, more craggy country than that at either side. Perhaps the greater alteration of the rocks may have made them harder and less readily subject to atmospheric wear. But, besides this, it seems possible that the crumpling of the prominent planes may have contributed to the same result,— water and other decomposing agents not having so many straight running channels to advance along. In this area the more micaceous and the more quartzose pebbly schist make about equally prominent crags, and as their colour is also much the same in the landscape, they cannot be confidently distinguished except on close approach.

On and near Cruach Tuirc the albite and greywacke schists within ¾ mile of the granite edge make a steep craggy slope on the south-east side of Glen Fine. The schists in question have, at equal distances from the base of the green beds, a common habit of forming rough rocky ground even far away from the granite, *e.g.*, on the two Beinn Lochains, Beinn Bheula, etc., and so it is doubtful to what extent we may attribute the roughness of Cruach Tuirc to the special mineral metamorphism in the vicinity of the granite. The schists near Achadunan make a smoother slope on the glen side than at Cruach Tuirc, but near the former place the foliation is in nearer agreement with the slope of the hill, and this of itself would make a considerable difference in the look of the slopes. Though Glen Fine is straight it does not always make the same angle with the foliation of the rocks traversed. Owing to the twist near the granite edge, the strike of foliation near Cruach Tuirc is nearly at right angles to the glen.

The granite weathers in large tabular blocks with rounded edges. It crumbles somewhat readily into a gravel-like soil which may form a covering of considerable thickness on the slopes below some of the crags. The approximate line between the granite and the schists is not easy to discern on first inspection. This is no doubt due partly to the alternation of granite and schists which occurs for a little breadth, there being no absolutely sharp line at which the main granite mass begins. The granite is crossed, too, by prominent regular joints which split the rock up into bands a few feet thick, and give the impression at first that we are dealing with bedded rocks. The general strike of the joints in the crags ₁₂ mile slightly east of north of Cruach Tuirc and in the river near the "F" of "Fine" is about north-east, and the dip to north-west at an angle of 28° but sometimes as low as 12°. As a matter of fact, of course, the bedding in the schists is nowhere dipping so steadily as these joints.

The Ardrishaig phyllites and the garnetiferous mica schists form together an unusually smooth area, the smoothest on the north-west side of "the anticline," and we think, too, smoother than that of the Dunoon phyllites on the south-east side. They are more shivery under the hammer than the Dunoon phyllites and are cut into more deeply by the burns, *e.g.*, by the burns that enter Loch Fine between Creag a' Phuill and St Catherine's. In the Ardrishaig phyllites the close intermixture of thin calcite layers and grains no doubt helps to this result. The agreement in direction also of the foliation and the chief hill slopes is closer than that in the Dunoon area. If the formation of the garnets and actinolites in the garnetiferous schist was originally attended with hardening of the schist, such as we commonly find in the neighbourhood of large igneous masses, it is clear that the extra hardness has now disappeared. Perhaps the shearing action which the garnets and actinolites have undergone since their formation would itself bring the schist back to a softer condition, supposing it had been hardened before then. It is curious that the same band of garnetiferous schist begins to form much rougher ground almost directly it reaches the north-west side of Loch Fine, on the slopes of the Brannie, Clachan Hill, and the hills further north-east. At the same time the garnets become larger and the actinolites more abundant. The dip being into the hill instead of down the hill would inevitably tend to produce the more rugged appearance, but the rock also becomes harder of itself as we go north-east, breaking less readily under the hammer and with more ring. It would seem that we are getting nearer the influence of some special altering and hardening agent, the effect of which in Cowal was either very much less or has been subsequently destroyed.

The hornblende schists and epidiorites usually form the more rocky exposures among the schists, and can often be distinguished by this means even at some distance. For examples we may refer to the fir-clad ridge on the east side of Loch Loskin (184) and the continuation of this band on the west side of this loch ; the outcrops on Kilmarnock Hill (194), near Blairmore and Strone Saul (173 and 183), Ardnagowan Cottage (141), Creggans, and St Catherine's ; the ridges between Loch Fine and the valley of Strathlachlan (151).

It is probable that the degree of roughness of the different outcrops is to some extent dependent on the amount of schistosity developed in the rock. The less schistose the rock, the more it projects. The more schistose it is, the more planes there are for the entry of water and other decomposing agents, and the more readily it weathers away. Some of the highly sheared chloritic bands on Loch Fine side are hardly to be distinguished in roughness from the other schists with which they are associated.

When the green beds are free from drift and peat, but not actually crag forming, they make rather greener and richer looking ground than the greywackes and phyllites. It is possible that they contain a rather greater proportion of lime and the alkalies, being presumably formed in greater proportion from old igneous rocks. For examples we may refer to the hill slopes south-east of Ardkinglas wood (134), and near Cairndow and Upper Clasheoin (126). In the south-west part of Cowal the green beds form somewhat lower and less rugged ground than the mica schists on their south-east side, and it is easy to follow the boundary between them.　　　　　　　　　　　　　　W. G., C. T. C., J. B. H.

The limestones of the Dunoon phyllite series never make swallow holes, nor, as far as we can call to mind, any green bands on the hillsides.

In some rare cases the Loch Tay limestone makes small incipient swallow holes, e.g., on the watershed a little more than $\frac{1}{2}$ mile slightly south of east of Cruach nan Capull (133), $\frac{1}{6}$ mile S.S.E. of Cruach nan Capull (152), doubtfully, and $\frac{3}{4}$ mile north of Cruach an Lochain (151). The part above the hornblende schist in the north part of Cowal makes steep banks with small limestone exposures near the top, and green grassy slopes below. The prominence of the bank is no doubt increased by the hornblende schist, which almost invariably accompanies the limestone. The best crags are those on the south side of the burn $\frac{1}{4}$ mile E.S.E. of the last " n " of " Sith an' t- Sluain " (152), 1 mile west of the "s" of "Socach Uachdarach" (141), and in Glendaruel. The limestone readily dissolves away in some positions near the surface and far in along joints, and gives rise to a soft ochreous residue, covering for a few inches or more the sounder heart of the rock. We have noticed this particularly 240 yards E.S.E. of Cruach nan Capull (152). Such ochreous residue, partly mixed with or colouring the local drift or soil cap, may be the first indication of the presence of the limestone.　　•　　C. T. C., J. B. H.

There is a sharp distinction between hillsides according as they slope with or opposite to the dip of the foliation ; the latter are much the steeper and more craggy. This is seen well on the sides of Glendaruel, on the north-west and south-east sides of the Bishop's Seat, on Creag Tharsuinn, Beinn Bheula, Cruach nan Capull (133), Stob an Eas, and many other hills. In Bishop's Seat, the difference is accentuated by the carry of the ice having been roughly at right angles to the hill crest, from the north-west to the south-east, so that the south-east side is, in accordance with the usual rule, more drift covered than the other.

As we steam down the Firth of Clyde, we cannot but be struck with the smooth aspect of the nearer Highlands, those near Dunoon, Blairmore, and Inellan, when compared with those in the distance, near the heads of Loch Long, Loch Goil, or in the direction of Loch Eck. More than one cause no doubt contributes to the contrast. The rougher areas are composed of the more altered and more crumpled anticline schists, the general rough weathering of which has already been spoken of. Besides this, it is only their more elevated and drift-free portions that can be seen in the distance, while in the nearer hills, the low drift covered slopes strike the eye at once, and in parts above the drift, it is to dip slopes chiefly that the observer is turned. It is quite possible, also, that the greater amount of glaciation to which the nearer and lower ground has been subjected as compared with the highest, may have tended to give them a smoother aspect.

The foliation strike is very steady in the district generally, and the main crag faces keep roughly on the same lines or parallel to one another. If we look from some commanding point on the north-west side of the "anticline," in a direction rather nearer east to west than the strike, we see bluff after bluff advancing with bold front to the south-east, like great waves one after the other, until in the far distance, they "go from less to less and vanish into light." Points of view which may be specially mentioned as bringing out this structure are those near the foot of Allt Glinne Mhoir (134), and on Beinn Dubhain looking towards Creag Tharsuinn. In the area between Ardmarnock and Portavaidue (192), some of the more gritty and pebbly bands form features or even crags for short distances. The features are accentuated by a series of small streams that run nearly along the strike.

But features which run with the general strike of foliation cannot be considered to represent bedding without actual examination of the rock on either side. As a matter of fact the great majority of features of this kind, which at a distance, in the early days of the Survey of the district, led one to expect they might indicate margins of beds, turned out on closer approach not to do so: they were due generally to crush lines, thin lamprophyre intrusions, or joint planes. The joint planes in this district are, however, not generally so prominent or so constant in direction, as to deceive one in this way. We have recorded them as unusually strong for a little space on the south side of the Coire Athaich burn (173) near Corrarsk; here they incline north-west, directly opposite to the foliation dip, at an angle of about 29°. About two miles slightly west of north of Corrarsk, there are prominent joints dipping in the opposite direction at about 30°. On the south-east face of Creag Tharsuinn (163) there are prominent joints dipping south-east, nearly with the hill slope.

Joints sloping nearly with the prominent foliation have at times acted as lines for the entry of lamprophyre intrusions (see p. 108). Such intrusions often form little vertical scars, weathering in retreat from the schist above and below, or narrow grassy ledges with steep schist crags on either side. At other times they have given rise to the formation of crags above them, e.g., in Creag an Fhithich (193). All these appearances are due to the more ready wearing away of the lamprophyres, compared with the schists in which they occur.

The basaltic dykes often form prominent features in the landscape. Owing to their superior hardness many of them stand up, as their name implies, in a wall-like way, particularly along coast lines, whether of the present day or of old beach times, and they have not unfrequently been mistaken for artificial structures. The wall-like form is often less marked inland, because, in cases where the dyke runs nearly with the contours, the hollow at one of the sides, that facing the upward slope of the locality, becomes gradually filled in by movements of the soil cap, etc. Other dykes resist weathering less than the surrounding rocks, and these form hollows and the lower parts of features. Why they should do so is not always clear, but the reason is probably often connected with the grain of the dyke rock, as suggested on p. 127, the coarser grained being more readily penetrated by water and other decomposing agents. Sufficient examples of these two classes of dykes are given in the place just cited. The more readily decomposed dykes often also form dry smooth banks and mounds clothed with short sweet grass, which form a striking contrast to the peaty and "benty" ground on either side. The broad east to west Dunoon Castle dyke affords good examples of this. We may suppose it due, partly to their porous character, and partly to their richness in lime and the alkalies.

Except in the north-east and south-west portions of Cowal, the basalt dykes are extremely abundant. The difference expresses itself somewhat markedly in the coast scenery of the Clyde and Loch Striven, as compared to that of Loch Long, Loch Goil, and the part of Loch Fine above Strachur; also, but to a less extent, on the hillsides adjacent to these lochs.

A short examination of any area where dykes and crush lines are abundant, e.g., the sides of Glendaruel and Glen Tarsan, will reveal what an important part these play in directing the course of the minor streams. The crush rocks and decomposing dykes are acted on more readily by denuding agents than the rocks at their sides; so they would form hollows in early times, and these would get gradually deepened by the waters running down the hill, until at last deep burns are formed. In some districts, we can hardly find a burn which is not running along either a dyke or a crush; and by a careful examination of the plain Ordnance Map before visiting the ground, we can usually make a good guess as to what burns, or parts of burns, are running along dykes or crushes, these being almost invariably straighter than the others. These straight burns are sometimes so deep and so steep sided, that one can hardly advance along them except in the driest seasons.

If we take position on some hillside on a fine clear day, and carefully scan the bare crags on the opposite side of the glen, we are struck with the immense number of small nicks and features, which traverse them in almost every direction. The crags on the north-west side of Glen Massan show good examples of these. We do not suppose that the generality of these are due to dykes or faults; in fact, in the particular instance alluded to, it is certain they are not; but even a strong joint or the thinnest crush line, without any appreciable throw, may be enough to give rise to them.

Some of the larger features coincide with lines of important fault, and, we have little doubt, may be attributed indirectly to them; either to their having thrown together rocks of different degrees of hardness, or to the softness of the crush rock itself. The boundary line of the Highland schists, as far as it occurs within this district, is a conspicuous example of this, and the courses of some of the other fault lines mentioned in ch. XV.–XVII. afford further illustrations. We may mention especially the north-west fault on the north-east side of the Saddle (133), the N.N.E. fault on the west shoulder of Ben Donich, the north to south fault ½ mile east of Lochgoilhead, the N.N.W. fault on the west side of Am Binnein (164), the Corran Lochan north to south fault, the Hell's Glen north-west fault, the Bernice Glen and Stuck burn faults

(163). None of the above mentioned are accompanied by dykes. At least an equal number could be mentioned accompanied by basalts, for in the area where basalts are abundant, nearly all the faults are also filled with them.

In considering the origin of some of the larger river valleys, such as those of Glen Massan, and Glendaruel, it does not seem that we can ascribe them to fault lines. There is often some breadth of obscure ground, free from rock exposures, in the bottoms of these glens, but the tracing of the beds on the glen sides is sufficient to show that no large faults exist along them. The direction of these glens depends perhaps on the slopes and ready direction of drainage of some old schist tableland, before it was carved out into separate hill shapes. It would seem that to the geologist's eye these valleys are not of very great age, or that at all events they have been greatly deepened in late times, for if they had existed at their present depth at the time of the intrusion of the tertiary dykes, the material of some of these could hardly fail to have been poured out on the surface and signs of it would still remain. (See p. 147.)

In the Loch Eck valley it is curious that a considerable fault (see p. 179) can be traced for some distance parallel to the loch, on its east side. At the north end of the loch there is a large fault (see p. 214) running from Ballimore to the alluvium near the first "a" of "Creag Bhaogh." The fault should, if continued south-east, run along the loch as far as Whistlefield. At Whistlefield there are crushes proceeding south-east from the loch in the direction of the pass over into Glen Finart. In a north-west direction the fault may continue across the low watershed between the Loch Eck valley and the valley that leads down from Ballemeanoch into Strachur Bay. This watershed is very low—I think only about 20 feet above the level of the Cur at Ballemeanoch—and is occupied partly by a peat moss and partly by drift with one or two small rock exposures near the north end of the moss.

The lowness of the watersheds separating some other of the important burns and glens, e.g., Glen Lean and Glen Tarsan, Glen Tarsan and Glen Massan, Lettermay burn and Leamhanin, Garrachra Glen and Glen Shellish, Caol Ghleann and Stralachlan River, is also notable. If we look on the sea lochs as representing old land valleys which are now submerged, we may add to the above the two Kyles of Bute : the deepest channel connecting these is that on the south side of the Burnt Islands, but this is only 3¼ fathoms near its centre $\left(\text{Admiralty Chart} \frac{\text{Sheet 2}}{2134} \text{DE}\right)$ the channel on the north side of these islands is 2 fathoms : the west Kyle is the natural continuation of Loch Riddon, and west of the Burnt Islands attains a depth of from 21 to 27 fathoms.

Perhaps we may suppose in the above cases that two adjacent streams have gradually denuded backwards towards one another, and thus gradually lowered the surface of the area between. We know of no special lines of weakness at the three first mentioned watersheds, but the last two are partly traversed by N.N.E. crushes.   C. T. C.

The upper part of Loch Fine between Otter Ferry and Ardno runs almost with the strike of the beds, lying in the comparatively easily eroded Ardrishaig phyllite series. A set of prominent N.N.E. faults come down to its south-east shore, e.g., the Coire No fault, the crushes accompanying the north-west graphite schist, the N.N.E. fault near Ardchyline, the faults ½ mile west of An Carr, but we have no reason to suppose that anyone of them continues along it for more than 3 to 4 miles —this is the distance the Coire No fault continues along the loch, and the alluvium at the head of it, before it reaches the north-west side of the valley, between Achadunan and Newton Hill. The Coire No fault (see p. 222) is the south-west extension of the great Tyndrum fault, the course of which has probably determined the line of drainage which has resulted in the excavation of the upper part of the valley of Glen Fine, within 1-inch Map 46. C. T. C., J. B. H.

The north-west fault near the Saddle may be supposed to be the same as that which enters Loch Goil near Drimsynie. The part of the loch between Stuckbeg and Drimsynie runs in much the same direction as the fault and perhaps the original formation of it was influenced by the fault. A little north of Stuckbeg there are several north to south faults coming down to the loch from the north. South of these, and after the Drim-synie-Saddle fault has reached the east side of the loch, there is a slight bend in the direction of the loch almost into parallelism with the north to south faults. The changed path continues as far as Carrick Castle, where one of the north to south faults comes out of the loch again.

Our former colleague, Mr H. M. Cadell, has called attention ("The Dumbartonshire Highlands." *Scottish Geographical Magazine*, vol. 2., p. 336) to the number of large valleys in Cowal and Dumbartonshire which run south-east or S.S.E., in directions often continued across watersheds, as that of the upper part of Loch Eck and Glen Finart, or across both low watersheds and valleys which run in different directions, as the valley of Loch Goil and the Gareloch, Coilessan burn and the Douglas water (1-inch map 38). He looks back to a time when perhaps the original water parting in this district between the east and west coasts of Scotland was a tract of ground in the neighbourhood of what is now Loch Fine. He says "Before Loch Lomond and Loch Long came into existence all the country east of Inverary, and northwards to the Arrochar Highlands and Glen Falloch, was probably drained by the Forth, and not as at present by the Clyde." He supposes the direction of original slope of the table-land out of which the valleys were carved was nearly south-east, somewhat oblique to the strike of the rocks. The different bands of soft rock were more easily worn away in a south-west direction than in a north-east direction, because the slope of the ground more nearly coincided with the former than the latter. Hence the general direction of some of the valleys is S.S.E., or even south, rather than south-east. It is further suggested that Loch Long is of comparatively recent origin compared to Loch Goil and the Gareloch, and is due to erosion by ice.

The well marked fringe of raised beaches along the coast cannot fail to attract attention. These have rendered it easy to make good level roads, and also formed convenient sites for the villas and gardens which abound along the Clyde, etc. But the frequent alteration of level indicated by them has had the effect of interrupting the cutting action of the sea, and setting it to work again at new levels, so that high cliffs have rarely been formed. This limitation of height is from a geological point of view somewhat to be regretted, as the coast sections are on a smaller scale than they would otherwise have been.

Owing to the "carry" of the ice having been down rather than across the more important glens and sea lochs one does not see so clearly as in some districts that the greater spreads of boulder clay are on the lee sides of the carry. An instance is afforded by the upper part of Glen Kin (174), but even here the extra smoothness of the north-west side is not solely due to the drift distribution, but also to the structure in the schists, the north-west side having a slope in the same direction as the dip of foliation.

As stated, p. 243, it is not safe to attribute all the smooth slopes without rock exposures to boulder clay or other drift. It is clear that even when there is no drift the rock on the hillsides has often been so planed and polished, that it now presents a smooth grass covered face and one would not suspect how near the surface it really is.

Moraines have quite a different influence on the landscape than the boulder clay. Instead of the smooth slopes of the latter we see a congregation of rather steep sided ridges and hummocks with very little order in their arrangement, and often strewn over with blocks lying in fantastic positions, the whole assemblage presenting a scene of disorder and dreariness.

The moraines in the south-west part of the area are not numerous and do not affect the scenery to any marked extent. In the north-east part the hills are higher and the rainfall is greater, and we may suppose that in glacial times the snowfall was also greater. The later glaciation in this part was on a much more extensive scale than in the south-west, but still hardly came up to that in some part of the Highlands. The main areas of the occurrence of moraines are detailed on pp. 240, 241, 256-259. Good examples are seen by the traveller on the coach roads between Lochgoilhead and St Catherine's, Arrochar and Cairndow, the head of Loch Eck and Strachur. Those by the side of the first-mentioned road are particularly rough and boulder strewn.

The torn crags at the head of landslips, and the huge banks and masses of fallen rock below them, make some of the wildest scenery in the district. The two big crags on the north side of Beinn an Lochain (134) are both the heads of landslips. So too are the crags on the north side of the glen, $\frac{3}{4}$ mile east and $\frac{5}{8}$ mile slightly south of west of Binnein an Fhidleir (134). The smaller but better defined slip on the south side of the same glen, about $\frac{1}{2}$ mile W.S.W. of the "l" of "Glen" is still more prominent, being so near the roadside. This is the slip that is said to have taken place about 200 years ago. One quarter of a mile north-east of Upper Clasheoin (126) another comes down almost to the raised beach, from about the 1250-foot contour.

The south-west side of Hell's Glen, from near the top of Stob Liath, is one great mass of tumbled blocks. The rock at which the coach horses stop to drink, out of "Moses' well," is one of these blocks, but not nearly so large as some. The bottom of the slipped material is nearly $\frac{1}{2}$ mile long. The sharp vertical face of An Stiobull (The Steeple) $\frac{1}{2}$ mile south-east of Lochgoilhead is the head of a slip.

The north-east side of Loch Long is particularly marked by landslips. On the south-east side of Cnoc Coinnich there is one, both sides of which can be traced to the loch side from between the 2250 and 2000 foot contours—a distance of $1\frac{1}{2}$ mile. Outside the area being described, the Brack and Ben Arthur (The Cobbler) are in a still more ruinous condition.

On the west side of Loch Eck there is a very rough slip on the north side of Clach Beinn. One can see the large tumbled blocks quite clearly from the steamer deck.

The above form only a small proportion of the whole lot of slips, and have been selected for mention merely because they are the most readily discerned from the common tourist routes. A detailed list of the others is given on p. 276. The causes which appear to have led to the formation of some of them are also mentioned.   C. T. C.

CHAPTER XXIII.

ECONOMIC RESOURCES.

The fault breccias, crush lines, and veins in the schists are rarely accompanied with useful mineral, and, as far as known, never in quantities to repay working. Three trial levels were driven by Mr Duncan, the former proprietor of Benmore, into crush lines near Blairmore burn, but met with such little success that they may well act as deterrent examples. The adits are engraved in the published 1-inch Map 29 : one occurs just on the north side of Blairmore burn, another on the hillside ¼ mile slightly east of north of this, and the third at the west side of the south branch of Stronchullin burn ¾ mile slightly south of the "S" of "Stronchullin." A fourth has been unfortunately engraved, owing to an error which escaped notice, a little north of that last mentioned. We have to thank Mr Wood, Mr Duncan's factor, for particulars about the trials, and for accompanying us into two of them after their abandonment. The level in Stronchullin was driven for 858 feet, along a crush line the whole way, and various short cross cuts were also made where subsidiary crushes seemed to be coming in. A small wheel for crushing the ore was erected near the mouth, but did not appear to have been ever used. In both the other levels small specimens of argentiferous galena were found here and there, about half an inch long ; the main crushes in each appear to be the same ; the more northerly level follows the crush for 104 feet ; the south level cuts the main crush at a distance of 225 feet from the mouth, but another at half this distance.

In the schist quarry ⅒ mile south of the "y" of "Trinity" (Dunoon) there is a thin vein, hading north-west, which contains small specks of galena mixed with quartz and calcite.

Veins of galena were formerly worked near Tyndrum (1-inch Map 46) along the N.N.E. disturbances which pass by there. It is interesting to observe that some of the N.N.E. crushes which continue into this area from Tyndrum are also accompanied with small strings of this mineral, though never in such quantity as to have induced trial of them. Three-quarters of a mile N.N.E. of Bridgend (141 and 152) a N.N.E. crush about 6 feet wide, hading west and accompanied with a felsite dyke (see p. 104), is said to contain strings of argentiferous galena. We did not observe any, but we think the information was correct. We noticed one or two thin strings of this mineral too, in the Loch Tay limestone in the upper part of Coire Nò (133).

In the north part of the area, not far from the edge of the granite, there are some galena veins which appear rather more promising than those yet mentioned. In a steep crag ½ mile slightly south of west of the Eagles' Fall, at intervals for a breadth of about 4 feet in the lower part of the scar, there are veins of barytes, nearly vertical, and running north-west. Some of these also contain a ferriferous carbonate, small specks of copper pyrites as large as small shot, and galena in isolated crystals or short ribs. The best of the veins noticed was 6 to 8 inches wide, and the ribs of galena in it were in places rather over 1 inch in breadth.

In the course of a streamlet (not marked in the Map), ⅜ mile slightly south of east of Achadunan (126) there is a vein of barytes about 1 foot thick, which contains little strings of galena ¼ inch wide. The direction is apparently north-east. This occurrence was kindly called to our attention by Mr Brodie, the chief game-keeper on the Ardkinglas estate.

In the green beds ⅔ mile W.S.W. of the "M" of "Garbb allt Mor" (126) there is a 1 foot east to west vein of calcite and ferriferous carbonate, and thinner parallel veins in which there are occasional specks of galena.

On the shore of Loch Long 120 yards north-east of Mark (153), there is a north to south crush with a ferriferous carbonate in a phyllitic mica schist. A few yards east of the crush there are flattish strings, about 1 inch thick, which cross the foliation of the schist. These are composed of quartz, specular iron, magnetite, pyrrhotine and chalcopyrite.

The pyritous schist in the Ardrishaig phyllite series on the shore of Loch Fine 70 yards N.N.W. of M'Phun's Cairn and south east of Creggans Point (141) is mentioned on p. 58. In the first locality small proportions of galena, blende, silver and gold also occur. The results of an assay of a sample are given on the page quoted. The occurrences seem quite an integral part of the schist series and not dependent on the proximity of any crush line. They are perhaps more promising than any other of the metalliferous veins in the district, and it might in certain circumstances be worth while to further explore them.

The occurrences of copper ore in Glendaruel, and the Kyles of Bute, are mentioned elsewhere, p. 81. They are interesting from their mode of occurrence, forming apparently an integral portion of the schist, but are on far too small a scale to be of economic value. The copper pyrites found here and there along veins and crush lines is also generally in very small specks. Three such occurrences have been already mentioned, one in the fault rock in the burn ¼ mile W.N.W. of the Ordnance Station 697 (184) (p. 173), another in the quarry in the Bull Rock greywacke on the south of Dunoon, p. 178, and another in the galena vein ½ mile slightly south of west of the Eagles' Fall. The most easterly crush rock in the burn ½ mile south-east of the "Badd" (184), also contains small specks of this mineral mixed with quartz and ferriferous carbonate.

Small quantities of ferriferous carbonate and specular iron are found along many of the fault lines and in thin veins. Of the former we may mention especially a vein about 1 foot thick that occurs near the head of the burn on the north-east side of Cruach na Capull (183) : in one place a good deal of iron pyrites is mixed with this. A thick rib of quartz and ferriferous carbonate with iron pyrites crosses the stream just below the junction of the three burns ¾ mile east of the Badd. Small quantities also occur in crushes in the following places,—⅝ mile north-east of Rashfield (174), ⅓ mile S.S.W. of Garrachoran (crush not engraved), ⅝ mile S.S.E. of the top of Cruach nau Capull (183), between Mark and Carraig nan Ròn (6 inches wide at a point ⅓ mile N.N.E. of the Carraig), in thin "flats," sometimes 6 inches thick, near the north-west crush ½ mile E.S.E. of the Saddle (153), in the north-east crush ½ mile north-east of the "c" of "Beinn Bhreac" (153), in the N.N.E. crush ¼ mile S.S.W. of Corriesyke (142), in the Cur 130 yards east of Socach (141), in Liogan by the "g," in the burn north of Criogan by the "g," in the N.N.E. crush in Dunans burn, in the N.N.E. crush ½ mile north-east of An Cruachan (162), and many other places.

Specular iron is seen in the veins mixed with quartz and other minerals on the shore near the Bull Rock, south of Dunoon, in the greywacke quarry near the Bull Rock, and in the veins with chalcopyrite, etc., 120 yards north-east of Mark. It is probably common in veins at a little depth under the surface, and the staining derived from it is the cause, we may suppose, of the deep red colour often seen in the crush lines of the district, *e.g.*, on the west side of Meall Buidhe (173), ½ mile south-east of Uig (173), and ⅛ mile south-west of the "e" of "Coire Athaich" (173).

In spite of the scantiness of these ores it is not impossible that in old times some of the better exposures among them may have been used by the clans for smelting down into claymores, etc. Here and there small slag heaps of old bloomeries can still be recognized. Two such occur near together on either side of the burn by the "e" of "Gleann Laoigh" (163) : there is plenty of wood charcoal lying about near the east heap. Others were noticed close to the shore ¼ mile north-east of Eilean Dearg (Loch Riddon), by the first "n" of "Glen Finart" (164), by the burn at the 400 foot contour E.S.E. of Leak (151), in the wood $\frac{1}{10}$ mile north of the "s" of "Creggans" (141), and ⅓ mile W.S.W. of the "g" of Gleann Laoigh."* But we cannot say that in any of these cases we could find any signs of old mining, or promising veins in the neighbourhood, so that it may be that ores foreign to the district have been imported, and smelted with the help of charcoal from the native wood. It is stated (W. I. MacAdam. *Trans. Geol. Soc. Edin.*, vol. iv., 1881, p. 95, "Notice of veins of Specular Iron Ore at Strachur, Argyleshire"), that so important was the oak wood that used to grow all over the lower ground, that, long after the use of native ores had been given up, red iron ore was shipped from Cumberland to Furnace (1 inch map 37) for smelting. And from this the name "Furnace" is no doubt derived. Mr MacAdam gives analyses of two specimens of specular iron ore from Strachur. The exact locality is not stated. Specimen 1 was picked ore, as pure and free from rock as possible. Specimen 2 was the ore as found in the veins. To these are added analyses of mica schist close to the veins (3), and 6 feet from the veins (4). The results are as follows :—

	1	2	3	4
Fe_2O_3	88·83	37·51	16·03	15·18
CaO	trace	0·28	0·38	0·36
MgO	trace	trace	0·98	0·84
SO_3	0·16	0·25	0·12	0·09
SiO_2, and Silicates	10·92	61·79	82·49	83·53
	99·91	99·83	100·00	100·00

Barytes sometimes occurs with the fault breccias, but not commonly. In the south part of the area the thickest rib noticed was in the crush crossing Gleann Laoigh burn about 50 yards north of where the dyke near the "g" of "Laoigh" leaves the burn. This is 4 inches wide. Small strings and plumose aggregates of it also occur in the crush rock on either side of the above dyke, at the side of the dyke ⅓ mile north-east of Meall an Fharaidh (173), and in the burn in Drynain Glen (164), south of the "a." In the north part of the district, near the granite, it is perhaps commoner than elsewhere. Some instances of its occurrence have been already mentioned in association with galena and copper pyrites.

A magnesian limestone in the Upper Old Red Sandstone near Toward was quarried by Messrs Merry and Cunningham, in 1886, for lining their basic Bessemer steel converters, but as the sample lot of some 30 tons showed 10·8 per cent. of siliceous matter they did not use any of it.

---

* Since the above was written Mr Charles Mackintosh, the borough surveyor of Dunoon, has mentioned to us two heaps apparently of the kind under description, which were met with in extending the Dunoon Waterworks. These are both near together about ¼ mile west of the bottom of the "D" of "Dunoon." Mr James Tod of Rashfield has also reminded us of a similar heap on the hillside $\frac{1}{15}$ mile E.N.E. of Rashfield (173 and 174).

T

For particulars of the locality quarried, and for the analyses kindly furnished by Mr Main of Messrs Merry and Cunningham, refer to p. 94. We were informed by Mr Main that the samples broken by hand made him think that a purer stone might yet be got in bulk, and that it was in contemplation to make arrangements for further trials on the west side of the burn—the opposite side to that from which the stone was quarried.

The Dunoon limestones do not appear to have been ever quarried except in one place, by the junction of the little burns ⅓ mile W.N.W. of the " D " of " Dunoon " (town). Even the quarries in the thicker and much better limestone of Glendaruel, etc., the Loch Tay limestone, which used to be extensively worked for lime burning, are now very generally disused. The amount of lime used for agricultural purposes is much less now than formerly, and is generally imported ready made. The old lime kiln on the roadside south-west of Glenlean was never used for local stone, but limestone was imported from Ireland, brought to the head of Loch Striven by ship, and thence carted to the kiln. Judging from the abundant chalk flints still noticed on walking over the old ploughed fields this source must once have been largely drawn on. C.T.C.

The Loch Tay limestone at the hamlet of Glenslnan (141) is still occasionally burnt into lime for agricultural use, at an adjacent kiln. Most of the limestone now quarried seems to be used for road metal. There are 2 quarries on the north-west side of the burn, and another, ½ mile above the hamlet, on the south-east side. The latter has evidently been long since abandoned. At the time of the Survey only the quarry nearest to the hamlet was being used. Mr W. I. Macadam (" On the Chemical Composition of certain Limestone Beds at Ballimore, Argyllshire." *Trans. Edin. Geol. Soc.*, vol. iv., 1883, p. 101) gives some analyses of certain bands in the limestone, and vein rocks going through it. Two of these analyses are quoted on p. 47. The main band contains 65·44% of carbonate of lime, a little iron and magnesia, and 29·63% of silica and silicates. Mr Macadam states that, beside the kiln at present used, " there are several old and disused ones, the formation of which is peculiar, they being so formed that a continuous process is impossible and thus the fires had to be lighted before each charge, the contents burned, and the whole cooled before removing the charge." There are several old quarries in the Loch Tay limestone south of Kilfinan (181) and it seems to have been formerly burnt for lime as there is an old limekiln near the roadside north of Drum. A big quarry has been opened in the limestone at Otter Ferry : this provides an excellent road metal and the roads in this district are almost entirely repaired from this source. There are various old limestone quarries, now abandoned, at the sides of Kilbridemore burn (162), which formerly were largely wrought for agricultural purposes by the farmers of Glendaruel.

Some of the old slate quarries in the Dunoon phyllite series and the adjoining phyllites are elsewhere mentioned, pp. 30, 32. None of these are now worked, as they cannot compete with the slates imported from other districts. The cheapness of freight at the present day has brought into such close proximity and rivalry districts which are geographically far apart, that only those best fitted for their respective products and industries can retain even the local market.

To state this is almost a truism, but still we were hardly prepared for the great extent to which some of the local products have consequently fallen into neglect. In spite of the numberless basaltic dykes of the district a good deal of the road metal used in the south-east part of Cowal is shipped from Gourock, and the cost is considered to be lessened by this means. The Gourock stone is readily distinguished from the various local whin stones, and is carted up for considerable distances from its landing places. Among the basaltic dykes which have been quarried we may mention the following,—those ⅓ mile south-west of the foot of Loch Eck, ⅙ mile south of Auchateggan (172), ¹⁄₁₂ mile east of Auchenbreck (172), ¼ mile slightly west of north of Ardtarag (173), on the raised beach ¼ mile slightly north of west of Toward cottages, at the landward margin of the 20-foot beach, ⅛ mile E.S.E. of Cluniter (195), 200 yards E.N.E. of Leanach (151), by the roadside near Drum (181) and Craignafeich (191 and 192), and on the west side of Loch Riddon south of upper Callow.

We cannot call to mind many instances of lamprophyres having been quarried in this area. We should suppose that many of them would make good metal, but only few of them crop out conveniently for use along the roadsides. The band in the limestone quarry nearest to the hamlet of Glensluan is quarried together with the limestone, and a mixture of the two rocks applied to the roads near. A band near Auchanaskioch (181) has also been quarried. A clay, apparently formed from a crushed decomposed lamprophyre, occurs in a line of fault ⅜ mile east of Socach (141), and was formerly known locally as a good pottery clay (see p. 109).

The sheet of felsite near St Catherine's has been quarried, both for road metal and field dykes, a little below the road near the top of the "h" of "Tighe Claddich" (133), and on the hill ⅛ mile south-east of the "s" of St Catherine's. Half a mile slightly north of east of Ardno (133) a felsite sheet has been quarried, and metal from it is carried over the watershed for use on the Hell's Glen road. The extensive quarries in the somewhat similar rock lower down Loch Fine, at Furnace and Crarae, are all on the north-west side of the loch and so outside Cowal. There are no exposures on the south-east side at all so extensive, or so conveniently situated for shipping. There is a fair sized quarry in the felsite south of Loch na Melldalloch (181), which is worked for the roads.

The granite has not been quarried anywhere. The nearest part of it is 3 miles from the head of Loch Fine. Some varieties would probably serve well for ornamental purposes, being rather like the Shap Fell granite of Westmoreland, but with even larger porphyritic felspars.

The hornblende schist in the fir-clad ridge east of Loch Loskin has been extensively quarried in several places, and supplies most of the road metal used in the town of Dunoon.

The north-west outcrop of epidiorite at St Catherine's has been largely quarried, and is said to have supplied most of the stone used in building Inveraray Castle. The rock is evidently liable to vary in character a good deal, both structurally and in chemical composition. The more schistose and fissile parts are not so suited for building purposes. The microscopical characters of the variety preferred are described on p. 64. The rock is called by Mr Teall a tremolite-chlorite-rock. It is said in the neighbourhood that the new quarry, close to the road leading south-east from the side of the inn, did not furnish such good stone as the old quarry in the garden on the west side of the ruined chapel of St Catherine's. Besides these quarries, there are others at the back of the raised beach south and south-west of the old chapel. The stone has a great local reputation. It is soft and easy to cut, but yet offers great resistance to atmospheric weathering. Many of the bridges in the neighbourhood are built of a rather schistose variety of it, or some similar stone, and the depths of the names cut into them, and the dates added, show how easy the stone is to cut and how long it retains tool marks. These varieties remind one somewhat of the stone of which the old crosses of Iona and Kilchoman (Islay) have been formed.

The somewhat similar epidiorite bands near Creggans (141), have been worked for building stone near the west end of the wood, and in some older quarries further inland. There are also small quarries for general purposes in the epidiorites south of Kilfinan (181).

Small quarries for road metal are not unfrequently opened out in schist outcrops which happen to be conveniently situated at the road side. One third of a mile N.N.W. of Inveronich an albite schist has been extensively quarried, partly for this purpose and partly for building stone.

The more massive schists, quartzose or pebbly beds, or the green beds, are generally used for building purposes in the schist area. Some of the latter may tool with comparative facility, but with the other rocks mentioned it is hard to prepare even an approximately smooth face, and so the outsides of buildings are usually covered with cement or fine concrete, and subsequently painted. For the same reason imported freestone is used for window sills, door lintels, etc. Most of the stone used in building Dunans Castle (162), was procured from a tough pebbly green bed in the north-west banks of Dunans burn ½ mile south-west of Caol-ghleann. It is said to have given great satisfaction. The building stone chiefly used in Dunoon comes from a quarry in a massive portion of the Bull Rock greywacke schist, a little north of the Bull Rock. There are various large quarries of pebbly schist within the phyllite series near the town. One of these lies near the "d" of "Castle Crawford," another in the wood ¼ mile north of the first "o" of "Dunoon," another just north-east of the same letter "o," another ¹⁄₁₀ mile south of the "y" of "Trinity." Of these four the second was the only one in active work at the time the Survey was being carried on. In the south-west part of Cowal the more gritty schists have been quarried for building purposes in several places, but not on a large scale; among the more important, are those of Ardlamont west of Point Farm, and between Corra and Craig Cottage, at Acharosson, north-east of Auchanaskioch, Kilbride, Auchagoyl, 400 yards north of Ardmarnock House, Tighnabruaich, etc.

The red sandstone near Inellan and Toward is generally too full of bits of schist and other rock to make good building stone, particularly as these bits are not firmly cemented in. It has been quarried close to Toward Church, together with a basalt dyke, and we think also at the back of the raised beach near Chapel Hall.

The main areas of peat are detailed on pp. 273-274. Nearly all these are some distance off the farms in the glens, and at the time of the Survey we think there were only 3 houses in the part east of Glendaruel where peat was in common use for fuel. These were the shepherd's house at the head of Glen Tarsan, Upper Duillater, the house near the "t" of "Duillater" (164), and Caol-ghleann (152). It is perhaps worth mentioning that in Ardrishaig, 1-inch Map 29, but just without Cowal, the peat that the whisky distillery used was at the time of the Survey imported from Ireland. Along the south-west coast of Cowal bordering on Loch Fine peat is pretty generally used, and there is abundance of it in the peat flats marked in the Map, and elsewhere.

The clay shell-bed which occurs on the shore between Newton and Strachur is stated (p. 261) to be sometimes used locally as a hearthstone, to whiten the hearths before the fire. Various other exposures of similar beds are mentioned on pp. 245-247. These beds are often formed of a fine stiff clay, fairly free from stones, and pale grey in colour. They might, perhaps, make good pottery clay.

The sand in the present beach at the head of the Holy Loch was at the time of the Survey being shipped for making the moulds for iron castings. The gravel in the beach of the west bay of Dunoon is largely used for the garden paths of the numerous villas adjoining. Excellent gravel is also obtained from the beaches of Kilfinan Bay and Otter Ferry. At Sand-hank (174) slightly more than ¼ mile south-east of the "k" of "Sand-bank" there is an extensive pit in the gravel and sand of the 100-foot beach. Pits in the same beach have also been opened out about ½ mile south of Gortanausaig (194) and near Blairmore Farm. There is a good section of fine sand within the 100-foot beach on the south side of Dunoon, ¹⁄₁₂ mile couth east of the "y" of " Trinity," and this sand has sometimes been used for mixing with lime for building purposes. In the south-west part of Cowal sand and gravel pits have been opened in the raised beaches at Upper Callow, several places near Millhouse and in the valley of Allt Osda, near Stillaig Farm, Port Leathan, Ardmarnock Ho., etc.

The reservoir that supplies the towns of Dunoon and Kirn with water is situated on the course of the burn that runs into the west bay of Dunoon : the centre of the embankment is about ¼ mile south-west of the "D" of "Dunoon" and the reservoir extends up the burn for about 330 yards. With the exception of some narrow burn channels the site is in drift,—the lower part stiff boulder clay and the upper part morainic. We understand that at present (1896) an additional reservoir is being constructed further up the burn. The reservoirs that supply Iuellan and Sandbank are along the burns ½ mile south-west of Garrow-choran Hill, and ½ mile south-east of Finbracken Hill respectively. Both these reservoirs and also the one at Dunoon have been constructed since the Ordnance Survey of the district was completed, and they are not shown on the 6-inch or 1-inch maps. We believe that a storage reservoir for the district of Blairmore and Strone is now being constructed.

There are two dams or reservoirs used at the Kames Powder Works (193), on the Creag-an-fhithich burn E.S.E. of Loch na Melidalloch. and the water of Loch Asgog is also made use of for the same purpose. The Tighnabruaich reservoir for the supply of water to the town is partly in boulder clay, on the stream called Allt Mor, west of Barr Liath, and there is another reservoir further down the same stream near Middle Ineus. There is another reservoir in red till on the moor ½ mile south of Kames Farm.

The amount of arable land is small, nearly all of the area being given over to sheep farming. The winter climate is generally mild and snow does not lie long except near the hill tops, but there is little feeding in the winter, and near the end of autumn the lambs of the preceding spring have to be sent away for wintering, or the death-rate among them is excessive. On some of the hills in the north-east part of the district there is very little heather, and even the sheep farmers sometimes express a wish there were rather more of it, for the sake of the mixed feeding it affords. In this part the grouse shooting is of little value. Red deer are often seen on the hills north of Kinglas Water and occasionally considerably further south, but these have wandered in from neighbouring forests on the north. There is no area from which sheep are excluded for the sake of the deer.

The patches of arable land are generally on the marine terraces or alluvial flats, or the expanses of boulder clay adjoining these areas. Among the larger patches we may mention the valley of the Echaig, the Lochgoilhead valley below Polchorkan, the lower part of Glen Finart, the valley between Ballimore and Strachur Bay, the valley between Bridgend and the head of Loch Eck, the head of Loch Riddon and the valley of Glendaruel, the sides of Ardyne burn below Knockdow (194), and the vicinity of Toward Point. The alluvial flats in Glendaruel, the Lochgoilhead valley, and at the head of Loch Eck are often flooded, and receive deposits of fine silt, which add greatly to their fertility. At Kilfinan there is a considerable area of land under cultivation extending from Kilfinan Bay to a little past the village of Kilfinan. The various raised beaches between Ballimore and Largiemore (171), and the river terraces of Stralachlan, are also under cultivation. On the south side of Kilfinan there is some tillage land between this place and Ardmarnock, and again to a larger extent on the south side of the road between Tighnabruaich, Millhouse, and Portavaidue, in the low lying ground about Ardlamont, and in the valleys of the streams running into Kilbride Bay and Asgog Bay.

At the time of the Survey there were only two crofters in the area east of Glendaruel, but it is not very long since the glens were thickly peopled with them, and their old ruined houses and field dykes can still be seen in some places, e.g., near the edge of the alluvium ¼ mile east of the "1" of "Glen Tarsan" (173), on the north-west side of Allt Glinne Mhoir (134) a little north of the " i " of "Mhoir," in Glen Shellish by the "n" and "s," in Glen Branter 100 yards north of the "R" of "Robuie." Remains of higher lying, more isolated dwellings remind us of times when it was customary to remove for the summer time to distant shielings and tend herds on the small bits of pasture near the mountain tops—as the Norwegians still resort to the "seters" with their goats. Such remains occur ½ mile south-east of the the Eagles' Fall (126), ⅜ mile E.N.E. of the "c" of "Beinn Bhreac" (153), in Glen Branter 100 yards east of the "G" of "Glen," in Garrachra Glen 100 yards south-east of the "h" of "Garrachra Glen" (163). In those days there were comparatively few sheep kept, but many Highland cattle. The large sheep farms are said to be a later introduction.

In the south-west part of Cowal, there are ruins of farm buildings at Inveryne on the north side of the Acharosson burn, and on the south side of the same burn are others at Upper Auchalick, Upper Auchanaskioch, and at another place ½ mile north-east of the latter. Glenan is now unoccupied, as also Creag-an-fhithich cottages and Stroue-duich, west of Kilbride. As examples of moorland which was once cultivated may be mentioned a tract 300 to 400 yards east of Kilbride church, and another on the moor ½ mile east of Auchagoyl cottages. There are also many old fences on the moor west of Stillaig Farm near the ruins of the old chapel.

Some of the older inhabitants tell you tales of these old times, and speak of them with regret. They were humble and simple days, but it was more possible then for the young people to find employment at home, and in comparatively healthy surroundings. Each little feature in the glen was well known and had a Gaelic name distinctive of its peculiarities, and gradually acquired a corner of affection in their hearts.

The population is now nearly all concentrated in the coast villages and towns. These have either sprung up or largely increased since crofter days, and afford a pleasant and convenient outlet for the holiday makers of Glasgow and the other large towns near. The summer population of the coast places is very much larger than that in the winter, many of the houses being closed at the latter time of year. In the case of Dunoon it is said (1894) that the winter population may be taken as 6000, while in the height of the summer it amounts to 35,000. But even the winter residents, including the business men, who in yearly increasing numbers go up to Glasgow or other towns day by day, probably considerably exceed in number the population of old crofter days.

W.G., C.T.C., J.B.H.

APPENDIX.

I.—General Remarks on the Petrography of the Cowal District. By Mr J. J. H. Teall, F.R.S.

The main interest of the Cowal district, so far as petrographical questions are concerned, centres in the fact that it is largely composed of sedimentary deposits which have been converted into crystalline schists by metamorphic agencies operating on a regional scale. It has been shown in the body of the memoir that the least altered rocks, consisting mainly of phyllites and schistose grits or greywackes, occur in the south-eastern portion of the district, and that, as we proceed towards the north-west, across the general strike of the formations, the metamorphism is seen to increase until we reach, near the centre of the "anticline," an extensive area of typical crystalline schist. No abrupt change can be observed; at the same time it may be regarded as certain that the rocks of the most highly altered area are not simply the metamorphosed representatives of those which occur in the least altered area: new petrographical elements are introduced into the complex as we advance towards the north-west.

Another point which has been clearly brought out by a study of the rocks in the field is that all portions of the district have been profoundly affected by earth movements of different types and of different ages. Structures produced by one set of movements have been subsequently modified or even obliterated by a later set.

In comparing the rocks of the most highly altered with those of the least altered areas the most striking feature, however, is not a difference in the apparent amount of movement, but in the nature and relative abundance of the minerals which have been developed *in situ*, or, in other words, of the authigenic constituents. An unaltered mechanical sediment consists wholly of allothigenic constituents, whereas a typical crystalline schist consists wholly of authigenic constituents. Between these two extremes every intermediate stage may be found, and many, though not all, of these occur in the district in question. The initial stages are absent, for the least altered rock belonging to the Highland series shows a certain amount of new mineral development.

It is well known that in studying the metamorphism of a complex series of stratified deposits the coarser grained sediments retain traces of their original character long after all such traces have disappeared from the finer grained deposits. It is also known that rocks which can be clearly recognised, in the field or in hand specimens, as of clastic origin, often so closely resemble crystalline schists in microscopic structure that if we had to depend on the microscope alone it would be impossible to form any definite conclusion as to their origin.[*]

These points are well illustrated in the Cowal district by a widely distributed rock to which the terms gneissose grit, biotite-gneiss, and granulitic biotite-gneiss have been applied during the progress of the Survey.[†] This rock is found not only in the district in question but in many other portions of the southern Highlands, along a belt of country stretching from the coast north of Stonehaven to the peninsula of Kintyre. Although distinctly gneissose in structure and composition it differs markedly from igneous gneisses in its relation to the other rocks with which it is associated, and frequently contains relics of original grains of quartz, often bluish in colour, and felspar. The main mass of the rock is a granulitic mosaic of quartz and felspar with which definite flakes of biotite and flakes or confused aggregates of white mica are associated.

[*] *L. Milch*, Beiträge zur Lehre von der Regional-metamorphose. *Neues Jahrbuch*. Beilage Band ix. (1894), p. 101.

[†] In the body of this memoir, however, they have been called simply schistose grits, or schistose greywackes. C.T.C.

A very common type of this rock is represented by a specimen from Fearna Bagh (3834). It is formed of fine-grained, irregular, quartzo-felspathic layers separated by thin dark streaks which have a tendency to run into each other, and so cut out

FIG. 54.— × 1.　Granulitic biotite gneiss.　Fearna Bagh

the broader granulitic folia of quartz and felspar. The above diagrammatic sketch of the cut surface of this specimen represents the relation of the dark laminæ to the light coloured folia.

In this particular specimen the original clastic character has been almost entirely obliterated. The rock tends to split along the dark laminæ, and the planes of schistosity are coated with plates of black mica. Under the microscope the light coloured folia are seen to be composed of quartz, felspar, a carbonate, and here and there a flake of biotite. Of these constituents quartz is by far the most abundant. It forms a mosaic of interlocking grains. The felspar and the carbonate are present only in small quantity, but they interlock with the individuals of the quartz mosaic in the same way as these do with each other. All the constituents are authimorphic, and most, if not all, are anthigenic. The plates of biotite are isolated, and are idiomorphic so far as the basal plane is concerned. One and the same plate may lie in two or three individuals of the quartz-mosaic. This fact is important because it has been suggested that the idiomorphism of the biotite is a distinctive character of gneissose rocks of igneous origin.

The dark laminæ are extremely rich in micas of two kinds. One is strongly pleochroic in light brown and deep brown or green tints ; the other is nearly, but not quite, colourless. The dark mica occurs in flakes which usually show definite extinction, but it is not so markedly idiomorphic as the plates which lie in the granulitic folia. The white or pale green mica sometimes shows definite extinction, but is more apt to form sericitic aggregates. Iron ores occur sparingly in the dark layers.

In addition to the constituents above described there are a few larger grains of quartz and felspar. These, like the other constituents, are authimorphic, but they are probably relics of original clastic (allothiclastic) grains which have not been completely absorbed into the granulitic mosaic ; in other words, they may be in part allothigenic. The felspars sometimes appear to have been fractured ; the different portions of what was probably an original grain being now separated by veins of granulitic material. It is clear, however, that the amount of allothigenic material in this particular specimen is extremely small.

Another specimen from Tom an Iasgaire (5295) belonging to the same type of rock has preserved its original character in a much more perfect manner. More or less deformed grains of blue quartz and pink felspar, often measuring ·5 mm. in diameter, are clearly recognisable in the hand specimen. The bed of grit has been bent into a fold, and, although all the grains have been deformed, those which occur near the axial plane have suffered less than those which occur in the limbs. The composition of the rock is the same as the last. The constituents are, for the most part, authimorphic, but the amount of allothigenic material is much greater. The specimen furnishes a striking illustration of the difference in behaviour of quartz and felspar under the influence of the deforming stresses. The former changes its shape by granulitisation, the latter by fracture. Under the microscope detached authiclastic fragments of felspar which clearly belonged to the same grain are often seen to be separated by veins of microcrystalline quartz. Folia of microcrystalline material and micas sweep round the lenticles or phacoids of quartz and felspar. The mosaic produced by the deformation of the large quartz grains is much coarser than that of the matrix.

The extensive series to which these two specimens belong consists of foliated crystalline rocks essentially composed of quartz, felspar, and one or two micas. They are, therefore, gneisses in the usual sense in which that term is employed. Where conspicuous traces of their clastic origin remain they may be termed gneissose grits; when all or nearly all such traces have disappeared they may be termed granulitic gneisses. They pass by insensible gradations into felspathic mica schists.

The rocks in the area of least metamorphism consist of phyllites and greenish schistose grits or greywackes. The latter are composed of more or less deformed grains of quartz and felspar in a matrix of micro-crystalline quartz, sericitic mica, and chlorite. The rock possesses under the microscope a marked flaser structure, and the larger grains of quartz frequently have "tails" of granulitic material which merge themselves in the micro-crystalline matrix. The streaks of sericitic mica wind round the larger constituents and thus produce the characteristic flaser structure.

The rocks grouped as "green beds" vary in character. Many specimens have the appearance of greenish, more or less schistose grits, others are finer in grain and more markedly schistose. A special feature is the almost constant occurrence of granular epidote. Chlorite and a dark green or brown biotite are usually present, and a white mica, in well developed flakes, is not uncommon. The coarser grained rocks are light in colour, and are mainly composed of quartz, felspar, and epidote.

The quartz and felspar appear in many cases to represent clastic fragments, but some of the felspar, especially in the finer grained schists, is albite and undoubtedly authigenic (3831). In some rocks large plates of authigenic white mica may be seen under the microscope lying at right angles to the plane of foliation. Rutile and tourmaline occur as accessory constituents in some specimens. The more gritty looking varieties have a marked flaser structure under the microscope, the larger grains apparently representing clastic quartz and felspar, the matrix being composed of the ubiquitous micro-crystalline mosaic with which more or less granular epidote is associated.

The rocks with authigenic albite, grouped as albite-schists, are sufficiently described in the body of the memoir and in the Appendix No. II. It is, therefore, only necessary, in this connection, to refer to other rocks of a similar character. Albito-gneisses are recorded as occurring at several localities along the northern border of the central zone of the eastern Alps. The rocks from the north-eastern prolongation of this border, south of Wiener Neustadt, have been described by Böhm.* He divides the rocks of the district, according to the dominant mineral, into mica-rocks, chlorite-rocks, and hornblende-rocks. The rocks containing albite belong to the first group which he further subdivides into albite-gneiss, granulitic albite-gneiss, mica-schist, epidote-mica-schist, and quartzite-schist. His detailed petrographical descriptions show that there are many points of resemblance between the rocks of the Wechselgebirge, and those of the Cowal district. Amongst these points may be mentioned the occurrence of a green mica, and epidote as well as that of albite. Böhm regards the rocks as a transitional group between the old crystalline schists and the true phyllites.

Rocks of a similar character occur in the upper part of the Ennsthal and in the Rädstadter Tauern.†

Rocks, termed albite-phyllites, have been described by the officers of the Geological Survey of Saxony as belonging to the lower portion of the phyllite-formation of the Erzgebirge.‡ They consist of a white potash mica referred to damourite, chlorite, and a variable quantity of quartz and albite. The albite occurs sporadically, as it does in the Cowal district, and where abundant the rock may be termed albite-gneiss. Dr Credner has examined hand specimens of the Cowal rocks, and states that they can be exactly matched by specimens from Saxony. The upper portion of the "phyllite-formation" of the Erzgebirge is referred in part to the Cambrian. It is less crystalline than the lower portion, and contains rutile needles. It is therefore the petrographical equivalent of the Dunoon phyllites.

Rocks closely allied to the albite-schists of the Cowal district occur in the Green Mountains of Massachusetts,§ where they are said to be of later date than quartzite containing Olenellus. The albite-schist of Hoosac Mountain overlies a metamorphosed grit or conglomerate containing blue quartz and having other points of resemblance to the gneissose grits of the Cowal district. The albites often give a porphyritic character to the microscopic section; but they contain the other minerals as inclusions. "The groundmass is composed of muscovite and biotite, or muscovite alone, chlorite, grains and aggregate lenses of quartz, magnetite in octahedra or grains, apatite, tourmaline, and rutile." ‖ The main difference between these rocks

* Ueber die Gesteine des Wechsels. *Tschermak's Min. u. Petro. Mitth.* v. 201.
† Von Foullon. *Jahrb. K. K. geol. Reichsanst*, 1883, xxxiii. 207 ; and 1884, xxxiv. 635.
‡ See K. Dalmer. *Erlauterungen zur geologischen Specialkarte*, Section Lossnitz.
§ Geology of the Green Mountains by Pumpelly, Wolff and Dale. *Monograph of the U. S. Geol. Survey*, xxiii. (1894). .
‖ *Op. cit.*, p. 61.

and those of Cowal appears to lie in the relation of the albite to the micaceous minerals. These have not been observed as inclusions in the Cowal schists.

It will thus be seen that a considerable amount of doubt remains as to the age and mode of origin of the different occurrences of albite-schist and gneiss. These remarkable rocks sometimes undoubtedly form a kind of transitional zone between normal crystalline schists and unquestionable sedimentary deposits ; and those who still believe that this transition in space corresponds to a transition in time refer the rocks in question to the "transitional period." The albite-gneisses of the Alps have been in part referred to the carboniferous and in part to this "transitional period" ; the albite-phyllites of Saxony are supposed to belong to a continuous series, the upper part of which is Cambrian, while the albite-schists of the Green Mountains are said to be Cambrian or Silurian.

The Cowal district furnishes no satisfactory evidence as to the age of the rocks in question, but it shows that they belong to a series of highly metamorphosed stratified deposits which cannot be sharply separated from another series, far less highly metamorphosed and consisting of phyllites or clay-slates, schistose grits, and limestones.

The rocks of the hornblende-schist and epidiorite group probably represent in all cases basic igneous intrusions. The type to which they belong is so well known and so widely distributed that it is unnecessary to refer to special localities where similar rocks are to be found. They were in existence before the movements which have so powerfully affected the district had commenced, or at any rate before these movements had ceased.

The granitic rocks were, on the other hand, intruded after the movements. They vary in character from granitites to quartz-diorites, and often contain clove-brown crystals of sphene which may be easily recognised in hand specimens. Basic patches richer in ferro-magnesian constituents and sphene than the main mass are not uncommon. Both in macroscopic and microscopic characters the specimens (6113-6116) bear a very close resemblance to rocks from the Galloway district and especially to those from the Criffel mass. They are also allied to rocks occurring in the neighbourhood of Ballachulish.

The hyperites (4743, 4744) are extremely interesting as furnishing another bond of union between the later intrusive rocks of the Cowal district and those of Galloway. The description of these rocks which has been given in the text may be compared with that of the Galloway hyperites already published in the Memoir on Sheet 5.

A word or two is necessary as to the use of the term lamprophyre. Professor Rosenbusch says[*] :—" I adopt Gümbel's term lamprophyre for a group of dyke-rocks which occur in folded districts, and which, in respect of mineralogical composition, belong in part hornblende and in part to dioritic types. They are characterised macroscopically by a fine grained, compact or porphyritic structure, by a grey or black colour when in a fresh condition, and by a great tendency to decomposition accompanied by a rich development of carbonates. In the porphyritic varieties, apart from certain rare exceptions and some transitional forms, the phenocrysts belong to ferriferous minerals of the mica-, augite-, and hornblende-families . . . . felspar rarely occurs as a phenocryst."

In the rocks of the Cowal district which answer in a general way to the above description the dioritic types prevail. Of the three minerals, augite, biotite, and hornblende, the first is the one most commonly found, although it is frequently not the most abundant. Unlike the common type of augite in dolerites and diabases it is colourless or nearly so in thin section, and shows a strong tendency to idiomorphism, frequently occurring in large phenocrysts. Hornblende and biotite occur separately or together, and seem more or less to replace each other. The former is often developed in long brown prisms. These three minerals occur in a felspathic matrix.

Plagioclase is seen to be the dominant felspar when the rocks are sufficiently well preserved to allow of the determination of the mineral. It may occur in large irregular grains, or in zoned and more or less idiomorphic individuals. In the latter case a little interstitial quartz may occasionally be detected (3417). When the darker rock contains pink patches (4012) these are largely composed of orthoclase, and it is probable that this mineral is frequently present, in small quantities, in many of the other rocks.

Names such as diorite, camptonite, augite-diorite, mica-dolerite or diabase, kersautite, and vogesite might be applied to different varieties ; but as these varieties cannot be systematically separated in the field, and as the state of preservation of the rocks is often such as to render their precise determination impossible, even with the aid of microscopic sections, it has been found necessary to group them together under the comprehensive but somewhat ill-defined term lamprophyre.

The porphyrites of the Cowal district are essentially similar to those of Galloway described in the Memoir on Sheet 5. They represent the porphyritic phase of a

[*] Mikroskopische Physiographie der massigen Gesteine. II. Edition 1887, p. 308.

hornblende-granitite or quartz-mica-diorite.   The ground-mass may be micro-granitic, felsitic, or even microlitic, as in some andesites.

In addition to the normal porphyrites, with numerous phenocrysts of plagioclase, biotite, and hornblende, there are many light coloured rocks for which the term felsite is retained.   These rocks are often so much altered that their precise determination is impossible.   In many cases they are merely altered porphyrites, but one specimen from the north-west side of Loch Fine, belonging to this group (4806), contains large Carlsbad trains of altered orthoclase similar in form and size to the crystals of sanidine in the Drachenfels trachyte.   It is probable, therefore, that some ortho-felsites are included in this group.   True quartz-porphyries appear to be absent, but the porphyrites occasionally contain a few scattered phenocrysts of corroded quartz, and such varieties must be classed as quartz-porphyries.

II. DESCRIPTION OF AN ALBITE SCHIST NEAR ARD A' CHAPUILL.
BY MR TEALL.

Slide 3418.   ½ mile E.N.E. of Ard a' Chapuill (182).

Macroscopically this rock consists of narrow wavy folia of white granulitic quartz and a grey highly micaceous schist.   The micaceous portions of the rock contain numerous crystals of a glassy-looking mineral, having well-marked cleavages.   These crystals measure on the average about 1 mm. across.   They give a porphyritic aspect to the rock.   Isolated grains, tested by Szabo's method, give a strong soda colouration to the flame of a Bunsen's burner ; but it is impossible to obtain fragments sufficiently large to yield perfectly satisfactory results by this method.   So far as they go, they point to a felspar rich in soda.

More satisfactory results are obtained by examining crushed grains under the microscope.   When mounted in hard balsam, and examined in parallel light, the refractive index of the mineral is seen to be almost exactly that of the balsam, which is about 1·536.   A crushed albite mounted in the same way shows precisely the same refractive phenomena.   In both cases the flakes which show the somewhat oblique emergence of a positive bisectrix, and which therefore contain approximately the greatest and mean axes of elasticity, nearly disappear when viewed by rays vibrating at right angles to the trace of the optic axial plane, but stand out distinctly when viewed by rays vibrating parallel to the trace of the plane.   From this it follows that the mean and least refractive indices of the mineral are the same as those of albite, or at any rate that the difference is less than 2 in the third place of decimals (refractive indices for albite—n_g 1·540, n_m 1·534, n_1 1·532).   Another point of agreement with albite is the extinction angle in flakes which show the oblique emergence of a positive bisectrix.   This varies from 19° to 21° in the positive direction.

An attempt was made to isolate the mineral, but it was found impossible to obtain it in a pure state, in consequence of the inclusions which are especially abundant in the central portions of the individuals.   The material analysed floated in a solution of sp. gr. 2·643.   It yielded the following result :—

SiO_2,	81·4
Al_2O_3, } Fe_2O_3, }	11·5
CaO,	0·5
MgO,	Slight traces.
Na_2O,	4·9
K_2O,	0·5
Ignition,	0·3
	99·1

It is evident from this that the substance analysed consisted essentially of a mixture of albite and quartz, in about equal proportions.   The analysis was merely undertaken for the purpose of verifying the results obtained by the other methods.

When the rock is examined in thin section under the microscope it is seen to consist of albite, quartz, white-mica, chlorite, magnetite, and tourmaline.

The albite occurs as more or less idiomorphic crystals, and also as grains with somewhat irregular boundaries.   Sometimes a patch of albite consists of four or five individuals.   The substance of the albite is water clear, and never shows a trace of alteration.   The cleavages are often well marked in the section.   The individuals are generally simple, and the sections are of fairly uniform dimensions in the different directions. Binary twins may occasionally be observed. Lamellar twinning is entirely absent. Magnetite occurs as inclusions, especially in the central portions of the individuals, but the grains are smaller and less numerous relatively to the enveloping matrix than those occurring in the groundmass of the schist. It is especially worthy of note that the foliation of the rock, as defined by the stream-like arrangement of the grains of magnetite, is seen in the central portions of the albite crystals. In addition to the magnetite, we find also numerous minute and nearly colourless prismatic microlites occurring as inclusions in the albite.

The quartz occurs in irregular grains of variable size, and is for the most part singularly free from inclusions.

White-mica is present both in large and small plates. There is a considerable amount of confusion in its mode of arrangement ; and plates may not unfrequently be seen having their broad flat surfaces oblique or even at right angles to the general plane of schistosity.

The chlorite is not a very important constituent of the rock. Sections at right angles to the flat surfaces of the scales are strongly pleochroic. The colour for rays vibrating at right angles to the basal plane is yellow, that for rays vibrating parallel to the basal plane is green. Cross sections polarise in the neutral tints of the first order, and the *minor* axis of depolarisation is always at right angles to the length of the section, *i.e.*, to the basal plane. Observations on the flakes of the chlorite from an allied rock show that it is uniaxial. It belongs therefore to the pennine group, and the double-refraction is positive.

Magnetite occurs abundantly in the form of opaque grains. That the grains are magnetite has been proved by isolation and treatment with acids and a weak magnet. Small grains of rutile are often associated with the magnetite.

Tourmaline occurs as an accessory constituent. It is present in the form of small prisms. The larger individuals are pleochroic in pale rose-colour and deep bluish-grey tints. They often lie with their longer axes at right angles to the plane of schistosity, as if they had been developed *in sitû* after the schistosity had been produced.

The minerals above described are not uniformly distributed throughout the rock. A section at right angles to the general trend of the foliation is seen to be made up of wavy lenticular bands which vary in composition. Thus we have folia which consist almost entirely of quartz, and others which consist mainly of large flakes of white-mica. These alternate with bands composed of smaller scales of white-mica, magnetite, chlorite, and a little granulitic quartz and felspar. It is in these bands that the albite occurs. The white-mica ends off abruptly where it comes against the albite, and never occurs as inclusions in that mineral. The streams of magnetite, on the other hand, are continued through the albite. The foliation planes between the individual crystals of albite are often puckered, and it is possible that this may be a result of the crystallisation of the albite. In the thin section the total space occupied by the albite is about equal to that occupied by the schistose groundmass in those bands in which the albite occurs. The foliation of the rock is not even and regular, but wavy and of the flaser type.

III. LIST OF MICROSCOPE SLIDES.

Schists S.E. of the Bull Rock Greywacke.
    2834. Near Toward schist-boundary (194).
    2835. Near schist-boundary S. of Inellan pier.
    4840. ¼ mile E. of Inellan Hill.
    4841. ½ mile W. of Toward schist-boundary.
Bull Rock Greywacke.
    3823. Bullwood, Dunoon.
    3824. Same locality.
Dunoon Phyllite Series.
    3825. Bullwood, Dunoon.
    3826. Same locality.
    3827. Same locality.
Schistose Grits and Greywackes N.W. of Dunoon Phyllites.
    3828. 1 mile E.S.E. of Lephinkill (172).
    3834. Fearna Bagh (182).
    4731. 1 mile N.N.W. of Ben Donich (134) : with tourmaline.
    5295. Tom an Iasgaire (152).
Pegmatites, with Albite, in Schistose Grit.
    4198. ¼ mile N.E. of Mark (153).
    7314. 50 yards S. of the Bull Rock (184).

" Green Beds."
 2830. 1 mile S.W. of Ardnadam Farm (184).
 2831. Burn ½ mile N. of Bishop's Seat (183).
 3791. 150 yards W. of Conchra (162).
 3792. 200 yards S.E. of Ardnadam Farm (174).
 3793. ½ mile S. of Ardnadam Farm.
 3794. ½ mile N.W. of Blairbuie (183): with (?) albites.
 3795. Tighnuilt, Inverchaolain (183).
 3829. Lephinkill (172).
 3830. Same locality : with garnets.
 3831. ⅔ mile slightly N. of E. of Lephinkill : with albites.
 3832. ¼ mile S.E. of Lephinkill : with albites.
 4726. ¼ mile N.W. of Mullach Coire a Chuir (142) : with albites.
 4727. 1 mile slightly S. of E. of Dundarave Castle (133) : with albites.
 4728. ½ mile S.E. of Cruach nam Mult (134).
 4729. Same locality.
Albite Schist.
 2957. ¼ mile N. of Cruach nam Cuilean (172) : pegmatite.
 3413. ½ mile E.N.E. of Ard a' Chapuill (182).
 3833. Nearly 1 mile E. of Lephinkill (172).
 3866. Coast ⅔ mile N.N.E. of Knap (164).
 4197. ¼ mile N.W. of Ben Reithe : junction of lamprophyre sheet with albite schist.
 4732. A little over ¼ mile N.N.E. of Ben Lochain (142) : thin band in albite schist.
 4733. Beinn Tharsuinn (142) : altered near basalt.
 4734. ¼ mile W.S.W. of Drimsynie (142) : altered near basalt.
 4738. 1½ mile slightly W. of S. of Carrick Castle (153) : altered near basalt.
 5214. Head of Stuck Burn (152).
 5215. Same locality.
 5432. ⅔ mile slightly N. of E. of Mid Letter (141).
 6101. River Fine a little over ½ mile N.W. of Chruac Tuirc (126): altered near granite.
 6102. Garbh Allt Mor ¼ mile S.E. of Achadunan (126) : with garnets.
 6111. ¼ mile N.W. of Lochan Mill Bhig (126): with garnets.
Loch Tay Limestone and Bands in it.
 4730. Not quite ¾ mile E.N.E. of Bathaich ban Cottage (134).
 5133. Not quite ¼ mile E.N.E. of Bathaich ban Cottage (134).
 5135. Not quite ¼ mile S. of Cruach nan Capull (141).
 5136. Not quite ½ mile S. of Cruach nan Capull (141).
 5213. 1 mile slightly E. of N. of Socach (141).
 5546. Same locality.
 5547. Same locality.
Garnetiferous Mica Schists.
 4739. ¼ mile W. of Bathaich ban Cottage (133 and 134).
 6103. ⅔ mile slightly S. of W. of the Eagle's Fall (126).
Graphitic Schists.
 5053. 1½ mile S.E. of St Catherine's pier.
 5054. ⅔ mile slightly N. of W. of Cruach nan Capull (133) : knotted schist.
 5055. Burn 1 mile slightly S. of E. of St Catherine's pier.
Ardrishaig Phyllite Series.
 5053. 1½ mile S.E. of St Catherine's pier.
 5054. ⅔ mile slightly N. of W. of Cruach nan Capull (133) : knotted schist.
 5055. Burn 1 mile slightly S. of E. of St Catherine's pier.
Andalusite Hornfels.
 6098. Top of Chruac Tuirc (126).
 6099. Near above locality.
 6112. ₇⁄₁₂ mile S.E. of Cruach Tuirc.
Quartz-hornblende Schist.
 6100. River Fine ⅔ mile slightly N. of W. of Cruach Tuirc.
Schist altered near Basalt.
 4737. 1 mile S.W. of Beinn Mhor (163).
Schist Crushed near Intrusion in Donich Burn.
 4735. Stream ½ mile E.N.E. of Ben Donich Lodge (142).
Hornblende Schist and Epidiorite Group.
 2832. 230 yards N.N.W. of Hunter's Quay (174).
 4000. Stralachlan (151).
 4003. Kilbride Chapel, Stralachlan : "eye" in the schist.
 4005. Kilbride Chapel, Stralachlan.
 4006. N. of Kilbride Chapel, Stralachlan.

4008. ½ mile N. of Stralachlan.
4740. ¼ mile E.N.E. of Cruach nam Mult (134).
5056. ₁₋₂ mile N.E. of Ardchyline (133).
5057. ₁₋₂ mile S.W. of St Catherine's pier (133).
5058. From spoil heap of quarry 50 yards N. of St Catherine's old chapel.
5059. ₁₋₂ mile S.W. of St Catherine's pier.
5060. 130 yards E. of St Catherine's pier.
5061. Block in low sea wall ¼ mile N.E. of St Catherine's.
5062. Loose block near St Catherine's.

Serpentine and Allied Rocks.

2833. S. of Inellan pier (195).
4004. Glendaruel (162).
4369. Near Kilbridemore (162): interior of mass.
4370. Same locality: dyke-like band in serpentine complex.
4837. ½ mile S.W. of Chapelton (194).
4838. Toward schist-boundary fault: ferriferous dolomite with patches of serpentine.
4839. Same locality: schistose crush rock of carbonates and talc.

Granitite.

6113. River Fine ₁₋₂ mile slightly S. of E. of Newton Hill (126).
6114. River Fine ½ mile S.E. of Newton Hill.
6115. River Fine ₁₋₂ mile E.S.E. of Newton Hill: with inclusion.
6116. Boulder by road ¾ mile N.E. of Achadunan.

Hornblende-porphyrite.

6109. ¼ mile slightly N. of W. of Cruach Tuire (126): middle of sheet.
6110. Same locality: 2 inch off top of sheet.
6117. ₇₋₆ mile S.W. of Cruach Tuire: altered near a crush.

Felsite.

4013. Acharosson Burn, Kilfinan (182).
4741. ⅝ mile S.S.W. of Stob an Eas (134).
5063. 230 yards slightly E. of S. of Ard na Slaite (133).
5066. ¼ mile S.E. of St Catherine's pier (133).

Hyperite.

4743. Donich Burn 1½ mile above Donich Lodge (142).
4744. ¾ mile N. of Ben Donich (134).

The (?) Intrusive Andesite near Inellan.

2837. Near Inellan Perch (195): the most northerly exposure of this rock.

Trachytic Dykes.

2960. N.N.E. dyke in Allt na Airidh Loisgte (burn ⅝ mile E. of Cruach Mhor, 162): spherulitic edge.
2961. ¼ mile S.E. of Dun Mor (172).
2962. 200 yards W. of Craigendaive (172): spherulitic edge.
3452. Judge's Chair, Dunans Burn, 162 (¼ mile S.E. of the "D" of "Dunans").
3921. 200 yards W. of Craigendaive (162): spherulitic edge.
3922. ¼ mile S.E. of Dun Mor (172).
    a. Next to selvage.
    b. Intermediate portion.
    c. Furthest from selvage.

Pebble of Igneous Rock from the Upper Old Red Conglomerate.

2836. ¼ mile W. of Toward Point (194).

Partially fused Biotite Schist from a Vitrified Fort.

5083. Island 1 mile N.W. of Colintraive.
5084. Same locality.

Lamprophyre Group.

2838. ½ mile S.S.W. of Cnoc Madaidh (173): with hornblende.
2839. 300 yards N. of the Craig (183): with malacolite and hornblende.
2840. 160 yards N.W. of Doire Dairach (¼ mile E.N.E. of Drynain, 164).
2958. ½ mile S.W. of Glenmassan (163).
2959. 1¼ mile S.S.E. of Craigendaive (172).
3417. Old Chapel, Cnoc Dubh (182): allied to kersantite.
3453. ⅜ mile S. of "I" of "Tom an Iasgaire" (¼ mile N.N.W. of the "P" of "Dornoch Point," 152): with hornblende.
3835. ¼ mile N.E. of Ben Bhreac (153): with hornblende.
4011. Stralachlan (151): kersantite.
4012. Barnacarry, Stralachlau: kersantite.
4015. Craig Lodge, Loch Riddon.
4197. ¼ mile N.W. of Ben Reithe (142): spherulitic: junction with albite schist.

4369. Kilbridemore Burn (162).
4742. ¼ mile slightly W. of S. of Bathaich ban Cottage (134): mica trap.
4745. ¾ mile N. of Ben Donich.
4746. A little over ¼ mile S. of Bathaich ban Cottage (134): phenocrysts of malacolite.
4752. The most W. of the two dykes ½ mile E.N.E. of Cruach nam Mult (134): augitic mica trap.
4755. ¾ mile N.N.E. of Stob an Eas (134): biotite dolerite.
5067. ⅞ mile S.W. of St Catherine's pier (133).
5064. 1¼ mile slightly E. of S. of St Catherine's pier (133).
8107. ¼ mile W.N.W. of the Eagle's Fall (126).

Camptonite Type.
4747. 100 yards S.E. of Bathaich ban Cottage (134).
4748. ¼ mile W.N.W. of Loch Restil outlet (134).
4749. ½ mile E. of Monovechadan (134).
4750. Not quite ½ mile S.S.E. of Bathaich ban Cottage (134).
4751. ¼ mile W. of Loch Restil outlet (134).
4753. Portion of upper sheet ¼ mile S.E. of Loch Restil head (131).
4754. Lower portion of lower sheet, same locality.
5065. ₇⁄₁₀ mile slightly N. of E. of St Catherine's (133).

Crushed or Foliated.
6104. ₁⁄₁₆ mile W.N.W. of Cruach Tuirc (126). Hornblende-mica-schist.
6105. ½ mile less 50 yards N.E. of Achadunan (126): calc-chlorite-schist.
6106. ¾ mile E.N.E. of Achadunan (126).
6108. ½ mile E.S.E. of Achadunan.

Basaltic Group.

Sith an' t-Sluain (152) Dolerite Boss.
5438. Interior of intrusion.
5439. Edge of intrusion.
5440. Felspar-augite vein.

Early E. to W. Dykes.
2841. *a.* The broad dyke in Easan Biorach (the stream ¾ mile S.W. of Bodach Bochd, 183).
    b. Red vein in above dyke.
2843. The broad dyke S. of Larach Hill (164).
4199. ½ mile E. of Donich Lodge (142).
4200. Reddish less basic band in above dyke.
4756. Not quite ¾ mile S. of Bathaich ban Cottage (134).

Later Dykes.
2842. By the big bend in Inverchapel Burn (164).
2844. The N.W. running dyke S. of Larach Hill (164): porphyritic olivine dolerite.
2845. 200 yards S. of Brackleymore school (194): porphyritic olivine dolerite.
2846. Same locality: companion dyke.
2847. Same locality: companion dyke: glassy basalt.
2848. ⅞ mile above foot of Invervegain Burn (183): edge of dyke (? basalt).
2849. ¾ mile W. of Toward Lighthouse (194): hard band in dyke in old beach: porphyritic basalt.
2850. Easan Biorach (stream ¾ mile S.W. of Bodach Bochd, 183): most N. of the N.W. running dykes: tachylite.
2851. ½ mile S. of the foot of Easan Biorach: perlitic tachylite.
2852. ¾ mile E. of Cloineter Hill (100 yards N. of the "i" of "Buthkollidur," 184): tachylite.
3268. ¼ mile slightly N. of E. of the Clachan of Glendaruel (172): part of a "core": andesitic glass.
3269. ¼ mile slightly N. of E. of the Clachan of Glendaruel: part of a "sheath": andesitic glass.
3270. N.E. corner of 172 N.E.ᵂ. (¼ mile S.W. of Cruach Mhor): andesite, tholeite type).
3271. ¾ mile S.E. of Lower Duillater (162): andesite, tholeite type.
4014. Near Glendaruel House (162): porphyritic dolerite.
4016. Otter—3¾ miles N.E. of—(171): pyroxene andesite glass.
5435. Not quite ¼ mile N.E. of Tigh na Criche (141): dolerite with zeolite amygdules.
5436. 100 yards W.S.W. of Mid Letter (141): dolerite with chlorophæite amygdules.
5437. 250 yards N.N.E. of M'Phun's Cairn (141): dolerite with olivine pseudomorphs.

IV. LIST OF GLACIAL FOSSILS.

The collections of glacial fossils from the different localities were made by Mr James Beunie of the Geological Survey, and we have been much indebted to the late Dr David Robertson, the veteran naturalist of Cumbræ, for their determination. A list of glacial fossils previously noted in the district is given by Messrs Crosskey and Robertson in their paper on the "Post-Tertiary Fossiliferous Beds of Scotland" (*Trans. Geol. Soc. Glasgow*, vol. v. p. 29, 1875). See also the *Catalogue of Western Scottish Fossils*, compiled by James Armstrong, John Young, M.D., and David Robertson, 1876.

We give first a list of the exact localities from which the collections were made, and then the list of fossils. In the last list there are ten columns at the right hand side, a column for each locality. The localities at which each fossil was collected are denoted by crosses in the columns belonging to these localities.

Localities.

1. ¼ mile above Ormidale on N.E. (left) bank of River Ruel, and 70 yards up the river above this locality, on same side of the river : about 5 miles N.N.W. of Colintraive.
2. By Springfield House (182), below high-water mark : 3½ miles N.N.W. of Colintraive.
3. Port an Eilein (182), below high-water mark : 2 miles N.W. of Colintraive.
4. N. side of Fearna Bagh : 1½ mile N.W. of Colintraive.
5. ½ mile S. of Fearn'ach (182) : 1½ mile N.W. of Colintraive.
6. ¼ mile N.W. of Colintraive pier.
7. Balnakailly Bay, N. end of Bute.
8. 80 yards N.E. of Toward Quay (194).
9. ¾ mile N. of Gairletter Point (174), a little below high-water mark.
10. ½ mile S. of Ardentinny (164), below high-water mark.

	Ormidale.	Springfield House.	Port an Eilein.	Fearna Bagh.	Fearn'ach.	Colintraive.	Balnakailly Bay.	Toward Quay.	Gairletter Point.	Ardentinny.
	1	2	3	4	5	6	7	8	9	10
Plantæ. Algæ.										
Melobesia, *Linné.*										
Sp.,						×				
Protozoa. Foraminifera.										
Miliolina.										
agglutinans, *D. Orb.,*	×					×				
secans, *D. Orb.,*						×		×		
oblonga, *Montague,*						×		×		
semimulum, *Linn.,*						×	×	×	×	×
subrotunda, *Montague,*							×			
fusa, *Brady,*							×			
bicornis, *Walker and Jacob,*							×			
Biloculina, *D. Orbigny.*										
ringens, *Lamk.,*								×	×	
Lagena, *Walker.*										
lævis, *Montague,*				×				×		
striata, *D. Orb.,*							×			
semistriata, *Will.,*					×					
,, var. apiculata, *Reuss.,*							×			×
distoma, *Parker and Jones,*							×			
marginata, *W. and B.,*							×			
acuticosta, *Reuss.,*				×			×		×	
lucida, *Williamson,*							×			
Cristellaria, *Lamarck.*										
rotulata, *Lamk.,*								×		

	Ormidale.	Springfield House.	Port an Eilein.	Fearna Bagh.	Fearn'ach.	Colintraive.	Balnakailly Bay.	Toward Quay.	Gairletter Point.	Ardentinny.
	1	2	3	4	5	6	7	8	9	10
Protozoa. Foraminifera—continued.										
Polymorphina, *D. Orbigny.*										
ovata, *D. Orb.*,	×									
lanceolata, *Reuss.*,			×					×	×	
augustion, *Egger*,							×			
Textularia, *Defrance.*										
agglutinans, *D. Orb.*,							×	×		
Bulimina, *D. Orbigny.*										
elegantissima, *D. Orb.*,							×			
marginata, *D. Orb.*,							×			
Bolivina, *D. Orbigny.*										
punctata, *D. Orb.*,							×			
Discorbina, *Parker and Jones.*										
rosacea, *D. Orb.*,								×		
globularis, *D. Orb.*,								×		
Cassidulina, *D. Orbigny.*										
crassa, *D. Orb.*,							×			
Truncatulina, *D. Orbigny.*										
lobatula, *Walker and Jacob*,								×		
Rotalia, *Lamarck.*										
Beccarii, *Linn.*,				×		×	×	×	✓	×
Polystomella, *Lamarck.*										
macella, *Fichtel and Möll*,								×		
crispa, *Linn.*,								✓		
striata-punctata, *Fichtel and Möll*,	×		×	×	×	×	×	×	×	×
Arctica, *Parker and Jones*,							×		×	×
Nonionina, *D. Orbigny.*										
depressula, *Walker and Jacob*,	×			×		×		×		×
orbicularis, *Brady*,			✓	×			×	×	×	×
stellaria, *D. Orb.*,										×
Spirobulina.										
limbata, *D. Orb.*,								×		
Cæcum.										
glabrum, *Mont.*,								×		
Echinodermata. Echinoidea.										
Echinus, *Linné.*										
Drobachiensis, *Müller*,	×		×	×		×	×		×	×
spines,					×			×		
Spatangidæ.										
spines,			×					×		
Annelida. Tubicola.										
Serpula, *Linné.*										
triquetra, *Martin*,							×			
vermicularis, *Ellis*,							×		×	
Spirorbis, *Lamarck.*										
communis,							×		×	×
Crustacea. Ostracoda.										
Cythere, *Müller.*										
pellucida, *Baird*,			×			×	×			×
viridis, *Müller*,							×			
lutea, *Müller*,	×		×	×		×	×	×	×	×
„ —varidis,							×			
„ —reniformis, *Baird*,								×		
albo-maculata, *Baird*,								×		

U

	Ormidale.	Springfield House.	Port an Eilein.	Fearna Bagh.	Fearns' ach.	Collntraive.	Balnakailly Bay.	Toward Quay.	Gairletter Point.	Ardentinny.
	1	2	3	4	5	6	7	8	9	10
Crustacea. Ostracoda—continued.										
Cythere, *Müller.*										
cuneiformis, *Brady*,..								×		
concinna, *Jones,*	×		×			×	×	×	×	×
angulata, *G. O. Sars,*	×			×		×	×	×	×	×
tuberculata, *G. O. Sars,*			×			×	×	×	×	×
Dunelmensis, *Norman,*	×									
confusa, *Norman and Brady,*			×			×	×	×		
cetosa, *Brady,*								×		
Cytheridea, *Bosquet.*										
papillosa, *Bosquet,*	×		×			×	×			×
punctillata, *Brady,*	×		×				×	×	×	×
Loxoconcha, *G. O. Sars.*										
impressa, *Baird,*								×		
tamarindus, *Jones,*	×		×	×		×				
Cytherura, *G. O. Sars.*										
nigrescens, *Baird,*	×					×				
similis, *G. O. Sars,*				×		×				
undata, *G. O. Sars,*	×					×			×	×
clathrata, *G. O. Sars,*						×				
Cythcropteron, *G. O. Sars.*										
latissimum, *Norman,*	×		×			×			×	v
declevis, *Norman,*										×
Sclerochilus, *G. O. Sars.*										
contortus, *Norman,*	×								×	
Polycope, *G. O. Sars.*										
orbicularis, *G. O. Sars,*										×
Cirripeda.										
Balanus, *Lister.*										
balanoides, *Linn.,*			×			×	×			
crenatus, *Brug.,*	×			×					×	
porcatus, *Da Costa,*	×								×	
Verruca, *Schumacher.*										
strömia, *Müller,*	×									
Mollusca. Conchifera.										
Anomia, *Linné.*										
ephippium, *Linn.,*	×		×	×		×				
" var. aculata, *Linn.,*	×									
Pecten, *Pliny.*										
Islandicus, *Mull.,*	×			×		×			×	×
Mytilus, *Linné.*										
edulis, *Linn.,*				×						
modiolus, *Linn.,*	×					×				
Nucula, *Lamarck.*										
tenuis, *Mont.,*										×
Leda, *Schumacher.*										
pygmœ, *Münst.,*	×									
Lucina, *Bruguière.*										
borealis, *Linn.,*		×								
Axinus, *Sowerby.*										
flexuosus, *Mont.,*	×			×		×	×			
Cardium, *Linné.*										
Sp.,									×	
fasciatum, *Mont.,*						×				
edule, *Linn.,*							×			
Cyprina, *Lamarck.*										
Islandica, *Linn.,*	×								×	×

	Ormidale.	Springfield House.	Port an Eilein.	Fearna Bagh.	Fearn'ach.	Colintraive.	Balnakailly Bay.	Toward Quay.	Gairletter Point.	Ardentinny.
	1	2	3	4	5	6	7	8	9	10
Mollusca Conchifera—continued.										
Astarte, *Sowerby.*										
sulcata, *Du Costa,*		•				×				×
,, var. elliptica, *Brown,*							×		×	×
compressa, *Mont.,*									×	×
Venus, *Linné.*										
exoleta, *Linn.,*		×								
fasciata, *Da Costa.*										
Tellina, *Linné.*										
calcaria, *Chemn.,*	×		×	×		×	×	×	×	×
Mya, *Linné.*										
Sp.,				×			×		•	
truncata, *Linn.,*	×	×				×			×	×
,, var. Uddevallensis,			×	×			×		×	×
Saxicava, *Fleurain de Bellevue.*										
rugosa, *Linn.,*	×					×	×		×	
Gasteropoda.										
Chiton, *Linné.*										
plates,						×				
Tectura, *Cuvier.*										
virginea, *Mull.,*	×			×		×		×		
Puncturella, *R. T. Lowe.*										
Noachina, *Linn.,*	×									
Trochus, *Rondeletius.*										
Sp.,				×						
helicinus,						×	×			×
Grœnlandicus,						×				
cinerarius, *Linn.,*						×				
Mölleria, *Jeffreys.*										
costalata,						×	×		×	
Lacuna, *Turton.*					√					
divaricata, *Fabr.,*	×					×			×	
Littorina, *Ferussac.*										
littorea, *Linn.,*	×		×	×		×	×			×
obtusata, *Linn.,*						×				
Rissoa, *Freminville.*										
striata, *Adams,*	×					×				
acreata,										×
Natica, *Adamson.*										
affinis, *Gmelin,*				×					×	
Grœnlandica, *Beck.,*	×									×
Purpura, *Bruguière.*										
lappillus, *Linn.,*				×						
Trophon, *De Montfort.*										
truncatus, *Ström.,*				×						
Pleurotoma, *Lamarck.*										
pyramidalis, *Ström.,*	×					×				×
turricula, *Mont.,*						×	×			

Since the above list was made up, Mr James Tod of Rashfield, Kilmun, has informed us of two additional exposures of glacial shell beds. One of these lies about ¼ mile east of Rashfield, near the school-house, and the other is in the bed of the river Echaig less than ⅓ mile above Rashfield. From the former locality Mr Tod forwarded some specimens, among which Mr James Bennie has determined the following:—Echinus, spines; Serpula vermicularis; Balanus, plate; Pecten Islandicus; Cyprina Islandica; Mya; Astarte; Leda; Littorina littorea; Littorina rudis; Purpura lapillus.

V. BIBLIOGRAPHY.

1800. JAMESON, ROBERT (Prof.).—Mineralogy of the Scottish Isles. 2 vols. 4to. Edinburgh.

1819. MACCULLOCH, JOHN, M.D.—A Description of the Western Islands of Scotland and the Isle of Man. London. 2 vols. 8vo, and 1 quarto vol. of Plates.

1820. BOUÉ, AMI. — Essai géologique sur L'Écosse. 8vo. Paris (no date, but probably 1820).

1823. SMITH, JAMES (of Jordanhill).—'On the Vitrified Forts on one of the Burnt Islands, Kyles of Bute,' Trans. Antiq. Soc. Edin. and Trans. Roy. Soc. Edin., vol. x. 79.

1839. SMITH, JAMES (of Jordanhill).—'On the Phenomena of the Elevated Marine Beds of the Basin of the Clyde,' Mem. Wern. Soc., vol. viii. p. . (Shells of Balnakaille Bay.)

1840. MACCULLOCH, JOHN, M.D.—A Geological Map of Scotland, with Memoir on do. (published several years after Macculloch's death).

1844. NICOL, JAMES (Prof.).—Guide to the Geology of Scotland. 8vo. Edinburgh.

1845. MACKAY, Rev. M.—New Statistical Account of Scotland, Argyllshire, Parishes of Dunoon and Kilmun, pp. 579–681. 8vo. Edin. and London.

1850. MACLAREN, CHARLES.—'On Glacier Moraines in Glen Messan, Argyllshire,' Rep. Brit. Ass. Trans. Sect., vol. xix. p. 90, and Edin. New Phil. Jour., vol. xlix. p. 330.

1851. HOPKINS, W.—'On the Granitic Blocks of the South Highlands of Scotland,' Quart. Jour. Geol. Soc., vol. viii. p. 20.

1852. SHARPE, D.—'On the Southern Border of the Highlands of Scotland,' Quart. Jour. Geol. Soc., vol. viii. p. 126.
　SHARPE, D.—'On the Arrangement of the Foliation and Cleavage of the Rocks of the North of Scotland,' Phil. Trans., vol. cxlii. p. 445.

1855. MACLAREN, CHARLES.—'Notices of Ancient Moraines in the Parishes of Strachur and Kilmun, Argyllshire,' Edin. New Phil. Jour., vol. i. New Series, p. 189, and Proc. Roy. Soc. Edin., iii. 279 (1850–1857).

1858. NICOL, JAMES (Prof.).—Geological Map of Scotland.

1859. BRYCE, JAMES.—The Geology of Clydesdale and Arran. 8vo. Glasgow.

1861. HARKNESS, Prof. R.—'On the Rocks of Portions of the Highlands of Scotland South of the Caledonian Canal and on their Equivalents in the North of Ireland,' Quart. Jour. Geol. Soc., vol. xvii. p. 256.
　JAMIESON, T. F.—'On the Structure of the South-West Highlands of Scotland,' Quart. Jour. Geol. Soc., vol. xvii. p. 133.

1862. JAMIESON, T. F.—'On the Iceworn Rocks of Scotland,' Quart. Jour. Geol. Soc., vol. xviii. p. 164.
　SMITH, JAMES (of Jordanhill).—Researches in Newer Pliocene and Post-Tertiary Geology. 8vo. Glasgow.
　MURCHISON, Sir R., and GEIKIE, A.—First Sketch of a New Geological Map of Scotland, with Explanatory Notes.

1863. NICOL, JAMES (Prof.).—'On the Geological Structure of the Southern Grampians,' Quart. Jour. Geol. Soc., vol. xix. p. 180.
　SORBY, H. C.—'On the Original Nature and Subsequent Alteration of Mica Schist,' Quart. Jour. Geol. Soc., vol. xix. p. 401.
　GEIKIE, A. (Sir).—'On the Phenomena of the Glacial Drift of Scotland,' Trans. Geol. Soc. Glasg., vol. i. part ii.

1865. GEIKIE, A. (Sir).—The Scenery of Scotland viewed in connection with its Physical Geology. 8vo. London and Cambridge. (2nd edition, 1887.)
　BRYCE, JAMES.—The Geology of Arran and Clydesdale. 8vo. Glasgow. New Edition of Work of 1859.
　CROSSKEY, Rev. H. W.—'Glacial Deposits of the Clyde District,' Trans. Geol. Soc. Glasg., vol. ii. p. 45. (Kyles of Bute.)

1868. ARGYLL, DUKE OF.—'On the Physical Geography of Argyllshire in connection with its Geological Structure,' Quart. Jour. Geol. Soc., vol. xxiv. p. 255.
　ARGYLL, DUKE OF.—'On Six Lake Basins in Argyllshire,' Quart. Jour. Geol. Soc., vol. xxiv. p. 508.
　HASWELL, JAMES.—'On Columnar Structure developed in Mica Schist from a Vitrified Fort in the Kyles of Bute,' Proc. Geol. Soc. Edin., vol. i. p. 229.

1871. YOUNG, J. WALLACE.—'Miscellaneous Notes on Chemical Geology,' Trans. Geol. Soc. Glasg., vol. iii. p. 28. (Read in May 1868.)
　BELL, DUGALD.—'On the Aspects of Clydesdale during the Glacial Period,' Trans. Geol. Soc. Glasg., vol. iv. p. 63.

874. BELL, DUGALD.—'Notes on the Glaciation of the West of Scotland with reference to some recently observed instances of Cross Striation,' *Trans. Geol. Soc. Glasg.*, vol. iv. p. 800.

1875. CROSSKEY, Rev. H. W., and ROBERTSON, D.—'The Post-Tertiary Fossiliferous Beds of Scotland' (Tighnabruaich, Kyles of Bute, and Loch Riddon), *Trans. Geol. Soc. Glasg.*, vol. v. p. 29.

GLEN, D. C.—'On a Magnetic Sand from East Bay, Rothesay,' *Trans. Geol. Soc. Glasg.*, vol. v. p. 158.

1876. ARMSTRONG, JAMES, YOUNG, JOHN, and ROBERTSON, DAVID.—Catalogue of the Western Scottish Fossils, with Introduction by Prof. Young, M.D., on the Geology and Palæontology of the District. 8vo. Glasgow.

1880. SORBY. H. C.—Anniversary Address to the London Geological Society, vol. xxxvi. p. 46.

1883. MACADAM, W. IVISON.—'Preliminary Notice of a Clay Shell-Bed between Newton and Strachur,' *Trans. Geol. Soc. Edin.*, vol. iv. p. 94. (Read in 1881.)

MACADAM, W. IVISON.—'Further Notice of the Tigh-na-Criche Shell-Bed, Loch Fyne,' *ibid.*, vol. iv. p. 232. (Read 1882.)

MACADAM, W. IVISON.—'Notice of Veins of Specular Iron Ore at Strachur, Argyllshire,' *ibid.*, iv. 95. (Read in 1881.)

MACADAM, W. IVISON.—'On the Chemical Composition of certain Limestone Rocks from Ballimore (Argyllshire),' *ibid.*, vol iv. p. 101. (Read in 1881.)

1886. CADELL, H. M.—'The Dumbartonshire Highlands,' *Scot. Geograph. Mag.*, vol. 2, p. 336.

ARGYLL, DUKE OF.—'Our Highland Mountains,' *Good Words* for 1886, pp. 32 and 119.

1888. GEIKIE, Sir ARCHIBALD.—'The History of Volcanic Action during the Tertiary Period in the British Isles,' *Trans. Roy. Soc. Edin.*, vol. xxxv. p. 21.

1889. ARGYLL, DUKE OF.—'On certain Bodies, apparently of Organic Origin, from a Quartzite Bed near Inveraray,' *Proc. Roy. Soc. Edin.*, vol. 21, p. 39.

1892. GEIKIE, Sir ARCHIBALD.—Geological Map of Scotland reduced chiefly from the Ordnance and Geological Surveys, with Explanatory Notes. Edin.

MILL, Dr HUGH ROBERT.—'The Clyde Sea Area,' *Trans. Roy. Soc. Edin.*, vol. xxxvi. p. 641. (Read May 1891.)

DAKYNS, J. R., and TEALL, J. J. H.—'On the Plutonic Rocks of Garabal Hill and Meall Breac,' *Quart. Jour. Geol. Soc.*, vol. xlviii. p. 104.

1892-93. MUNRO, Dr ROBERT.—'Notes on Crannogs or Lake-dwellings in Argyllshire,' *Pro. Soc. Antiq. Scotland*, vol. xv. pp. 205–222. (Notices some in Loch Askaig, near Tighnabruaich.)

1894. MURRAY, JOHN, and IRVINE, ROBERT.—'On the Manganese Oxides and Manganese Nodules in Marine Deposits,' *Trans. Roy. Soc. Edin.*, vol. xxxvii. p. 721.

1895. BELL, DUGALD.—'On the Origin of certain Granite Boulders in the Clyde Valley,' *Trans. Geol. Soc. Glasg.*, vol. x. p. 16. (Read in 1892.)

1896. In February, before Glasgow Geol. Soc.

MACNAIR, PETER, read a paper on 'The Altered Basic Rocks of the Highlands as Exemplified by the Sill of Hornblende Schist underlying the Loch Tay Limestone.' (Reported in *Glasgow Herald* for Feb. 18, 1896.)

ANDERSON, JAMES.—'Evidences of the Most Recent Glaciers in the Firth of Clyde District,' *Trans. Geol. Soc. Glasg.*, vol. x. p. 198. (Read Oct. 1894.)

W. G.

INDEX

X

NEILL AND COMPANY, PRINTERS, EDINBURGH.

DESCRIPTION OF PLATES I.–V.

These are reproductions of Photographs of Natural Rock Exposures taken by R. Lunn.

PLATE I.

Schistose limestone with folding of the prominent foliation planes. Half a mile north of the Bull Rock, Dunoon. Observer looking south-west. See p. 14. A basalt dyke is seen in the background and a large erratic block from the "anticline" grits.

PLATE II.

Folded phyllite with quartz veins. East Bay of Dunoon. Observer looking west. The axes of the fold shade S.S.E.

PLATE III.

Contorted mica schist. North side of Fearna Bagh, Colintraive. Observer looking east. The prominent folds have axes hading north-west at a low angle. They are probably of "anticline" age.

PLATE IV.

Schistose greywacke with folded strain slips. North side of Fearna Bagh, Colintraive. Observer looking east. The folds which affect the strain slips have axes hading north-west, and are probably of "anticline" age. See p. 26.

PLATE V.

Contorted mica schist in cliff by the roadside a little south of Coylet Inn, Loch Eck. Observer looking east. The prominent folds have axes hading north-west, and are probably of "anticline" age.

To face Plate I.

PLATE 1

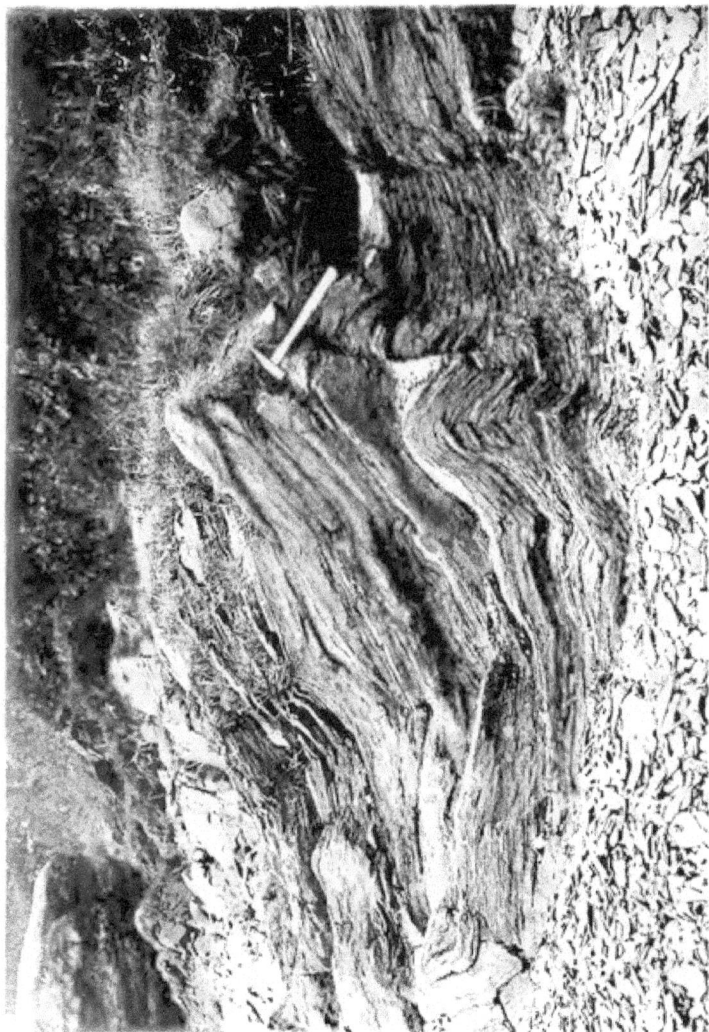

Schistose limestone with folding of the prominent foliation joints. Hilt at angle N. on the Boit Rock Traverse.

Observer looking South West.

PLATE II

Folded psillite with quartz veins. East face of Duncan Quarry, looking West.

PLATE III

Contorted mica schist. N. side of Fianna Bagh, Colonsare. Observer looking East

PLATE IV

Solution gneissoide with folded strain slips N ale of Future Bagh, Coimbatore Oblique looking East

PLATE V

Contorted mica schist in cliff by the roadside a little S. of Eight Inn, Lewis Esk. Observer looking East

PLATE VI.

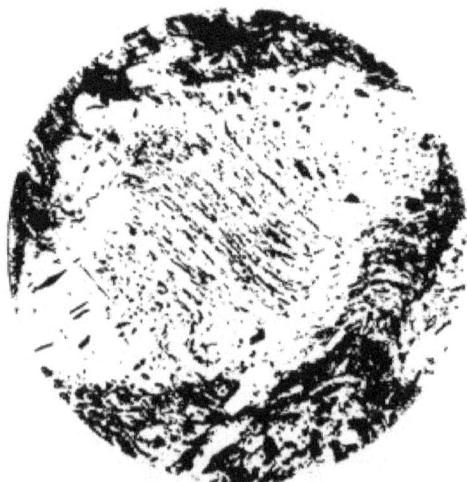

1

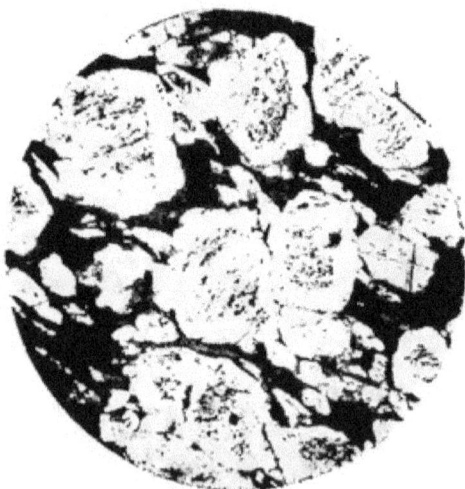

2

MICROSCOPIC SECTIONS OF ROCKS.

Reproduced from Photographs taken by J. J. H. Teall.

DESCRIPTION OF PLATE VI.

By J. J. H. Teall, F.R.S.

FIG. 1.— × 35. Natural light. Slide 3418. Albite schist ½ mile E.N.E. of Ard a' Chapuill (182). See p. 40 and Appendix I.

Section of a large albite crystal containing inclusions. The lines of inclusions are curved near the margins, but the crystal has not been strained. The ground-mass is mainly formed of white mica, chlorite, and iron ores.

FIG. 2.— × 35. Natural light. Slide 4733. Albite schist of Beinn Tharsuinn (142), somewhat altered by basalt dyke.

Numerous albite-crystals in a matrix similar to the above. The marginal portions of the albite-crystals are almost entirely free from inclusions.

To face Plate VI.

DESCRIPTION OF PLATE VII.

By J. J. H. Teall, F.R.S.

Fig. 3.— × 35. Natural light. Albite schist from head of Stuck Burn (152). Similar to the figures on Plate I., but containing also large plates of white mica, two of which are seen in cross-section in the lower half of the figure.

Fig. 4.— × 35. Natural light. "Green bed" with albite and epidote, 1 mile slightly south of east of Dunderave Castle (133). See p. 44.

The minerals in the portion of the slide represented in this figure are chlorite, biotite (scarce), granular epidote, quartz and albite. The biotite cannot be distinguished from the chlorite: nor the albite from the quartz.

To face Plate VII.

MICROSCOPIC SECTIONS OF ROCKS.

Reproduced from Photographs taken by J. J. H. Teall.

NOTES ON PLATES VIII. AND IX.

In the Index of Plate VIII. the tablets for albite schist, schistose limestone, and graphitic schist, have, through a misunderstanding, been placed in close connection. These schists have, however, no closer relationship to one another than to the other schists.

In Plate IX. the shelves of marine erosion and the blown sand are shown by the same pattern as the alluvia. The width of this pattern is slightly exaggerated along some of the sea margins.

To face Plate VIII

SKETCH MAP
OF THE
GEOLOGY OF COWAL.

By C. T. CLOUGH, M.A.

SCALE 4 MILES = 1 INCH.

INDEX.

Upper Old Red Sandstone rocks.

Phyllites and phyllitic mica schists, and
horizons, with a considerable proportion of these.

Albite schists where mapped.

Schistose Limestone.

Graphitic schist, with dark limestone
in places.

Schistose grit, greywacke or quartzite, or
horizons with a considerable proportion of these.

"Green Beds."

Serpentine.

Epidiorite, Hornblende schist and re-
lated Chlorite schist.

Granite.

Hypersite.

Non-schistose dykes and sills omitted.

Faults.

Portions left plain in the
area described, consist
chiefly of alternations

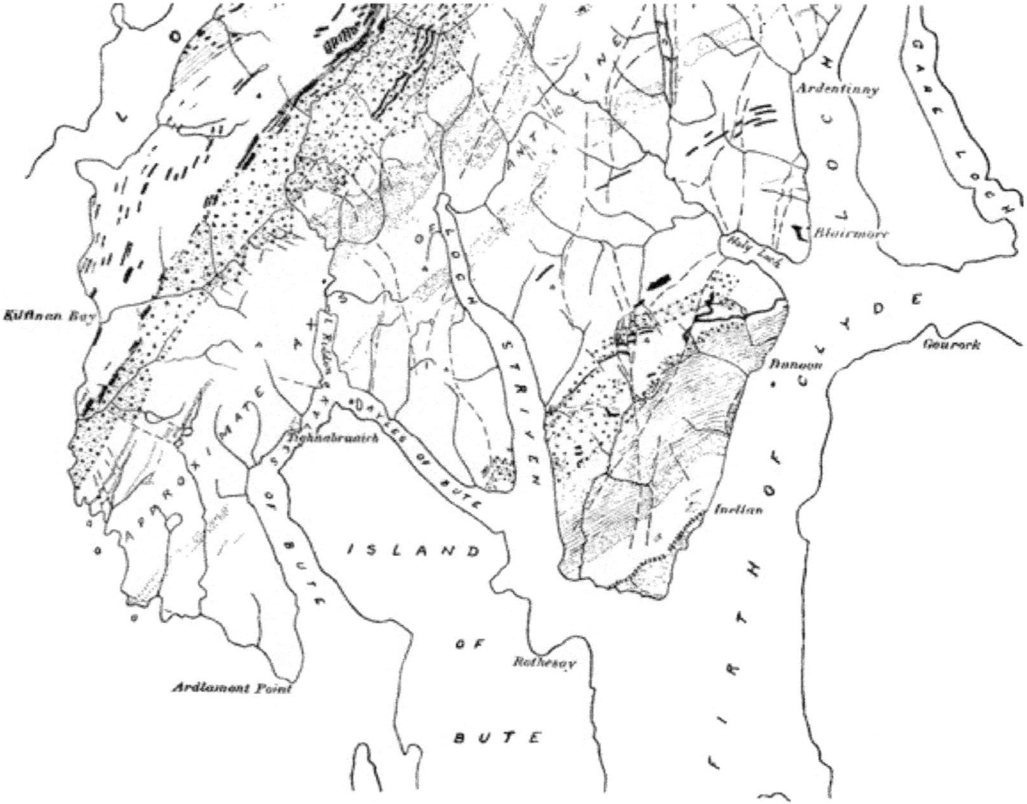
GARE LOCH
LOCH
Ardentinny
Blairmore
Holy Loch
FIRTH OF CLYDE
Gourock
Dunoon
Innellan
LOCH STRIVEN
KYLES OF BUTE
Tighnabruaich
L. Riddon
Kilfinan Bay
Ardlamont Point
ISLAND OF BUTE
Rothesay

PLATE IX.

SKETCH MAP
OF
GLACIAL AND POST-GLACIAL
GEOLOGY OF COWAL.

By C. T. CLOUGH, M.A.

SCALE 4 MILES =1 INCH.

INDEX.

Landslips.

Marine and freshwater alluvia and peat in basin-shaped hollows.

Boulder clay and sandy drift without definite moraine shapes.

Drift with well defined moraines.

Glacial striations: arrow shows supposed direction of ice-flow.

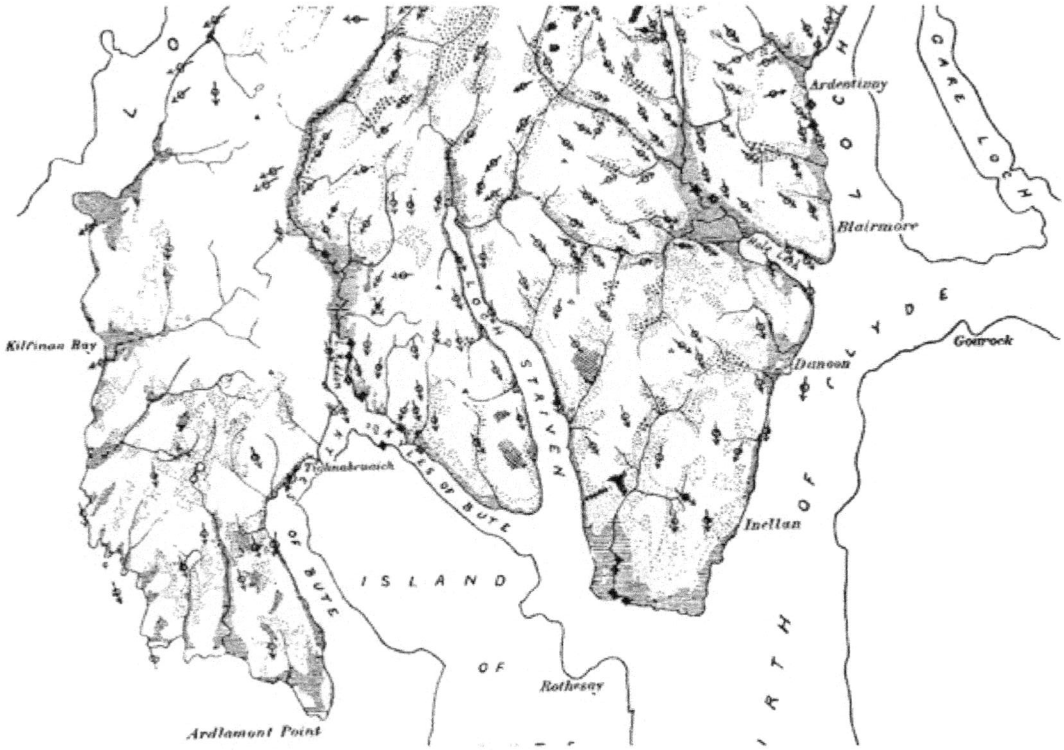

GARE LOCH
LOCH
Ardentinny
Blairmore
Holy Loch
CLYDE
Gourock
Dunoon
FIRTH OF
Inellan
LOCH STRIVEN
KYLES OF BUTE
Tighnabruaich
ISLAND OF BUTE
Rothesay
Kilfinan Bay
Ardlamont Point

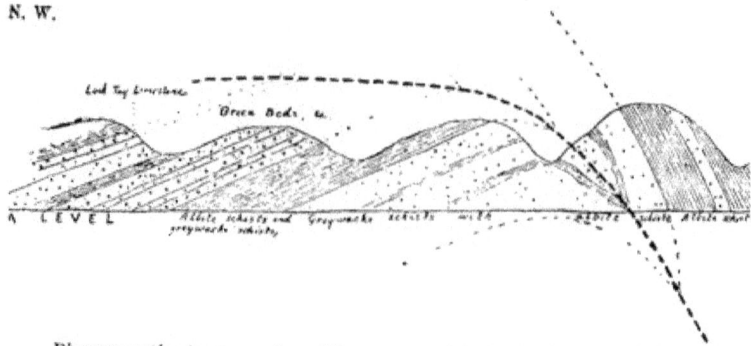

Diagram section to represent possible structure of the "anticline" between Inellan and
syncline is drawn in an interrupted dar[

S. E.

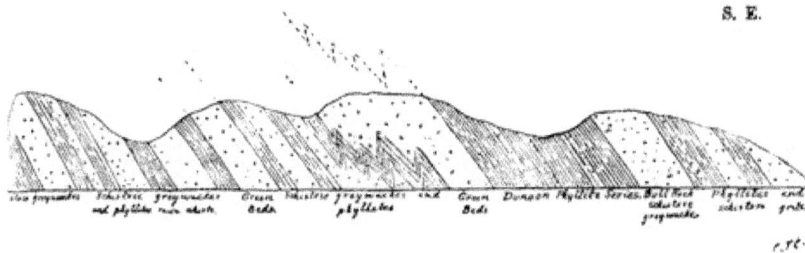

New greywacke Schistose greywacke Green Schistose greywacke and Green Dunoon Phyllite Series Bell Rock Phyllites and
and phyllites some schists Beds phyllites Beds schists schistose grits
greywacke

aruel. Not drawn to scale, and all faults and minor folds omitted. The axis of the supposed
· It is folded by the action of the " anticline."

www.ingramcontent.com/pod-product-compliance
Lightning Source LLC
Chambersburg PA
CBHW021939220326
41599CB00011BA/923